Lipid-Protein Interactions

Volume 2

LIPID-PROTEIN INTERACTIONS

Volume 2

Edited by
PATRICIA C. JOST
and
O. HAYES GRIFFITH

*Institute of Molecular Biology
and Department of Chemistry
University of Oregon, Eugene, Oregon*

1807 1982
175 YEARS OF PUBLISHING

A Wiley-Interscience Publication
JOHN WILEY & SONS
New York • **Chichester** • **Brisbane** • **Toronto** • **Singapore**

NMU LIBRARY

Library of Congress Cataloging in Publication Data:

Main entry under title:

Lipid-protein interactions.

  "A Wiley-Interscience publication."
  Includes indexes.
  1. Lipoproteins.   2. Protein binding.
3. Membranes (Biology)   4. Phospholipids.
I. Jost, Patricia C.   II. Griffith, O. Hayes.

QP552.L5L53 v. 2   574.19′24   81-16157
ISBN 0-471-06457-2 (v. 1)   AACR2
ISBN 0-471-06456-4 (v. 2)

Printed in the United States of America

10 9 8 7 6 5 4 3 2 1

Dedication

To the memory of Jaakko R. Brotherus, our young Finnish colleague, whose untimely death on June 18, 1981, cut short a promising research career. He was a beloved friend as well as a talented scientist and we shall miss him.

Contributing Authors
Volumes 1 and 2 (Current Addresses)

Banaszak, Leonard J., Department of Biological Chemistry, Division of Biology and Biomedical Sciences, Washington University, St. Louis, Missouri 63110 U.S.A.

Boggs, Joan M., Department of Biochemistry, Research Institute, The Hospital for Sick Children, Toronto, Ontario, CANADA M5G 1X8

Brotherus, Jaakko R., Department of Medical Chemistry, University of Helsinki, Siltavuorengpenger 10, 00170 Helsinki 17, FINLAND (Deceased)

Brown, James R., Clayton Foundation Biochemical Institute, Department of Chemistry, University of Texas at Austin, Austin, Texas 78712 U.S.A.

De Haas, Gerard H., Biochemical Laboratory, State University of Utrecht, Padualaan 8, 3508 TB Utrecht, THE NETHERLANDS

Edelstein, Celina, Departments of Medicine and Biochemistry, University of Chicago, Pritzker School of Medicine, Chicago, Illinois 60637 U.S.A.

Griffith, O. Hayes, Institute of Molecular Biology and Department of Chemistry, University of Oregon, Eugene, Oregon 97403 U.S.A.

Jost, Patricia C., Institute of Molecular Biology and Department of Chemistry, University of Oregon, Eugene, Oregon 97403 U.S.A.

Khorana, H. Gobind, Departments of Biology and Chemistry, Massachusetts Institute of Technology, 77 Massachusetts Avenue, Cambridge, Massachusetts 02139 U.S.A.

Marsh, Derek, Max-Planck-Institut für biophysikalische Chemie, Göttingen, Federal Republic of GERMANY

Matthews, Brian W., Institute of Molecular Biology and Department of Physics, University of Oregon, Eugene, Oregon 97403 U.S.A.

Moscarello, Mario A., Department of Biochemistry, Research Institute, The Hospital for Sick Children, Toronto, Ontario, CANADA M5G 1X8

Papahadjopoulos, Demetrios, Cancer Research Institute, School of Medicine, University of California, San Francisco, California 94143 U.S.A.

Radhakrishnan, Ramachandran, Departments of Biology and Chemistry, Massachusetts Institute of Technology, 77 Massachusetts Avenue, Cambridgree, Massachusetts 02139 U.S.A.

Reynolds, Jacqueline A., Department of Physiology and Whitehead Medical Research Institute, Duke University Medical Center, Durham, North Carolina 27710 U.S.A.

Robson, Robert J., Chevron Research Co., 576 Standard Avenue, Richmond, California 94802 U.S.A.

Ross, Alonzo H., Departments of Biology and Chemistry, Massachusetts Institute of Technology, 77 Massachusetts Avenue, Cambridge, Massachusetts 02139 U.S.A.

Ross, Joe M., Department of Biological Chemistry, Division of Biology and Biomedical Sciences, Washington University, St. Louis, Missouri 63110 U.S.A.

Scanu, Angelo M., Departments of Medicine and Biochemistry, University of Chicago, Pritzker School of Medicine, Chicago, Illinois 60637 U.S.A.

Seelig, Anna, Biocenter of the University of Basel, Klingelbergstrasse 70, CH-4056 Basel, SWITZERLAND

Seelig, Joachim, Biocenter of the University of Basel, Klingelbergstrasse 70, CH-4056 Basel, SWITZERLAND

Shen, Betty W., Department of Biophysics, University of Chicago, Chicago, Illinois 60637 U.S.A.

Shockley, Paula, Clayton Foundation Biochemical Institute, Department of Chemistry, University of Texas at Austin, Austin, Texas 78712 U.S.A.

Silvius, John R., Department of Biochemistry, McGill University, Montréal PQ, CANADA H36 1Y6

Tamm, Lukas, Biocenter of the University of Basel, Klingelbergstrasse 70, CH-4056 Basel, SWITZERLAND

Takagaki, Yotaroh, Departments of Biology and Chemistry, Massachusetts Institute of Technology, 77 Massachusetts Avenue, Cambridge, Massachusetts 02139 U.S.A.

Volwerk, Johannes J., Institute of Molecular Biology, University of Oregon, Eugene, Oregon 97403 U.S.A.

Watts, Anthony, Department of Biochemistry, University of Oxford, South Parks Road, Oxford OX1 3QU, U.K.

Wirtz, Karel W. A., Laboratory of Biochemistry, State University of Utrecht, Transitorium 3, Padualaan 8, NL-3508 TB Utrecht, THE NETHERLANDS

Wrenn, Richard F., Department of Biological Chemistry and Department of Electrical Engineering, Division of Biology and Biomedical Sciences, Washington University, St. Louis, Missouri 63110 U.S.A.

Preface

Lipid-protein associations are central to understanding structure-function relationships in such diverse fields as ion transport, oxidative phosphorylation, photosynthesis, cell recognition, other problems in membrane biochemistry, soluble enzyme interactions with lipid substrates, lipid-requiring enzymes, lung surfactant, phospholipid transfer proteins, and serum lipoproteins.

At the molecular level, there must be common unifying principles governing the hydrophobic, ionic, and steric contributions that determine the molecular arrangement of lipids interacting with proteins. Progress in determining their relative roles is likely to occur in any of a number of seemingly unrelated fields. This active area of biochemical research is now approaching the state where critical reviews of major developments in the field will be useful to a large number of investigators. At present there is no single source that focuses on the problem of lipid and protein interactions. It is the aim of these volumes to bring together advances in the molecular biology of protein associations with lipid, so that an investigator in any of these fields will be aware of fundamental developments.

The philosophy has been to incorporate basic reference material, including sequences, tabular sources of relevant data, and basic theoretical treatments where appropriate. Our hope is that, in addition to providing in-depth current reviews, these volumes will serve as a compact reference source for investigators of diverse backgrounds and interests.

Lipid-Protein Interactions is organized into two initial volumes published simultaneously. One deals with water soluble lipid-protein complexes, and the other focuses on membrane systems. Also included is a thorough compilation of data on the thermotropic behavior of phospholipids that is relevant to many of the experimental approaches to lipid-protein interactions in both water-soluble and membrane systems.

Volume 1 covers selected watersoluble lipid-protein complexes, and contains six chapters including a discussion of the high resolution X-diffraction studies on the chorophyll-binding protein showing the interaction of the phytol chains with the protein backbone (Brian Matthews), the proposed structure of the fatty acid and drug binding sites of serum albumins (James R. Brown and Paula Shockley), the properties of phospholipase A_2 and its association with organized lipid-water interfaces (Johannes J. Volwerk and Gerard de Haas), the properties and characterization of the phospholipid transfer proteins (Karel W. A. Wirtz), lipid-protein associations in the microcrystalline yolk complex (Leonard J. Banaszak, Joe M. Ross, and Richard Wrenn), and the physico-chemical aspects of the serum high density lipoproteins (Angelo M. Scanu, Celina Edelstein, and Betty W. Shen).

This volume, *Volume 2*, is concerned with lipid-protein interactions in membranes and begins with the structural organization of myelin and lipid-protein interactions of both the small water-insoluble protein, lipophilin, and the water-soluble highly basic protein. This chapter, by Joan M. Boggs, Mario A. Moscarello, and Demetrios Papahadjopoulos, brings together the data from different disciplines to examine the role of the proteins in the unique organization of the myelin membrane.

Chapter 2, by Derek Marsh and Anthony Watts, discusses the structure of membrane proteins and describes the results of spin labeling experiments in investigating the lipid dynamics and average lipid composition at the lipid-protein interface. This reporter group technique has the fortunate attributes of both motionally dependent lineshapes and a time-scale that can resolve different lipid environments. Nuclear magnetic resonance has a longer time-scale, complementing the shorter time-scale techniques. The application and potential of nuclear magnetic resonance in the study of lipid-protein interactions in membranes is explored by Joachim Seelig, Anna Seelig, and Lukas Tamm in Chapter 3.

Photoreactive groups on lipid molecules show great promise in identifying polypeptide segments of the folded membrane protein that are exposed to the immediate lipid environment. The general properties of various classes of useful photoreactive species, some synthetic procedures, and the results of photochemical cross-linking studies on several membrane proteins are discussed in Chapter 4 by Robert J. Robson, Ramachandran Radhakrishnan, Alonzo H. Ross, Yotaroh Takagaki, and H. Gobind Khorana.

Chapter 5 by Jacqueline A. Reynolds provides a quantitative treatment of the interaction of proteins with amphiphiles such as detergents. Detergents are required in the isolation and characterization of hydrophobic membrane proteins, and the thermodynamic treatment for these interactions is fundamental for both the water-soluble lipid complexes and hydrophobic interactions with solubilized membrane proteins. In the intact membrane continuum, the formalism for multiple binding equilibria between lipid and proteins differs from the water-soluble systems both because of the experimentally accessible parameters and the concentrated nature of membranes where lipid, rather than water, is the solvent. This special case, including the derivation of relevant expressions and their application to membrane proteins in lipid bilayers, is treated in Chapter 6 by Jaakko R. Brotherus, O. Hayes Griffith, and Patricia C. Jost.

The development and characterization of defined lipids has been crucial for studies of lipid-protein associations in both water-soluble and membranous systems. In the final chapter John R. Silvius presents a discussion and the first extensive tabulation of the available data describing the thermotropic behavior of pure lipid systems and binary lipid mixtures, and the effect of proteins on lipid phase behavior in reconstituted systems. These data are a basic tool in designing experimental approaches to many important problems in the field of lipid-protein interactions.

We owe a debt to Debra A. McMillen, who provided invaluable assistance in the organizational and editorial work on these volumes, to our colleagues for their patience, and to Dr. Stanley F. Kudzin, who served as our editor at Wiley-Interscience and who proposed the idea for these volumes.

PATRICIA C. JOST
O. HAYES GRIFFITH

Eugene, Oregon
October 1981

Contents

One

Structural Organization
of Myelin—Role of Lipid-Protein
Interactions Determined
in Model Systems

JOAN M. BOGGS, MARIO A. MOSCARELLO, AND DEMETRIOS PAPAHADJOPOULOS

Department of Biochemistry, Research Institute, The Hospital for Sick Children, Toronto, Ontario, Canada (J.M.B. and M.A.M.)

Cancer Research Institute, School of Medicine, University of California, San Francisco, California, U.S.A. (D.P.)

ABBREVIATIONS

BNPS-skatole, 2-(2-nitrophenylsulfenyl)-3-methyl-3′-bromoindolenine-skatole
CD, circular dichroism
[^{3}H]DIDS, 4,4′-diisothiocyano-2,2′-ditritiostilbene disulfonic acid
DLPC, dilauroylphosphatidylcholine
DMPC, dimyristoylphosphatidylcholine
DPPC, dipalmitoylphosphatidylcholine
DPPG, dipalmitoylphosphatidylglycerol
DSC, differential scanning calorimetry
DSPC, distearoylphosphatidylcholine
EAE, experimental allergic encephalomyelitis
ESR, electron spin resonance
IgG, immunoglobulin G
NMR, nuclear magnetic resonance
N-2, original designation for human myelin lipophilin [Gagnon et al. (1971)]
PA, phosphatidic acid
PC, phosphatidylcholine
PE, phosphatidylethanolamine
PG, phosphatidylglycerol
PS, phosphatidylserine
P7, rat brain myelin P7 proteolipid fraction

I. INTRODUCTION

The membranous system of the myelin sheath is composed of lipids and proteins. The structural organization of the membrane appears to follow the principles summarized for membranes in general by Singer (1974). In contrast to other membranes that are unilamellar (i.e., consisting of a single lipid bilayer into which various proteins are embedded), the myelin sheath is multilamellar (i.e., consisting of 5–20 lamellae that surround axons greater than 1 μm in diameter). The lamellae of compact myelin are directly continuous with the plasma membrane of the oligo-dendroglial cell, which is considered to be the myelin forming and sustaining cell of the central nervous system (Davison & Peters, 1970). Each oligodendroglial cell sends out several processes (30–50) and may myelinate a number of closely adjacent axons (Peters & Vaughn, 1970). Clearly, the synthetic capacity of these cells is prodigious. Norton and Poduslo (1973) have calculated that the oligodendroglial cell produces three times its own weight of membrane per day during the period of active myelination.

To understand the structural organization of the myelin sheath it is essential to study two fundamental problems: (1) the nature of protein-lipid, protein-protein, and lipid-lipid interactions in the myelin membrane (the unilamellar structure), and (2) the nature of the interactions that are necessary to maintain the unique multilamellar structure of myelin. This chapter addresses these two major areas.

The structural organization of the myelin membrane is discussed with special emphasis on the protein-lipid interactions. Although X-ray diffraction, electron microscopy, and labeling studies of intact and isolated myelin have provided abundant information concerning the structure of myelin, the role of the proteins in myelin can be better understood by studying their effects on lipid organization and their conformational properties in lipid vesicles containing the purified proteins. All the lipid components have been isolated and the two major proteins, the basic and proteolipid proteins (which constitute 80% of the protein in myelin), have been purified (see Table 1 for the lipid and protein composition of myelin). The chemical and physical properties of the proteins have been studied in some detail (Moscarello, 1976). The primary and secondary structure of basic protein is known, and some information on the tertiary structure of both proteins is available. Since these proteins

TABLE 1   Composition of Normal Human Myelin

	% Dry Weight
Protein	24.0
Basic protein	5.4
Proteolipid	
Lipophilin	7.2
Other LH-20 fractions	4.2
Thioethanol soluble	0.6
Others (Wolfgram, DM20, glycoproteins)	6.6

Lipid[a]	Mol % Total Lipid
Cholesterol	40.9
Cerebroside	15.6
Cerebroside sulfate	4.1
Phosphatidylcholine	10.9
Ethanolamine phospholipids	13.6
Serine phospholipids	5.1
Sphingomyelin	4.7
Others	5.1

[a]Lipid composition from Rumsby (1978).

represent examples of an extrinsic type (basic protein) and an intrinsic type (pro-teolipid), they can also be used as models for other membrane proteins to help understand the structure and role of these proteins in different membranes.

Reconstitution studies involving the incorporation of each of the proteins into lipid vesicles are well advanced. A detailed picture of protein-lipid interactions from a number of physical techniques including circular dichroism (CD), freeze-fracture electron microscopy, differential scanning calorimetry (DSC), electron spin resonance (ESR), nuclear magnetic resonance (NMR), and liquid diffraction X-ray analysis have contributed to the models presented below. In preparation for a discussion of the results obtained by these different methods, we begin by describing the structure of myelin and the properties of isolated myelin proteins.

II. STUDIES ON INTACT AND ISOLATED MYELIN

The oligodendroglial cell plasma membrane wraps around the nerve axon such that the intracellular membrane surfaces come together excluding the cytoplasm, and the extracellular surfaces come together excluding extracellular fluid as shown in Fig. 1. (For recent reviews, see Braun, 1977; Rumsby & Crang, 1977.) Freeze-fracture electron microscopy and X-ray diffraction studies indicate that the cyto-plasmic surfaces in myelin are more closely and more tightly apposed than the extracellular surfaces (McIntosh & Robertson, 1976; Caspar & Kirshner, 1971). Exposure of myelin to hypotonic solutions causes swelling and partial separation of the lamellae at the extracellular surfaces but not at the cytoplasmic surfaces (McIntosh & Robertson, 1976). Evidence has been reported indicating that myelin basic protein may be located at the cytoplasmic side (Adams et al., 1971; Herndon et al., 1973; Golds & Braun, 1976; Poduslo & Braun, 1975; Crang & Rumsby, 1977). Labeling studies both of isolated myelin (Crang & Rumsby, 1977; Wood et al, 1977) and of intact spinal cord (Poduslo & Braun, 1975) have been performed in order to determine the location of the proteins. Nonpenetrating reagents that should label only proteins in the outermost extracellular layer bind to proteolipid to a greater extent than to basic protein, even though basic protein is more reactive in its isolated form than proteolipid. This indicates that the proteolipid has some sites exposed to the aqueous phase on the extracellular surface while basic protein either is not exposed on the extracellular surface or its sites are occluded. However, most of the protein in myelin is not accessible to nonpenetrating reagents. Labeling of intact spinal cord with a penetrating reagent, salicylaldehyde, on the other hand, resulted in greater labeling of basic protein than proteolipid, suggesting that basic protein is located on the intracellular surface rather than the extracellular surface (Golds & Braun, 1976). There is also freeze-fracture evidence to indicate that the intramembranous particles in myelin from the central nervous system, which are undoubtedly the proteolipid proteins (Boggs & Moscarello, 1978a), cleave primarily with one fracture face. This has been claimed to be the cytoplasmic side of the bilayer (Schnapps & Mugnaini, 1975; 1976), which would indicate tighter asso-ciation with this side. On the other hand, Pinto da Silva and Miller (1975) observed

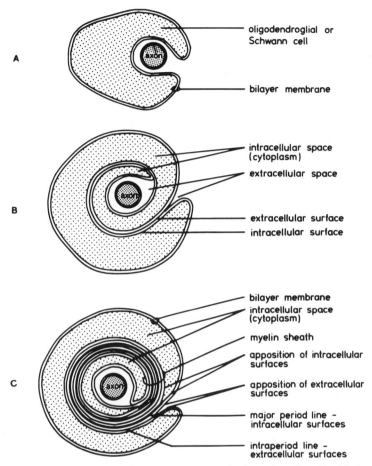

Figure 1. Schematic diagrams illustrating the stages in the formation of myelin. (A) Axon (dense stippling) surrounded by a Schwann cell (peripheral nervous system) or oligodendroglial cell (central nervous system). Cytoplasm is indicated by less dense stippling. (B) Intermediate stage showing apposition of extracellular surfaces of bilayer plasma membrane. (C) Compact myelin formed by apposition of both extracellular surfaces and intracellular surfaces of plasma membrane. The intracellular surfaces form the more dense major period line; the extracellular surfaces form the intraperiod line seen by X-ray diffraction.

that in myelin impregnated with glycerol before freeze-fracture, the particles aggregate into linear arrays that are propagated radially across many lamellae and that seem to keep the lamellae closely apposed, suggesting interlamellar interactions between the protein particles at both the extracellular and cytoplasmic surfaces.

X-Ray diffraction studies show that the repeat distance of myelin is 150–160 Å, the thickness of two bilayers, indicating that the bilayers are asymmetric (Schmitt et al., 1941; Worthington & Blaurock, 1969; Blaurock, 1971; Caspar & Kirshner, 1971). The electron density profiles indicate that both the polar head group region and the hydrocarbon region are different on the two sides of the bilayer (Caspar

& Kirshner, 1971). The asymmetry in the hydrocarbon region was attributed to a higher cholesterol concentration on the extracellular side, while the asymmetry of the polar head group region was attributed to protein. It was considered unlikely that protein contributed to the asymmetry in the hydrocarbon region, since the electron density indicated that protein occupied less than 10% of the space in the bilayer interior. However, as pointed out by Rumsby and Crang (1977), even if all the proteolipid protein were embedded in the bilayer, it would occupy only 9% of the bilayer interior. Melchior et al. (1979) have recently shown that the asymmetry in the hydrocarbon region is retained in myelin of the peripheral nervous system that has been compacted by Ca^{2+}, an interaction that causes displacement of intramembranous protein particles from the compacted area. This suggests that the asymmetry is not due to intrinsic protein. The asymmetry disappears slowly with time in compacted myelin, indicating that a rearrangement of lipid and/or protein may occur.

III. CHEMICAL AND PHYSICAL PROPERTIES OF ISOLATED MYELIN PROTEINS

A. Basic Protein of Myelin

1. Primary Structure

This protein is known by several names including "basic protein" (Kies, 1965), a name favored by us, "A_1 protein" (Eylar et al., 1971a), and "encephalitogenic protein", the latter because of its potent disease-producing activity. It comprises about 30% of the total myelin protein and can be extracted readily from whole brain in a variety of acidic buffers or from isolated human myelin by dilute mineral acid (0.2 N H_2SO_4) (Lowden et al., 1966). The ease of extraction into aqueous solvents classifies this membrane protein as the extrinsic type, which can be studied readily by conventional methods of protein chemistry.

The demonstration of the encephalitogenic activity of whole brain homogenates led to a flurry of activity in many laboratories to isolate the active principle. One of the proteins, basic protein, was easy to prepare, and many independent laboratories focused their attention on this material. These included Roboz-Einstein et al. (1962), Lowden et al. (1966), Tomasi and Korngurth (1967), Eylar et al. (1969), and Carnegie (1971).

The isolation from bovine spinal cord by Roboz-Einstein et al. (1962) involved a preliminary extraction in acetone–petroleum ether followed by isolation of the protein with 5–10% KCl. In other procedures the tissue was defatted by chloroform-methanol (Tomasi & Korngurth, 1967; Kies et al., 1965; Nakao et al., 1966; Wolfgram, 1965; Carnegie et al., 1967; Kibler & Shapira, 1968) prior to extraction of the protein into aqueous media. Basic protein has been isolated from purified human central nervous system myelin by dilute mineral acid (Lowden et al., 1966), which completely inhibits enzymatic activities.

The isolated material was shown to have potent disease-producing acitvity. It induced experimental allergic encephalomyelitis (EAE) in the guinea pig, which has been used as a biological assay for this material. Eylar (1972) reported later that less than 5 μg of the purified protein induced disease in guinea pigs. The basic protein was considered to represent all the encephalitogenic activity of brain homogenates, and thus a concerted effort was made to elucidate the primary structure in the hope that this would provide an explanation in molecular terms for the potent biological activity.

The problem of the elucidation of the primary structure was solved by studies from several laboratories. The publication of the amino acid composition following hydrolysis in constant boiling HCl revealed that cystine was not present in the molecule (Lowden, 1966). Consequently in subsequent studies, it was not necessary to reduce and alkylate cystines, since the molecule consisted of a single polypeptide chain, without disulfide linkages.

Eylar et al. (1971a) isolated 27 tryptic and peptic peptides to establish the complete sequence of the 170 amino acid residues of the basic protein isolated from bovine spinal cord. The complete amino acid sequence of the human central nervous system myelin basic protein was published by Carnegie (1971) and is similar to the bovine sequence with minor modifications. Indeed the sequence of basic protein has so far been found to be highly conserved, as shown in Fig. 2. The protein is not amphipathic, nor is there any general periodicity of basic residues. Instead they are distributed throughout the molecule at random.

Several unusual features of the primary structure are noteworthy. Also, a number of posttranslational modifications occur at various sites in the sequence. The arginine residue at position 107 is present as both a monomethylated and a dimethylated derivative. In the vicinity of the methylated arginine and near the middle of the sequence is a proline-rich region of sequence Pro-Arg-Thr-Pro-Pro-Pro. The presence of these three prolines may induce a bend in the molecule, as deduced from model building. Such a structure is in agreement with several physical studies to be mentioned later. A tri-proline sequence is not commonly found in proteins but has been reported in rabbit immunoglobin G (IgG), in which it forms a hinge region (Smyth & Utsumi, 1967).

Although basic protein is not a glycoprotein it has been shown to function as an acceptor of N-acetyl-D-galactosamine from UDP-N-acetyl-D-galactosamine in the presence of an enzyme from bovine submaxillary gland (Hagopian & Eylar, 1969). The principal site of glycosylation was reported as Thr-98, forming an O-glycosidic linkage, although other sites were glycosylated also. The ability of basic protein to accept N-acetylgalactosamine gives it a unique property as the only known natural acceptor of carbohydrate (Sturgess et al., 1978). Possibly basic protein is synthesized as a glycoprotein and the carbohydrate is subsequently removed prior to compaction of myelin. The protein in glycosylated form has not been isolated so far.

Another interesting feature of this protein is the demonstration of microheterogeneity at alkaline pH values. Microheterogeneity was described first by Martenson et al. (1969). Electrophoresis at alkaline pH values in polyacrylamide gels showed

B – Bovine   **H** – Human   **Rs** – Rat small   **Gp** – Guinea pig

Figure 2. Amino acid sequences of myelin basic proteins from human, bovine, rat, and guinea pig central nervous system myelin. The human sequence is shown in full with changes in the bovine and guinea pig above and rat below. The guinea pig protein has only been sequenced from residue 43–88. Substitutions, insertions (–) or deletions (–) are shown. Various posttranslational modifications that have been found in the isolated protein or can be made *in vitro*, as described in the text, are indicated. Residues 67–75 are encephalitogenic in the rabbit, 76–86 in the Lewis rat, 114–122 in the guinea pig, and 154–167 in the monkey and rabbit, as indicated. From Carnegie & Dunkley (1975); Martenson et al. (1977).

four or five distinct bands, whereas a single band was observed at acid pH values. These bands were partially separated on carboxymethyl cellulose columns at alkaline pH's (Deibler & Martenson, 1973). All components isolated showed similar molecular weights by gel filtration and similar biological activity. No difference was found in the amount of monomethyl- and dimethylarginine at position 107, which was considered to be responsible for the microheterogeneity by Baldwin and Carnegie (1971).

The basis of microheterogeneity was investigated further by Chou et al. (1976). They separated several components by ion-exchange chromatography in glycine buffer pH 10.4 containing 2 M urea, but three major components were obtained. The amino acid composition and tryptic peptide maps of two of the components were identical, and the carboxy terminal sequences Ala-Arg-Arg were identical. They numbered their components 4, 5, and 6 and showed the following differences in structure. Component 4 was partially phosphorylated (50%) at Thr-98 and Ser-165. Component 5 was partially deamidated at glutamine residues 103 and 147. Modified glutamine residues were present also in component 4, accounting for about 15% of this component. These investigators concluded that component 6 was native unmodified basic protein. The effect of these modifications was to alter the net charge on the molecule. Thus, component 4 differed from 6 by a net negative charge of 2 and component 5 by a net negative charge of 1. Since the modifications were nonrandom, they concluded that specific enzymes were responsible. Although not considered to be a major modification, the loss of the carboxy terminal Arg may also contribute to the microheterogeneity. It now appears likely that the microheterogeneity is even more extensive than reported above. The functional significance of the microheterogeneity remains an intriguing problem.

A number of studies have shown that basic protein is an excellent substrate for phosphorylation *in vitro*. The sites were different from those reported above (Carnegie et al., 1973; Miyamoto & Kakiuchi, 1974), when myelin basic protein was phosphorylated by a cyclic AMP-dependent kinase from brain. Miyamoto and Kakiuchi (1974) observed that 0.2 mol of phosphorus per mol of basic protein was transferred and the phosphorus was equally divided between phosphoserine and phosphothreonine. Under appropriate conditions, the amount of phosphorus added could be increased to 4 mol/mol basic protein. Therefore there appeared to be at least four potential sites. Using a protein kinase from muscle, Carnegie (1973) reported that Ser-12 and Ser-110 were phosphorylated primarily. An endogenous protein kinase was reported in myelin which phosphorylated basic protein principally at Ser and Thr (Miyamoto & Kakiuchi, 1974; Steck & Appel, 1974).

The conclusion from these various studies appears to be that basic protein in myelin is phosphorylated to variable extents. Since four potential sites of phosphorylation are present, differences in the degree of phosphorylation can give rise to a number of components with slightly different negative charges. This mechanism can introduce a much more extensive microheterogeneity than previously had been considered. No role for the phosphate has been postulated, but it could affect the conformation of certain segments that would interfere with the way these segments of the protein interact with the phospholipid bilayer. Posttranslational modifications of the sequence of basic protein are indicated in Fig. 2.

2. Secondary and Tertiary Structures

The absence of cystine from the primary structure of basic protein eliminated the possibility that disulfide bonds could stabilize the conformation. However the presence of the Pro-Pro-Pro region near the center of the polypeptide chain confers a high probability to the formation of a bend in the molecule, which might bring the N-terminus and the C-terminus close together.

Measurements of the size and shape of bovine basic protein in solution were first reported by Eylar and Thompson (1969). On the basis of viscosity studies, they reported an axial ratio of 1 : 10 and concluded that the shape was that of a prolate ellipsoid. However, Chao and Einstein (1970) determined the axial ratio to be 1 : 17, in contrast to the results of Eylar and Thompson. Epand et al. (1974) calculated the radius of gyration from low angle X-ray scattering and found it to be 39 ± 2 Å, which corresponded well with the prolate ellipsoid structure proposed earlier with an axial ratio of 1 : 10. Further confirmation of the axial ratio came from sedimentation velocity analyses, which yielded a calculated ratio of 1 : 12. The structure proposed by Epand et al. had the dimensions of 150 × 15 Å, and they concluded that the molecule possessed considerable structure although no specific α-helical or β-structure. In contrast, X-ray scattering results reported by Krigbaum and Hsu (1975) were consistent with a flexible coil model. Natural abundance ^{13}C-NMR studies by Chapman and Moore (1976; 1978) supported a structure that embodied an open structure flexible at the ends of the molecule but with considerable secondary structure near the center, in accord with theoretical stabilization of this part of the molecule by the Pro-Pro-Pro region. This conclusion is consistent with a later study by Deber et al. (1978) in which myelin basic protein from human brain was specifically enriched with [^{13}C]methyl iodide at the two Met residues 20 and 167. Enrichment of the methionines on the sulfur atom places the ^{13}C carbon a considerable distance away from the backbone of the peptide chain. The motion of the methyl group was not restricted, and it was concluded that the two ends of the molecule contain little structure. The concept of a specific secondary/tertiary structure of bovine myelin basic protein was supported by ^{1}H-NMR studies of the protein and constituent peptides by Littlemore (1978) on the basis of chemical shift heterogeneities for His, Tyr, Met, Thr, and Ile.

The absence of significant α-helical or β-structure has been confirmed in several laboratories. Optical rotatory dispersion studies by Eylar and Thompson (1969), Chao and Einstein (1970), and Palmer and Dawson (1969) were in agreement that little helical structure was present in the molecule. Circular dichroism spectra in 0.2 M acetic acid and 0.1 M sodium hydroxide were similar to each other but differed from the spectrum in 6 M guanidine hydrochloride, a denaturing solvent. An extensive study of the CD properties of human myelin basic protein at pH values 1.68–12.1 was reported by Moscarello et al. (1974). At pH 1.68 and 25°C ellipticity values observed at 220 nm indicated little or no helical structure. Heating the solution to 85°C resulted in a decreased ellipticity at 220 nm, indicating an increase in helical content. The ellipticity at 200 nm was less at 85°C than at 25°C, consistent with a more ordered structure at 85°C. At pH 11.3, the spectra showed little evidence of structure throughout the scan. Heating to 85°C decreased the ellipticity that was especially marked at 200 nm. At pH 12.1 and 25°C some helical structure was present at 220 nm, and it increased after heating to 85°C, in agreement with the spectra obtained after heating at pH 1.68. Since heating proteins is expected to reduce the helical content, basic protein was unusual in that heating to 85°C induced more structure.

Further evidence for the presence of secondary structure was obtained from studies of surface tension originally used to study conformational transitions in poly-L-lysine solutions (Neumann et al., 1973). The temperature dependence of the surface tension was studied at pH values from 2.2 to 12.5. The following features were observed: (1) the surface tension decreased with increasing pH due to an increased concentration of protein hydrophobic groups at the surface, possibly indicating a conformational transition; (2) there was no evidence of temperature-induced conformational transitions at intermediate pH values; and (3) a low temperature transition (40°C) was observed at low pH (2.2), whereas a high temperature transition (80–85°C) was observed at high pH.

The lack of changes in secondary structure at pH values 5.6–9.2 were confirmed by CD studies (Smith, 1977a) and proton NMR studies (Liebes et al., 1975). In the latter report, spin-spin relaxation studies showed little change over the pH range 5.6–9.2. In addition they confirmed the radius of gyration reported earlier by Epand et al. (1974).

A specific secondary/tertiary structure is implied by the immunochemical studies of Whitaker et al. (1977), who showed that antibody to peptide 43–88 reacted well with the peptide but showed little or no reaction with intact basic protein. The authors concluded that peptide 43–88 or a portion of this peptide was shielded and must have been buried in a specific conformation of basic protein.

It is clear from the discussion above that myelin basic protein is devoid of α-helical or β-structure in the isolated, water-soluble form. However, it is not devoid of secondary structure. A complete solution to this problem of secondary/tertiary structure will require X-ray studies on the crystallized protein.

3. Encephalitogenic Activity

The potent encephalitogenic activity of the basic protein in the guinea pig led to the rapid elucidation of the primary sequence. Further studies showing that tryptic or peptic digestion liberated a nonapeptide consisting of residues 114–123, Phe-Ser-Trp-Gly-Ala-Glu-Gly-Gln-Lys (Arg), which was also encephalitogenic in the guinea pig, stimulated the search to define the minimum requirements for encephalitogenic activity. Using synthetic peptides it was shown that disease induction in the guinea pig is absolutely dependent on the presence of virtually all the residues in the peptide (Westall et al., 1971; Eylar et al., 1970; Lamoureux, 1972; Burnett & Eylar, 1971; Hashim, 1978). This region of the sequence is highly conserved in mammalian basic proteins. However, in the chicken the Gln is replaced by His, making chicken basic protein at least 100 times less active than mammalian proteins (Eylar et al., 1971b).

Sensitization of guinea pigs to the nonapeptide results in sensitized lymph node cells that respond to both the peptide and the intact basic protein, indicating that the Trp region is exposed and recognized in the intact molecule (Spitler et al., 1972). Other peptides (Bergstrand, 1972) are immunogenic, resulting in cell-mediated immunity but not EAE, suggesting that these regions of the protein are not accessible in intact myelin.

Species variation occurs with respect to the encephalitogenic site. Peptide 43–89 elicited EAE in rabbits (Kibler & Shapira, 1968) but not in guinea pigs or monkeys. The major monkey determinant has been recently localized to residues 132–169 (Jackson et al., 1972), which could be relevant to the disease-inducing site in humans. Peptide 1–42 is also weakly active in the monkey (Brostoff et al., 1974), but the Trp peptide is not active (Jackson et al., 1972). The encephalitogenic site in the rat was elucidated in extensive studies by Hashim and colleagues to be peptide 74–81 of guinea pig protein (Hashim, 1977; Hashim et al., 1978, 1979), which lacks a Gly-His segment present in bovine basic protein (Fig. 2). The presence of Gly-His in a longer peptide, containing residues 74–81, results in loss of activity, which may mean that the conformation necessary for disease in the rat is disrupted by the Gly-His segment. This suggests that the conformation of this region in intact bovine basic protein is different from that of guinea pig basic protein. The disease-inducing sites for various species are summarized (underscored) in Fig. 2. The species variability in encephalitogenic determinants suggests that in different species, different regions of the protein may be exposed in myelin for recognition by sensitized cells.

B. The Proteolipid Proteins

1. Primary Structure

Proteolipid, described originally by Folch and Lees (1951) as a proteinaceous material soluble in chloroform-methanol and insoluble in water, has been recently defined as a family of different proteins soluble in chloroform-methanol (Lees et al., 1978). Although this definition is useful, it cannot be applied too rigidly. Myelin, which contains several proteins including the basic protein described above, will dissolve completely in chloroform-methanol due to the large amount of lipid present in this membrane. This group of proteins may be characterized better by the fact that they can be isolated and purified in organic solvents, so that the protein in the absence of lipid is still soluble in chloroform-methanol. They constitute a major portion of the total protein of myelin (about 50%) and are found in plant, animal, and bacterial cells (Folch-Pi & Stoffyn, 1972). A homogeneous protein can be isolated from myelin proteolipid by chromatography on Sephadex LH-20 in acidified chloroform-methanol (Gagnon et al., 1971) and has been termed lipophilin because of its avidity for lipids (Moscarello, 1976). Other similar proteins that are fractionated by this column have not yet been characterized.

The chemistry of this group of proteins is much less advanced than that of the basic protein of myelin. This is due largely to solubility problems, so that the chemistry which has been used so successfully to unravel the sequences of many soluble proteins has yielded only minimal results in the case of proteolipid. In addition, the presence of an "impervious hydrophobic core" (Cockle et al., 1978a) in the molecule limits digestion of large portions of the protein by specific enzymes. Even when peptides have been isolated, they have not yielded to automatic sequence analysis because they are so hydrophobic that they are readily washed out of the cup of the sequenator. In spite of these difficulties, some progress has been made and this is discussed below.

The amino acid composition has been reported by several laboratories. The compositions reported for bovine white matter proteolipid (Folch & Lees, 1951) human myelin lipophilin [originally referred to as N-2 (Gagnon et al., 1971)], and

TABLE 2   Amino Acid Composition (Residues per 100) of Bovine White Matter Proteolipid, Lipophilin, and Rat Brain P7

Amino Acid	Bovine White Matter Proteolipid[a]	Human Myelin Lipophilin[b]	Rat Brain P7[c]
Aspartic	4.2	5.4	5.5
Threonine	8.5	8.2	8.5
Serine	8.5	5.9	6.0
Glutamic	6.0	6.4	6.5
Proline	2.8	3.1	3.0
Glycine	10.3	10.4	11.0
Alanine	12.5	10.6	12.0
Valine	6.9	6.8	8.5
Cystine	4.0	2.9	4.0
Methionine	1.9	1.6	1.5
Isoleucine	4.9	4.9	6.0
Leucine	11.1	11.2	12.0
Tyrosine	4.6	4.2	5.0
Phenylalanine	7.9	8.3	8.5
Lysine	4.3	4.7	5.0
Histidine	1.8	2.6	2.5
Arginine	2.6	2.9	3.0

[a]Folch & Lees (1951).
[b]Gagnon et al. (1971).
[c]Nussbaum et al. (1974a).

rat brain P7 (Nussbaum et al., 1974a) are shown in Table 2. Small differences in overall amino acid composition are noted in aspartic, serine, alanine, cystine, and histidine but are probably not significant.

The purity of the human lipophilin preparation of Gagnon et al. (1971) was monitored by end group analysis (in which a single N-terminal glycine, and a single C-terminal phenylalanine were found), polyacrylamide gel electrophoresis at two pH values, and sedimentation and equilibrium ultracentrifugation. It is evident from the composition that about 60–65% of the residues are apolar, accounting for its limited solubility in aqueous media. Assuming a molecular weight of about 25,000, of the 10 cysteines present only 1 or 2 were titratable. Thus the molecule must contain three or four disulfide bonds, which probably contribute to the intractability of the hydrophobic core (Cockle et al., 1980).

Sequence data have been difficult to obtain so that only partial sequences are available. Nussbaum et al. (1974a, 1974b) and Jollés et al. (1977, 1979) have published the most extensive sequence data to date. Jollés et al. (1979) have reported on a 31 amino acid N-terminal residue, a 3 amino acid C-terminal residue, and some cyanogen bromide cleavage products. For bovine brain proteolipid four cyanogen bromide fragments were isolated: CN-1, the N-terminal sequence, the first 23 residues were reported; CN-2, 18 residues reported; CN-3, 13 residues; CN-4, 6 residues believed to be the C-terminal sequence, a total of 60 residues. With a molecular weight in the vicinity of 25,000, it is clear that the largest portion of the molecule is yet to be sequenced. Partial sequence data have been reported by Chan and Lees (1978) for bovine white matter proteolipid. Human proteolipid (Nussbaum et al., 1974b) and lipophilin (Gagnon, 1976; Boggs & Moscarello, 1978a) have also been partially sequenced. The available data are summarized in Table 3.

An interesting posttranslational modification of these proteins is the presence of about 2% covalently attached fatty acid. The presence of this fatty acid was reported by Wood et al. (1971) and Gagnon et al. (1971). The fatty acids were mainly palmitic and oleic and were not a part of a complex lipid molecule. Analyses for glycerol in either acid or alkaline hydrolysates of the protein were negative. Neither phosphorus nor sphingosine was detected. All attempts to extract the fatty acid by extensive Soxhlet extraction in a variety of solvents or using the charcoal method of Chen (1967) for the removal of fatty acids from albumin were negative. It was concluded that the fatty acid was probably covalently bound.

A similar conclusion was expressed by Folch-Pi and Stoffyn (1972) concerning bovine proteolipids. They found 2.2–3.0% fatty acids in their preparations, which was not removed by extensive dialysis against chloroform-methanol. Choline, sphingosine, inositol, and hexosamine were not detected and glycerol was less than 0.03%. The fatty acids were converted to their methyl esters after alkaline hydrolysis. Subsequent reaction with diazomethane and sodium borohydride converted them to their corresponding alcohols. Folch-Pi and Stoffyn concluded that the fatty

TABLE 3   Amino Acid Sequence Data for Various Brain Proteolipids

Rat

(1)   Gly-Leu-Leu-Glu-Cys-Cys-Ala-Arg-Cys-Leu-Val-Gly-Ala-Pro-Phe-Ala-X-Leu-
          Val-Ala-Thr-Gly-Leu-X-Phe-Phe-Gly-Val-Ala-Leu-Phe

Bovine

(2)   Gly-Leu-Leu-Glx-Cys-Cys-Ala-Arg-X-Leu-Val-Gly-Ala-Pro-Phe

(3)   CN-1   Gly-Leu-Leu-Glu-Cys-Cys-Ala-Arg-Cys-Leu-Val-Gly-Ala-Pro-Phe-Ala-
                   Ser-Leu-Val-Ala-Thr-Gly-Leu

          CN-2   Tyr-Gly-Val-Leu-Pro-Trp-Asn-Ala-Phe-Pro-Gly-Lys-Val-X-Gly-Ser-Asn-
                   Leu

          CN-3   Ile-Ala-Ala-Thr-Tyr-Asn-Phe-Ala-Val-Leu-Lys-Leu-Met

          CN-4   Gly-Arg-Gly-Thr-Lys-Phe

Human

(4)   Gly-Leu-Leu-Glu-Cys-Cys-Ala-Arg-Cys-Leu-Val-Gly-Ala-Pro-Phe-Ala

(5)   Gly-Leu-Leu-Glu-Cys-Cys-Ala-X-X-Leu-Val-Gly-Ala-Pro-Phe-Ala-X-Leu-Val-
          Ala-X-Gly-Leu

(1) Nussbaum et al. (1974a, 1974b); Jollés et al. (1974).
(2) Lees et al. (1978).
(3) Jollés et al. (1979).
(4) Nussbaum et al. (1974b).
(5) Gagnon (1976); Boggs & Moscarello (1978a).

acids were esterified to the peptide chain, probably with a hydroxy amino acid such as threonine or serine.

The presence of fatty acids directly linked to proteins has been shown to occur in other membrane proteins as well. MacLennan et al. (1973) have isolated a proteolipid from sarcoplasmic reticulum Ca^{2+}-ATPase complex and have reported 1–2% fatty acid in their material. Covalently bound fatty acids have been reported also in murein lipoprotein of *E. coli* (Hantke & Braun, 1973).

Such covalently bound fatty acids would increase the overall hydrophobicity of the protein and may play some role in the orientation of the protein in the lipid bilayer. Interestingly, removal of the fatty acid with hydroxylamine from lipophilin rendered the protein even less soluble in aqueous media (Cockle et al., 1980).

2. Secondary and Tertiary Structures

It is not surprising that hydrophobic membrane proteins have a great tendency to aggregate in aqueous media. This has been observed for bovine proteolipid, for which a number of sedimentation values were obtained by ultracentrifugation (Folch-Pi & Stoffyn, 1972). The sedimentation boundaries were not symmetrical and more than one component was present (Nicot et al., 1973). Lipophilin from human myelin was shown to exist in three water-soluble forms, A, B, and C, which depended on the method used for preparation (Moscarello et al., 1973; Cockle et al., 1978b). The molecular weights were 86,000, 500,000 (Moscarello et al., 1973), and 79,000 (Cockle et al., 1978b), respectively. Since the molecular weights in 98% formic acid or 1% SDS were 25,000–30,000, the A and C forms were probably trimers while the B form was a large aggregate.

Circular dichroism studies of the three aqueous forms showed that form C and form A, respectively, contained 74 and 66% α-helical structure. Form B contained 53% α-helical structure and 29% β-structure (Moscarello et al., 1973; Cockle et al., 1978b).

Form A was readily converted to B by adopting the appropriate dialysis procedure, so that this protein showed considerable conformational flexibility (Moscarello et al., 1973). A highly α-helical structure was found for lipophilin in chloroform-methanol. Sherman and Folch-Pi (1970) reported optical rotary dispersion studies on the bovine proteolipid, which they found to be highly α-helical in chloroform-methanol but of low helical content (16–40%) when dialyzed into water.

The subunit molecular weight was found to be between 25,000 and 30,000 in 98% formic acid, 0.5% SDS, pH 7.4, in phosphate buffer (Moscarello et al., 1973) or in 2-chloroethanol (Cockle et al., 1978b) for lipophilin, by equilibrium ultracentrifugation. The bovine proteolipid showed multiple bands on SDS–polyacrylamide gels with a major band in the 25,000–30,000 molecular weight range (Chan & Lees, 1978). Nicot et al. (1973) were able to isolate two major components from the bovine proteolipid by preparative electrophoresis. The molecular weights were 25,100 and 20,700 respectively. Amino acid analysis showed some differences in composition, notably Thr, Glu, Ser, and Leu. Because their material was isolated from bovine brain white matter, Nicot et al. were not able to rule out contamination of their proteolipid by non-myelin proteolipid. A homogeneous proteolipid P7 has been isolated from rat brain myelin by Nussbaum et al. (1974a). On preparative gel electrophoresis they obtained three proteolipids of which P7 accounted for about 60%. They reported a total of 219 residues, corresponding to a molecular weight of about 25,000, in agreement with most laboratories. A monomeric molecular weight of 5000 reported by Chan and Lees (1974) has not been confirmed. From the number of peptides they obtained from tryptic digests of their protein, Chan and Lees (1978) later confirmed a molecular weight of 25,000–30,000.

3. Encephalitogenic Activity

With the isolation of basic protein, a potent encephalitogen, from central nervous system white matter, the disease-inducing activity of whole brain homogenate was attributed to the presence of basic protein. Until recently, it has been considered to be the only encephalitogen of the central nervous system, although the possible presence of other antigens was not ruled out.

In an early report, Waksman et al. (1954) described the induction of EAE with large doses of proteolipids. Contamination of this preparation with basic protein was considered to be responsible for the activity observed. Recently lipophilin has been demonstrated to have potent encephalitogenic activity and represents a second neural antigen. Hashim et al. (1980) showed that injection of 50 μg of either form A (or C) or B induced EAE in guinea pigs. The development of clinical signs was correlated with the presence of histological lesions characterized by a diffuse cellular infiltration different from the lesions induced by basic protein. A more pronounced delayed type hypersensitivity reaction, indicating cellular response, was elicited with lipophilin than with basic protein, indicating that contamination of the lipophilin preparation with basic protein did not account for the data. In a number of experiments, attempts to isolate basic protein from lipophilin failed; therefore the encephalitogenic activity was considered to be due to the lipophilin molecule itself.

IV. LIPID-PROTEIN INTERACTIONS

A. Proteolipid Protein

1. Reconstitution into Vesicles

The proteolipid apoprotein lipophilin has been incorporated into lipid vesicles by several methods: (1) by dissolving the lipid and protein in chloroform–methanol–water, 10 : 5 : 1, evaporating the solvent, and dispersing the dry material in buffer (Papahadjopoulos et al., 1975a), (2) by adding the water-soluble form to preformed unilamellar (Braun & Radin, 1969; Papahadjopoulos et al., 1973; Vail et al., 1974) or multilamellar vesicles, or (3) by dissolving the lipid and protein in 2-chloroethanol followed by dialysis against buffer (Boggs et al., 1976).

The protein can be readily incorporated into neutral or acidic lipids by methods

(1) and (3), and it forms intramembranous particles approximately 100 Å in diameter (Papahadjopoulos et al., 1975a; Boggs et al., 1976) that can be seen by freeze-fracture electron microscopy. Dialysis from 2-chloroethanol results in the most homogeneous incorporation of protein, judging from sucrose density centrifugation and the freeze-fracture appearance (Boggs et al., 1976). The water-soluble forms of the protein bind to a significant extent only to acidic lipids (method 2). The protein causes precipitation of unilamellar vesicles prepared from acidic lipids but not from neutral lipids, although some binding to neutral lipids has been demonstrated by sucrose density gradient centrifugation (Braun & Radin, 1969). Freeze-fracture electron microscopy has demonstrated the presence of intramembranous particles in the bilayer of acidic lipid vesicles to which the water-soluble form was added (Vail et al., 1974) (Fig. 3). No such particles can be seen when the water-soluble form is added to neutral lipid vesicles (W. J. Vail, J. M. Boggs, & M. A. Moscarello, unpublished observation). The water-soluble form has also been shown to have a high affinity for monolayers of acidic lipids, causing expansion of the monolayer (Papahadjopoulos et al. 1973), and a low affinity for neutral lipids, with the surprising exception of cholesterol (London et al., 1974). It has also been shown to prevent Ca^{2+} binding to acidic lipid monolayers (Minassian-Saraga et al., 1973). Thus electrostatic interaction plays an important role for binding of the water-soluble form to lipids, although once bound it may then become embedded in the hydrocarbon region, producing the intramembranous particles that can be seen by freeze-fracture electron microscopy.

Use of a low salt concentration (10 mM NaCl), when preparing vesicles of the neutral lipid phosphatidylcholine (PC) containing lipophilin by dialysis from 2-chloroethanol, resulted in less precipitation of the vesicles than when a high salt concentration (100 mM NaCl) was used. At the low salt concentration, the suspension became clearer with increasing protein concentration in the vesicles, and at 40% protein content (% of total weight) single layered vesicles of diameter 2000 Å were obtained (Boggs et al., 1976). This may be due to the charge that the protein gives the bilayer, causing interlamellar repulsion. Higher salt concentration would neutralize this charge and allow formation of multibilayers. Vesicles prepared from negatively charged lipids by this method formed clear suspensions, which were not easily sedimented. However, in the presence of protein, vesicles of acidic lipids precipitate even at 10 mM NaCl, probably because of aggregation and fusion of the vesicles due to interlamellar interactions.

2. Conformational Properties and Orientation in Vesicles

The single layered vesicles formed at low salt and high protein concentration with PC are particularly useful for CD measurements, for which samples giving low light scattering are necessary, and for labeling studies requiring single layered vesicles. Lipophilin incorporated into PC vesicles was found to have a high α-helical content (70–75%) by CD spectroscopy (Cockle et al, 1978b), as anticipated for intrinsic membrane proteins (Singer, 1971). The water-soluble trimeric form C of the protein, obtained by dialysis from 2-chloroethanol into distilled water, also

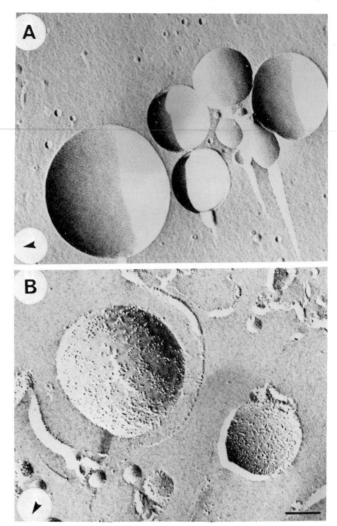

Figure 3. Freeze-fracture electron micrograph of (A) PS vesicles and (B) PS vesicles containing 46% of the water-soluble form A of lipophilin. Total magnification 75,000 ×, bar 0.1 μm. From Vail et al. (1974).

has a similar α-helical content, indicating that self-association via hydrophobic interactions as well as interaction with lipid is compatible with retention of α-helical conformation. Incorporation into vesicles, however, increased the average length of the α-helical segments relative to the protein in water and partially protected the protein from thermal denaturation. It was not possible to study the CD spectrum of the water-soluble forms of the protein when bound to lipid vesicles, since these forms bind only to acidic lipids and cause vesicle aggregation and precipitation. However, it was possible to study the effect of lysolecithin on the three water-soluble forms, and it was found to increase the total helical content and the average length of the helical segments of all three forms (Cockle et al., 1978b).

The appearance of intramembranous particles in the freeze-fracture electron micrographs suggested that the protein was embedded to some extent into the bilayer. To determine whether the protein spanned the bilayer completely, the protein in single layered vesicles of diameter 2000 Å (prepared by dialysis against 10 mM NaCl) was labeled with a covalent reagent (Wood & Moscarello, 1977; Wood et al., 1980) [³H]DIDS (4,4'-diisothiocyano-2,2'-ditritiostilbene disulfonic acid), which does not penetrate erythrocyte membranes (Cabantchik, 1974). However, because of the possibility that the reagent might penetrate the lipid-protein vesicles, the vesicles were also labeled with a large molecular weight (MW 250,000) dextran dialdehyde (Wood et al., 1980). After labeling, the labeled (and cross-linked) protein was separated from unlabeled protein by solubilization of the un-labeled protein in acidified chloroform-methanol. However, since these vesicles were formed by dialysis from 2-chloroethanol, the protein should be present on both sides of the bilayer. If it penetrates only partially into the bilayer, the protein on the inner monolayer would not be labeled. Nearly all of the protein was labeled by both reagents, suggesting that it completely spanned the bilayer. Since the orientation of the protein toward the inside or outside of the vesicles should be random, this result also indicates that the protein particle is exposed to the aqueous phase on both sides of the bilayer. Results of an X-ray diffraction study of these vesicles, using liquid diffraction techniques to identify the scattering domains of the protein, were in agreement with this model and suggested that lipophilin forms a spherical particle of 108 Å that completely spans the bilayer (Brady et al., 1979).

3. The Effect of Proteolipid on Lipid Organization and Formation of Boundary Lipid

A number of intrinsic membrane proteins have been reported to restrict the motion of lipid spin labels as originally observed by Jost et al. (1973a) for cytochrome oxidase. This lipid, which is restricted in motion on the ESR time scale, is probably localized at the interface between the protein and the bulk lipid, surrounding the protein and/or occupying the interstitial spaces between the polypeptide chains of the protein, and has been termed "boundary lipid" (Jost et al., 1973a). It is not surprising that lipophilin, with its very hydrophobic character and transmembranous location, also has boundary lipid. We have carried out a number of studies to investigate the properties of the boundary lipid of this protein more fully.

It was shown by ESR spectroscopy that lipophilin restricts the fatty acid motion of a population of the lipid to a frequency less than 10^7–10^8 s^{-1} while having a minor ordering effect on the bulk lipid as monitored by the motion of fatty acid spin labels (Boggs et al., 1976). The component characteristic of restricted motion could be first seen at a lipid to protein molar ratio of 150 : 1 and increased with increasing protein content in the vesicles. However, NMR spectroscopy, which is sensitive to a motional frequency slower than 10^5 s^{-1} does not detect the presence of a component with restricted motion in vesicles containing lipophilin (or a number of other intrinsic proteins) (Rice et al., 1979). This indicates that if spin labels are providing qualitatively accurate information, the boundary lipid is not long-lived and must be exchanging with the bulk lipid at a rate greater than 10^5 s^{-1} but less than 10^7 s^{-1}. For a discussion of the time scales in ESR and NMR, see Chapter 3, Section V.

The spatial orientation and degree of *gauche-trans* isomerizations of the boundary lipid cannot be deduced from the ESR spectrum of the component with restricted motion in isotropic samples. Lipid at the interface between the irregular surface of the protein and the regular liquid crystalline bulk lipid would be expected to be spatially disordered relative to the bulk lipid. Using cytochrome c oxidase samples oriented on a planar surface, Jost et al. (1973b) found no orientation dependence for the boundary lipid component of the spectrum, indicating a low degree of spatial order. ^{2}H-NMR spectra of vesicles containing intrinsic proteins including lipophilin and cytochrome oxidase were characteristic of even more motion than liquid crystalline phase lipid (Rice et al., 1979). Several interpretations are compatible with the ESR data, indicating restricted motion and low spatial order. The motion may be slowed down but the amplitude of the motion may be larger. Alternatively the boundary lipid may undergo more *gauche-trans* isomerizations but at a rate that is slow on the ESR time scale (Rice et al., 1979; Jost & Griffith, 1980).

Papahadjopoulos et al. (1975a) had shown earlier by differential scanning calorimetry (DSC) that lipophilin decreased the enthalpy of the gel to liquid crystalline phase transition without affecting the phase transition temperature. The size of the cooperative unit was only slightly decreased (Papahadjopoulos et al., 1975b), and the premelt transition was retained up to protein concentrations of 35% (Papahadjopoulos et al., 1975b; Boggs & Moscarello, 1978b). The decrease in enthalpy (ΔH) therefore suggested that a proportion of the lipid, probably the boundary lipid, did not participate in the gel to liquid crystalline phase transition. By plotting ΔH against the molar ratio of protein to lipid in the vesicles and extrapolating to ΔH = 0, an estimate of the amount of lipid that does not go through the phase transition could be obtained. This was found to be 15–30 molecules of lipid per monomer of lipophilin (Papahadjopoulos et al., 1975a; Boggs & Moscarello, 1978b; Boggs et al., 1980a).

The oligomeric state of lipophilin in the bilayer is not known. X-ray diffraction measurements (Brady et al., 1979) have shown that the protein forms a particle of 108 Å diameter in the bilayer. It can be estimated that a particle of this size could accomodate ~70 molecules of lipid around its perimeter. The boundary lipid estimate obtained by DSC of 15–30 molecules of phospholipid per monomer of lipophilin is consistent with an oligomeric association of 2–4 monomers (Boggs & Moscarello, 1978b). A model of lipophilin spanning the bilayer with its disordered layer of boundary lipid conforming to the irregular surface of the protein is shown in Fig. 4.

Lipophilin increased the permeability of vesicles to Na$^+$ and this increase was not inhibited by cholesterol as it was for other proteins (Papahadjopoulos et al., 1973). This suggested that the increase in permeability may not be through a disturbance of the lamellar portion of the bilayer but rather through weakened

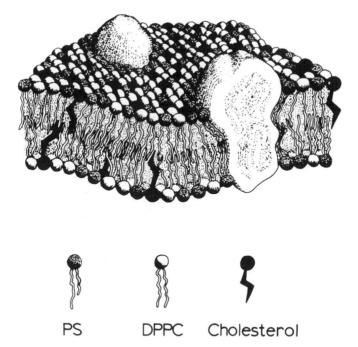

PS   DPPC   Cholesterol

Figure 4. Schematic representation of lipophilin spanning the lipid bilayer. A boundary layer of lipid surrounds the protein, with fatty acid chains disordered and conforming to the irregular surface of protein. In a mixture of lipids, acidic lipids such as PS are found preferentially in the boundary layer while cholesterol is probably partially excluded. From Boggs & Moscarello (1978a).

interactions at the lamellar lipid-boundary lipid interface or possibly through the protein itself. The water-soluble form of the bovine proteolipid apoprotein was recently reported to induce a voltage-dependent conductance in black lipid membranes (Ting-Beall et al., 1979).

The tryptophan fluorescence spectrum of lipophilin in dimyristoylphosphatidylcholine (DMPC) vesicles indicates that the tryptophan is in a hydrophobic environment and insensitive to the lipid phase transition (Cockle et al., 1978c). It was suggested that the phase transition may not be transmitted to the protein through the boundary layer. However, the tryptophan of the protein in solution was also in a hydrophobic environment, and it is possible that the tryptophan is buried too deeply in the interior of the protein to sense the lipid phase transition.

The decrease in enthalpy without change in phase transition temperature observed for lipophilin (from human myelin proteolipid) in vesicles by Papahadjopoulos et al. (1975a), and in a separate study by Boggs and Moscarello (1978b), was not in agreement with a DSC study by Curatolo et al. (1977) on bovine white matter proteolipid apoprotein. The latter study indicated that this protein incorporated into DMPC vesicles induced a new transition 2°C higher than that of the pure lipid. At a molar ratio of lipid to protein of 142 : 1 all of the lipid melted at the higher temperature, although with a reduced enthalpy. Curatolo et al. concluded that this higher melting lipid represented boundary lipid. Thus, this study concluded that the boundary lipid went through its own phase transition and at a value of 142 molecules of lipid per monomer of protein, included several layers of lipid surrounding the protein. The increase in T_c and decrease in ΔH suggested either that the boundary lipid was more ordered above its own phase transition and/or that it was more disordered than unaffected lipid below its phase transition. This possibility was not supported by the ^{2}H-NMR studies (also carried out on this protein) (Rice et al., 1979). However, Curatolo et al. (1978) reported Raman spectroscopic data that showed an increased *trans-gauche* ratio of the lipid fatty acid chains above the phase transition and a decreased *trans-gauche* ratio below it, in the presence of the protein. This was not confirmed in a later Raman spectroscopic study of this system by Lavialle and Levin (1980).

Curatolo et al. (1977) suggested that the discrepancy in DSC results with lipophilin (Papahadjopoulos et al., 1975a; Boggs & Moscarello, 1978b) and bovine proteolipid was due to the heating rates used. However, this has been ruled out. A second possibility might be chemical differences in the two proteins (although as discussed in Section II. B. 1 all properties studied indicate that the proteins are very similar), or to the method of delipidation of the protein, or the method of preparation of the vesicles. Lipophilin is fractionated and delipidated by chromatography on Sephadex LH-20 in acidified chloroform-methanol and has been shown to be a pure protein (Gagnon et al., 1971). In Curatolo's study, the proteolipid is delipidated according to the procedure of Folch-Pi and Stoffyn (1972) by dialysis against acidified chloroform-methanol but is not fractionated. Vesicles containing lipophilin were prepared by methods (1) (Papahadjopoulos et al., 1975a) and (3) (Boggs et al., 1976), whereas in Curatolo's study, chloroform-methanol solutions of the lipid and protein were evaporated directly into the DSC pans and hydrated with distilled water (Curatolo et al., 1977).

The effects of bovine white matter proteolipid and human myelin proteolipid on the phase transition of DMPC were compared in a recent study (Boggs et al., 1980a). The effects of methodological differences in delipidation of the protein and preparation of the vesicles were also investigated. The behavior of bovine proteolipid was similar to the behavior of lipophilin, regardless of the method of delipidation or preparation of the vesicles. Neither protein induced a higher melting peak. Incomplete delipidation resulted in a broadening of the DSC endotherm but did not shift the transition temperature.

However, when breakdown of DMPC to lyso-DMPC occurred (after several days storage of the vesicles), a new peak or shoulder on the original peak, approximately 3°C higher than the DMPC transition, appeared in the DSC scans. This new transition was observed in all the samples including the sample without protein, and the intensity increased with increasing protein content in the vesicles. Addition of as little as 6 mol % of the decomposition products of DMPC (i.e., equimolar amounts of lyso-DMPC and myristic acid to the vesicles) resulted in the appearance of a similar higher melting peak. The intensity of this new transition also increased with increasing amounts of protein (Boggs et al., 1980a). The effect of lyso-PC and myristic acid in increasing the lipid phase transition temperature had been reported previously (Papahadjopoulos et al., 1976a; Usher et al., 1978). Decomposition of phosphatidylcholine during the course of X-ray diffraction measurements on lipophilin in vesicles also was found to be the cause of an increase in lipid phase transition temperature (Rand et al., 1976). However, the increased effect of these compounds with increasing concentration of protein was not expected. The effect of the added compounds was similar to the effect of decomposition and also to the effect observed by Curatolo et al. (1977), not only in the appearance of a new peak somewhat higher than that of DMPC but also in the increased height of this peak with increasing protein concentration.

It has been suggested that motional restriction of fatty acid spin labels and decrease in enthalpy of the phase transition is due to lipid entrapment between protein aggregates, when proteins are squeezed into aggregates below the phase transition (Chapman et al., 1979). Lipophilin was reported to remain randomly dispersed in gel phase lipid as detected by freeze-fracture electron microscopy (Papahadjopoulos et al., 1975a). However, it does not have a noticeable broadening effect on the phase transition and does not abolish the premelt until the concentration in the vesicles reaches 35% (wt %). This corresponds to a mol ratio of lipid to protein monomer of 66 : 1, of which 25 are in the boundary layer and 41 are free. It could be argued that at this ratio of lipid to protein there might not be a sufficiently large lipid domain to allow for the high cooperativity of melting observed unless the protein were aggregated. The size of the cooperative unit for a lipid melting with a premelt transition and a fairly sharp main transition has been estimated to be 70–230 molecules (Papahadjopoulos et al., 1975b; Hinz & Sturtevant, 1972). If lipophilin is organized as an oligomer, there would be more free lipid. For example, if it is a trimer, the mole ratio of lipid to trimer at a concentration of 35% (wt %) is 200 : 1, of which 75 are in the boundary layer, leaving 125 molecules of free lipid per trimer. Thus retention of the premelt and the sharp main transition is compatible with a random distribution of protein trimers or oligomers in the vesicles at concentrations up to 35% (wt %).

The freeze-fracture appearance of these vesicles above and below the phase transition of DMPC has been reinvestigated recently (Boggs et al., 1980a). The protein was always randomly located above the phase transition (Fig. 3 and Papahadjopoulos et al., 1975a; Boggs et al., 1976; 1980a; Brady et al., 1979), but if the samples were left in the cold for a prolonged time, large areas of aggregated protein were seen below the phase transition. In freshly prepared samples, however, the protein was for the most part randomly distributed below the phase transition as well, although clusters of several particles did occur (Fig. 5A). Clustering of particles occurred especially at temperatures between the premelt and the main transition. At these temperatures, the ridge pattern of the lipid can be seen. As shown in Fig. 5B, the protein particles are excluded from the areas of regular ridge pattern and located at discontinuities in the fracture plane of the bilayer, where the ridge pattern changed direction. It is difficult to say whether these discontinuities occurred to a greater extent because of the presence of the protein or whether the protein migrated to the discontinuities. However, such discontinuities also exist in the ridge pattern of the pure lipid. At higher protein concentrations, the protein

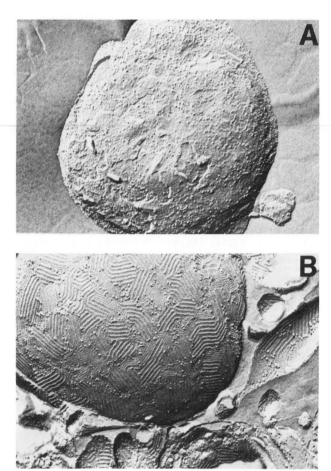

Figure 5. Freeze-fracture electron micrographs of vesicles of DMPC with (A) 21% lipophilin quenched from 4°C, below the premelt and main transition temperatures (magnification 81,900 ×), and (B) 17.5% lipophilin quenched from 19°C, between the premelt and main transition temperatures.

decreased the height of the premelt transition, distorted the ridge pattern, and resulted in less clustering of the particles (Boggs et al., 1980a).

It cannot be ruled out that the decrease in ΔH observed by DSC, showing that some lipid is withdrawn from the phase transition, is due to entrapment of lipid between particles of protein below the phase transition. Therefore the effect of lipophilin on the enthalpy of DMPC has also been measured from cooling as well as heating scans (Boggs et al., 1980a). The amount of lipid withdrawn from the phase transition from the liquid crystalline to the gel state on cooling was exactly the same as the amount withdrawn from the transition from the gel state to the liquid crystalline state on heating. This supports the notion that the withdrawal of lipid from the phase transition is not due to entrapment and that even though the protein becomes clustered below the phase transition, it is randomly dispersed above the phase transition and carries its boundary lipid with it.

The dimensions of the lipid fatty acid binding sites of lipophilin and the adaptability of the protein to bilayers of different thicknesses were investigated by measuring the dependence of the amount of lipid withdrawn from the phase transition on fatty acid chain length using phosphatidylcholine with fatty acids of lengths C12, C14, C16, and C18. The effect of a *trans* double bond was also studied using C18:1$_{tr}$ (Boggs & Moscarello, 1978b). The amount of boundary lipid was found to be similar, 21–25 molecules per monomer of lipophilin, for fatty acid chains of length 14–18, but somewhat less, 16 molecules, for C12 and C18:1$_{tr}$. No significant preferential interaction was observed with any particular lipid when the protein was incorporated into a mixture of these lipids. This was true for cocrystallizing and noncocrystallizing mixtures with the possible exception of dilauroylphosphatidylcholine (DLPC)–distearoylphosphatidylcholine (DSPC), for which the protein decreased the enthalpy of the DSPC peak more than the DLPC peak. With this exception, in a mixture of lipids the protein bound preferentially neither to a fatty acid of a particular chain length nor to the lower melting lipid to a significant extent. This is in contrast to cholesterol (de Kruijff et al., 1974) and gramicidin (Chapman et al., 1977) both of which, when added to a mixture of phosphatidylcholines, decreased the enthalpy of the lower melting lipid first.

A study of the effect of unsaturated fatty acids on the phase transition temperature of DMPC in the presence and absence of the bovine proteolipid apoprotein showed that unsaturated fatty acids lowered the phase transition temperature of DMPC, and this effect was reduced in the presence of the protein (Verma et al., 1980). It was concluded that in mixtures of unsaturated fatty acids and PC containing saturated fatty acids, the unsaturated fatty acids (with *cis* double bonds) bind preferentially to the protein, decreasing their concentration in the bulk lipid. This protein has been found to increase the T_c of egg PC by 17°C (Curatolo et al., 1978), and it was implied that the protein may bind more of the more unsaturated phosphatidylcholines relative to the more saturated ones. However, the results with unsaturated fatty acids do not necessarily mean that the protein will bind preferentially to phospholipids containing *cis*-unsaturated fatty acids. Our results with lipophilin in a mixture of a saturated fatty acid (myristic acid), lyso-DMPC, and a lipid containing saturated fatty acids (DMPC) suggest that the free fatty acid and/or the lyso-DMPC may be partially excluded from the boundary layer even though all components in the mixture contain saturated fatty acids. Thus the binding of free fatty acid and phospholipid to the protein may differ.

4. Preferential Binding of Acidic Lipids

Although the binding of boundary lipid to intrinsic membrane proteins is probably primarily through hydrophobic interactions, there is evidence that lipophilin can also interact electrostatically with acidic lipids as mentioned earlier. The proteolipid is difficult to delipidate; acidified chloroform-methanol must be used and the lipids that are most firmly bound are acidic lipids (Folch-Pi & Stoffyn, 1972). When the proteolipid was dialyzed against neutral chloroform-methanol, some acidic lipid, mostly phosphatidylserine (PS), remained bound to the protein, and the amount was equivalent to 35% of the total basic amino acids in the protein, or approximately 5 molecules of lipid per monomer of protein (Uda & Nakazawa, 1973). Braun and Radin (1969) extracted the complex formed between the water-soluble protein and PS vesicles with neutral chloroform-methanol and found that a portion of the PS corresponding to 30% of the basic amino acids could not be removed.

This suggests that the proteolipid may bind acidic lipids *in situ*. Although the amount of boundary lipid (estimated from the decrease in enthalpy) was found to be identical for vesicles of either a neutral lipid, dipalmitoylphosphatidylcholine (DPPC) or an acidic lipid, dipalmitoylphosphatidylglycerol (DPPG) (Papahadjopoulos et al., 1975a), it was considered possible that the protein might bind the acidic lipid preferentially when allowed to interact with a mixed bilayer containing both acidic and neutral lipids. This was demonstrated using DSC, utilizing the observation that some lipid is removed from participation in the phase transition (Boggs et al., 1977a). Vesicles containing binary combinations of PS and PC, phosphatidic acid (PA) and PC, or phosphatidylglycerol (PG) and PC were prepared with and without lipophilin. The lipids in the absence of the protein were demonstrated to be fairly randomly mixed, since for each mixture only a single peak, at temperatures intermediate between the two pure lipids, was observed. Lipophilin caused a shift in the transition temperature of the mixture in the direction of the PC component, indicating that it had caused an enrichment of the PC content of the bulk lipid, presumably through preferential binding of the acidic lipid. This was the case regardless of whether the acidic lipid was the higher or lower melting component in the mixture. A phase diagram was constructed for the PS-PC mixtures from which the amount of PS in the boundary layer could be estimated. This was found to be approximately 15 molecules of PS per monomer of protein at an equimolar ratio of PS to PC. Some PC was also found in the boundary layer but the ratio of PS to PC in the boundary layer was higher than the initial ratio in the vesicles. The results described above indicating that 5 molecules of acidic lipid per monomer of protein cannot be removed by extraction with neutral chloroform/ methanol suggest that a portion of the acidic lipid is bound even more tightly than the rest.

These results suggested that many of the 14 basic amino acid residues can be involved in electrostatic interaction with the acidic lipid polar head groups in the boundary layer, whereas the lipid fatty acid chains are motionally restricted by hydrophobic interactions with the protein. The protein has been only partially sequenced (Table 3), but so far there is no evidence that it is an amphipathic protein. If the basic residues are located randomly in the sequence, the protein may be able to take on a tertiary conformation in the vesicles such that the basic residues are located at the bilayer-aqueous interface, available for interaction with the acidic lipid polar head groups. This is the simplest model that can explain the observed effect, and it assumes that there are no gross distortions of the bilayer structure induced by the protein.

The strong interaction of the water-soluble form of bovine proteolipid with cholesterol monolayers found by London et al. (1974) is surprising in view of the

high affinity of both this protein and lipophilin for acidic lipids and the low affinity of the water-soluble forms for neutral lipids. This result has not been substantiated by a recent study of the effect of the water-soluble form of lipophilin on the surface tension of lipid monolayers. Although only neutral lipids were used, the protein was found to interact with sphingomyelin to some extent, with cerebroside slightly, and with cholesterol not at all (Katona et al., 1978).

This discrepancy may be due to the different methods of preparation or delipidation of the protein. When a steroid spin label was used to monitor the amount of cholesterol that might be in the boundary layer (Boggs et al., 1976), an immobilized component was present in the spectrum of vesicles containing lipophilin. The amount of this component did not appear to be as great as that when a fatty acid spin label was used, suggesting that steroids do not bind to the protein quite as well as fatty acids. However, spectral resolution and integration of the immobilized component is necessary to substantiate this conclusion. A more flexible molecule such as a fatty acid may be required to interact with the irregular surface of the protein. Other intrinsic proteins have been reported to exclude cholesterol from the protein-lipid interface (Bieri & Wallach, 1975; Warren et al., 1975). Preferential binding of acidic phospholipids and partial exclusion of cholesterol at the protein-lipid interface of lipophilin appear to be reasonable conclusions that are represented in the model in Fig. 4.

B. Basic Protein

1. *Electrostatic Interactions with Lipids*

The high affinity of basic protein for acidic lipids was first demonstrated by Palmer and Dawson (1969), using a biphasic solvent system to measure complex formation. Acetylation of some of the lysines in the protein decreased the number of lipid molecules bound by a proportionate amount that was consistent with the electrostatic nature of the interaction (Steck et al., 1976). A strong interaction with acidic lipids was also demonstrated using lipid monolayers (Demel et al., 1973) and vesicles (Gould & London, 1972).

This protein can be incorporated into vesicles by methods (2) and (3) discussed for lipophilin (Palmer & Dawson, 1979; Gould & London, 1972; Papahadjopoulos et al., 1973; Papahadjopoulos et al., 1975b; Boggs & Moscarello, 1978c), although its water solubility makes method (2)—addition in aqueous solution to preformed unilamellar vesicles or to multilamellar vesicles—the most useful method. Addition of the protein to unilamellar vesicles causes vesicle aggregation, increases the light scattering of PC vesicles (Smith, 1977b), and precipitates vesicles of acidic lipids. It may also cause fusion into multilamellar vesicles, although this has been examined by electron microscopy only for PC (Smith, 1977b) and phosphatidylethanolamine (PE) (Stollery & Vail, 1977), as shown in Fig. 6 for PE. Addition to multilamellar vesicles retained the multilayered structure as shown by X-ray diffraction and electron microscopy (Brady et al., 1980; J. M. Boggs, I. R. Clement & M. A. Moscarello, unpublished results).

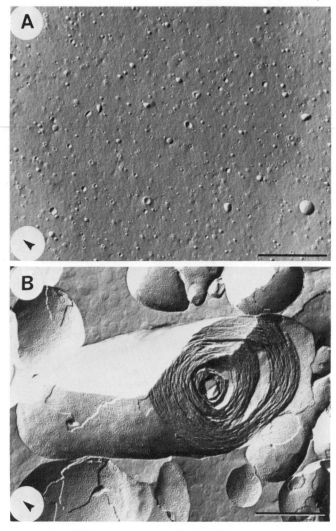

Figure 6. Freeze-fracture electron micrographs of small unilamellar vesicles of PE (*A*) before and (*B*) after addition of myelin basic protein and incubation at 37°C. J. G. Stollery & W. J. Vail, unpublished micrograph.

The interaction with PC and PE is weaker than with acidic lipids, and indeed no complex formation occurred with these two lipids in a biphasic solvent system (Palmer & Dawson, 1969). The basic protein caused precipitation of PE vesicles, however, but not of PC vesicles. The interaction with PC, which can be demonstrated by measurement of the surface pressure of monolayers (Demel et al., 1973) as well as of the light scattering of vesicles (Smith, 1977b), was most significant at higher pH values where the protein begins to aggregate. At pH 8.9, PC vesicles bind 18.1% basic protein (wt %), whereas at pH 6.4 they bind only 2.5% protein

(Smith, 1977b). PE vesicles bind a maximum of 30% protein at pH 7.4, although other acidic lipids bind up to at least 50% (Boggs & Moscarello, 1978c).

Basic protein has 31 positively charged residues (Lys, Arg) and 11 negatively charged residues (Asp, Glu) per mol of protein at physiological pH. Using a biphasic solvent system, it was demonstrated that the molar ratio of protein to monovalent negatively charged lipid in the water-insoluble complex was 1 : 23 (Palmer & Dawson, 1969) or in another study, 1 : 17 (Steck et al., 1976). This may represent the amount of lipid necessary to neutralize the net positive charge of the protein but may not indicate how many basic residues can actually be involved in electrostatic binding to lipids in bilayer form. This number was determined recently to be 26 ± 5 by measuring inhibition of Mn^{2+} binding to the lipid by the protein (Boggs et al., 1981a). The free Mn^{2+} concentration was measured from the intensity of its ESR spectrum. Thus nearly all the 31 positively charged residues can bind to negatively charged lipids. Four of the basic residues in the protein occur adjacent to acidic residues in the sequence, and intramolecular ionic interactions may occur for these four residues rather than interaction with lipid.

2. Hydrophobic Interactions with Lipid

Although basic protein is water soluble, it has 52% apolar amino acids, including 24% hydrophobic amino acids. These are indicated in the sequence shown in Fig. 7 and are arranged in segments of five to nine amino acids long, as originally pointed out by Eylar et al. (1971a). There is abundant evidence that these hydrophobic segments can interact with the lipid fatty acid chains either by partially penetrating into the bilayer as drawn in Fig. 8, or by deforming and expanding the bilayer such that the fatty acid chains are exposed to the hydrophobic regions of the protein as drawn in Fig. 9.

The protein causes an increase in the permeability of lipid vesicles (Gould & London, 1972; Papahadjopoulos et al., 1973, 1975b), increases the surface pressure or expands the area of lipid monolayers (Papahadjopoulos et al., 1973; Demel et al., 1973) and decreases the phase transition temperature, enthalpy, and cooperativity of melting of some lipids (Papahadjopoulos et al., 1975b; Boggs & Moscarello, 1978c; Boggs et al., 1980b, 1981a). In contrast, a completely basic protein such as the synthetic polypeptide polylysine has little effect on permeability (Papahadjopoulos et al., 1975b) or monolayer packing (Demel et al., 1973), but increases the phase transition temperature and enthalpy of melting (Papahadjopoulos et al., 1975b). Basic protein was reported in one study to increase the phase transition temperature of DLPG (Verkleij et al., 1974). This lipid does not form stable bilayers, as indicated by its inability to trap Na^+ (van Dijck et al., 1975), unlike PG with longer fatty acid chains. Hence basic protein may stabilize DLPG, resulting in an increase in its phase transition temperature while it decreases the phase transition temperature of DMPG and DPPG.

Specific regions of the protein are protected from tryptic hydrolysis when it is bound to lipid vesicles or monolayers, suggesting that these regions are sequestered from the aqueous phase by penetration into the bilayer (London & Vossenberg,

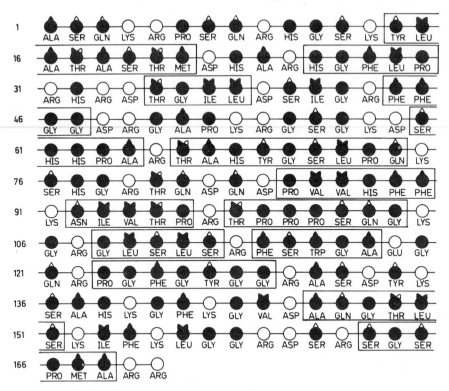

Figure 7. Amino acid sequence of human myelin basic protein indicating amino acids that are neutral (●) or charged (○) at pH 7.4. The presence of hydrophobic groups (▲) or OH groups (△) is also indicated. [Gln and Asn were incorrectly indicated as charged amino acids in our earlier publication of this sequence (Boggs & Moscarello, 1978a).] Boxes enclose regions of four or more amino acids uninterrupted by charged amino acids.

1973; London et al., 1973). A study of the effect of the protein on fatty acid spin labels with the nitroxide at different positions along the chain showed that it increases the order parameter of PA near the polar head group region while having little effect in the interior of the bilayer (Boggs & Moscarello, 1978c). This is consistent with the concept of partial penetration of some regions of the sequence into the bilayer. On the other hand, polylysine decreases the mobility of fatty acids even in the interior of the bilayer (Boggs et al., 1981a). Observation of the ^{13}C-NMR spectrum of PG indicated that basic protein may also restrict the motion of the glycerol head group, possibly through steric hindrance and/or hydrogen bonding with the glycerol hydroxyls, although the possibility that this effect is due to aggregation and slower motion of the vesicles has not yet been ruled out (Stollery et al., 1980a).

The models of basic protein with hydrophobic segments penetrating into or expanding the bilayer (Figs. 8 and 9) are the simplest models consistent with the effects of the protein on lipid organization. However, other models are also possible; for example, the protein may take on a tertiary conformation much different from

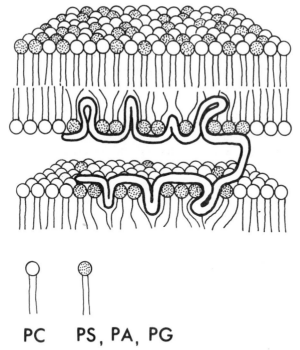

PC   PS, PA, PG

Figure 8. Schematic representation of myelin basic protein bound to lipid bilayer. Protein is shown bent at Pro-Pro-Pro region in center of sequence, with each half interacting with adjacent bilayers. However, it is probably just as likely that both halves of the protein interact with the same bilayer. Hydrophobic segments are shown penetrating into the lipid bilayers. In a mixture of lipids, acidic lipids such as PS, PA, and PG are separated out and bound to the protein as indicated.

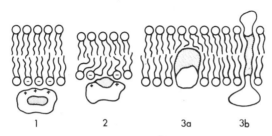

Figure 9. Schematic representation of different types of lipid-protein interaction: (1) electrostatic only, (2) electrostatic plus hydrophobic (shaded) resulting in deformation of the bilayer, (3a,b) hydrophobic interaction. The interaction of basic protein with lipid may be of type 2 while lipophilin may be of type 3. From Papahadjopoulos et al. (1975b).

its conformation in solution, or the protein may greatly distort the lipid bilayer. The protein might interact as shown with two bilayers or with only one bilayer, as discussed later.

Measurements of tryptophan fluorescence of basic protein in the presence and absence of sodium dodecylsulfate (SDS) indicated that the tryptophan, which is in a hydrophobic segment of the protein, moved into a less polar environment when the protein was bound to SDS (Jones & Rumsby, 1975). Similar results were found for the protein bound to lipid vesicles (J. G. Stollery, R. M. Epand, J. M. Boggs, & M. A. Moscarello, unpublished results). Interaction of the protein with SDS also broadened the ¹H-NMR resonance lines of the aliphatic and aromatic residues, suggesting hydrophobic interaction of these residues with the detergent micelle (Liebes et al., 1976).

Penetration of some segments of the protein into the bilayer would require some folding of the linear sequence. As mentioned earlier, basic protein appears to have a hairpinlike structure with dimensions 15×150 Å and an open or random conformation. Some folding of the protein must occur even in solution, since an extended sequence of 70 and 100 amino acids on each side of the tri-proline sequence should have a length of 245–350 Å. Calculation of the probability of β-bends in the molecule indicates a high probability $(>10^{-4})$ at many sites throughout the molecule, with a particularly high probability $(>4 \times 10^{-4})$ at seven sites in the molecule (Epand et al., 1974). NMR spectroscopy was indicative of molecular folding in the region of residues 85–116, since the phenylalanine and tryptophan resonances were broadened (Littlemore, 1978; Chapman et al., 1977).

Tryptophan fluorescence indicated that this residue was in a polar environment (Jones & Rumsby, 1975), although protected to some degree from quenching of its fluorescence (Feinstein & Felsenfed, 1975). Dimerization and association of this protein into water-soluble oligomers occurred at pH greater than 6 (Smith, 1980) and may be due to hydrophobic interaction. X-Ray diffraction studies, using liquid diffraction techniques, of the protein bound to PG vesicles suggested that its dimensions were similar to those in solution (G. W. Brady & M. A. Moscarello, unpublished results). However, CD measurements showed that some conformational changes did occur upon binding of the protein to detergents (Anthony & Moscarello, 1971; Smith, 1977a; Liebes et al., 1976) or lipid vesicles (Keniry & Smith, 1979). Analysis of the CD spectrum of the protein bound to PS and PA vesicles revealed 20% α-helical structure, 12% β-structure, and 45% random structure (Keniry & Smith, 1979), while the protein in solution has no α or β character. In contrast, interaction with PC vesicles did not result in the formation of a significant amount of α-helical or β-structures.

Comparison of the behavior of various large fragments of basic protein with the intact protein can help to understand which regions of the protein are involved in the interaction with lipids and the importance of the intact tertiary structure of the protein to these interactions. For example, the protein can be conveniently cleaved at the single tryptophanyl residue at position 116 with 2-(2-nitrophenylsulfenyl)-3-methyl-3'-bromoindolenine-skatole yielding a large N-terminal fragment (residues 1–116) and a smaller C-terminal fragment (residues 117–170). Several studies with these two fragments have led to the conclusion that their behavior was quite different from that of the intact protein (London et al., 1973; Jones & Rumsby, 1977), even though the proportion of positively and negatively charged apolar and hydrophobic amino acids is similar in the two fragments at pH 7.4. The large fragment had a substantially greater effect on the surface pressure of a lipid monolayer than the small fragment, when compared on a molar basis. The regions of the protein protected from tryptic hydrolysis were all in the N-terminal portion of the protein (London et al., 1973). Therefore it was concluded that the principal sites for hydrophobic interaction were in the N-terminal portion. However, it has been recently shown that when compared on the basis of equal moles of amino acid rather than equal moles of peptide, the large fragment and small fragment had an effect qualitatively similar to that of the intact protein on the lipid phase transition temperature and fatty acid mobility, although the effect of the small fragment was quantitatively not quite as great (Boggs et al., 1981a). Furthermore if the effects of the two fragments on surface pressure of monolayers described above (data from London et al., 1973) are compared on the basis of equal moles of amino acids, the two fragments have nearly the same effect on surface pressure as well. Thus both portions of the protein can participate in hydrophobic interactions with the lipid bilayer as shown in Fig. 8.

α-Helical structure was induced in both fragments upon binding to lipid, indicating that intact structure around the tryptophan is not required (J. G. Stollery, R. M. Epand, J. M. Boggs, & M. A. Moscarello, unpublished results). However, cleavage at the Phe-Phe linkage (residues 89 and 90) produced fragments whose ellipticity did not change significantly upon interaction with lipid, even though both fragments bound to the vesicles, suggesting that a major structure forming region was disrupted by this cleavage (Keniry & Smith, 1979).

The ability of the two fragments formed by BNPS-skatole cleavage to interact electrostatically with lipids has been studied also. Using a biphasic solvent system (Jones & Rumsby, 1977), it was found that only the smaller C-terminal fragment and the intact protein formed water-insoluble complexes with lipid and were carried into the lower organic phase. The larger fragment remained in the upper water-methanol phase and did not carry any lipid into the upper phase, leading to the conclusion that the sites for ionic interaction with lipid were localized in the C-terminal third. However the positively charged residues are distributed evenly throughout the molecule, 20 in the N-terminal two thirds, and 11 in the C-terminal one third. The intact protein inhibited Mn^{2+} from binding to 26 ± 5 phospholipid molecules as mentioned earlier, suggesting that almost all the positively charged residues bind to lipid. Inhibition of Mn^{2+} binding by the two fragments was also investigated, and it was found that the large fragment inhibited Mn^{2+} from binding to 17 ± 3 lipid molecules and the small fragment inhibited Mn^{2+} from binding to 11 ± 1.5 molecules of lipid (Boggs et al., 1981a). It appears then that both portions of the molecule can interact hydrophobically and electrostatically with lipid vesicles to similar extents, as might be expected from the random and equal distribution of hydrophobic and positively charged amino acids throughout the sequence. There is no unique sequence of amino acids in the molecule that allows hydrophobic interaction with the bilayer and its consequent effects on lipid organization. Cleavage at the tryptophan does not seem to disrupt any tertiary structure that is involved in these interactions with lipid, although cleavage at the Phe-Phe linkage may do so.

3. Dependence of Hydrophobic Interactions of Protein on the Type of Lipid and Its Physical State

There is evidence that the degree of hydrophobic interaction of basic protein with the lipid bilayer depends on both the phase state of the lipid (gel or liquid crystalline phase) and the type of lipid. Papahadjopoulos et al. (1975b) showed that the protein decreased the phase transition temperature of DPPG only when it was allowed to equilibrate with lipid in the liquid crystalline phase. Prolonged cooling reversed the effect. It was shown subsequently that the protein had a much greater effect on cooling scans of the phase transition than on heating scans, even when the sample was cooled rapidly and the heating scan immediately repeated, indicating that even brief cooling partially reverses the effect (Boggs et al., 1978c). These results suggested that the protein can interact hydrophobically with the expanded liquid crystalline phase and that this interaction is partially reversed upon freezing of the lipid. The protein also induced an exothermic transition in the heating scan just before the main endothermic transition indicating a transition from an unstable to a more stable phase (Boggs et al., 1980b). A clue to the nature of these phases comes from examining the effect of the protein on fatty acid motion in the gel phase.

As mentioned earlier, when lipids are in the liquid crystalline phase, basic protein increases the order parameter of fatty acid segments near the polar head group to a greater extent than the terminal methyl portions. The protein has a greater effect on the terminal methyl portions of PG than of PA, however. Furthermore, when DPPG is the gel phase, basic protein has a very pronounced immobilizing effect on the terminal methyl portions of the fatty acids and a much smaller immobilizing effect on the fatty acid segments near the polar head group (which are already fairly immobilized in the gel phase) (Boggs & Moscarello, 1978c; Boggs et al., 1980a). This might be explained by interdigitation of the fatty acid chains of DPPG in the gel state in the presence of basic protein (Boggs et al., 1981a; 1981c; 1982; Boggs & Moscarello, 1982). When the lipid freezes, the chains cannot pack together in a very stable arrangement due to interference from the hydrophobic segments of the protein. If the sample is heated to a temperature a few degrees below the normal lipid T_c, the chains slip into a more stable state by interdigitating. This gives rise to the exothermic transition seen by DSC. During this process the hydrophobic segments may be frozen out of the bilayer to some extent, since the sample then melts at almost the same temperature as the pure lipid T_c. In the liquid crystalline phase when the lipid is expanded, increased hydrophobic interaction of the protein can occur, and there is less interdigitation of the fatty acid chains and less effect of the protein on fatty acid motion in the interior of the bilayer.

Basic protein also decreased the phase transition temperature of DMPA, but the effect was only slightly greater on cooling than heating, suggesting that the hydrophobic interaction with this lipid is not reversed on freezing. The protein has only a slight immobilizing effect on DMPA in the gel phase, in contrast to its effect on DPPG. It also does not induce an exothermic transition in the DSC heating scan of DMPA (Boggs et al., 1980b).

Thus, the hydrophobic interaction of basic protein with lipids depends on the phase state for at least one lipid, DPPG, but not for DMPA. Differences in behavior even greater than between these two are seen with other acidic lipids. Variable effects on different lipids were first reported by London and colleagues (Demel et al., 1973; London & Vossenberg, 1973; London et al., 1973). These investigators found that the effect of the protein on the surface pressure of lipid monolayers at pH 5 decreased in the order cardiolipin > myelin acidic lipid extract > cerebroside sulfate >> total myelin lipid extract = PS = PE >> neutral lipids (Demel et al., 1973), suggesting that the degree of hydrophobic interaction decreases in the same order. Consistent with this interpretation, binding to PS vesicles did not protect the protein from enzymatic hydrolysis, although cerebroside sulfate did (London & Vossenberg, 1973). However, another study by this group indicated that, although the protein expanded monolayers of different lipids to a varying extent, the protein was partially protected from enzymatic hydrolysis by all the lipids (cerebroside sulfate, acidic lipid extract, PS, cerebroside, and PC). Tryptic hydrolysis caused loss of only about half the radioactive sites of the protein from the monolayer and loss of half its effect on the surface pressure (London et al., 1973). This study was carried out at pH 8, however.

Basic protein did not decrease the phase transition temperature or affect the fatty acid motion of all lipids to the extent of DPPG and DMPA described above. The effect on the phase transition temperature and fatty acid motion decreased in the order PG $\simeq$ PA > PS > cerebroside sulfate $\simeq$ PE (Boggs & Moscarello, 1978c; Stollery et al., 1980b). These results are consistent with the conclusion of London and colleagues that the hydrophobic interaction varies with different lipids. The order of degree of hydrophobic interaction was found in our studies (at pH 7.4) to be PG $\simeq$ PA > PS > cerebroside sulfate > PE.

The concentration of protein necessary to observe these effects on lipid organization is quite high, since the techniques used above observe unperturbed lipid as well as perturbed lipid. The results suggest that the conformation or environment of the protein varies with the phase state and type of lipid, and it is important to study the protein directly. This has been done by enriching the protein with ^{13}C or covalently spin labeling it by alkylation of the two methioninyl residues at positions 21 and 167 with [^{13}C]methyl iodide (Deber et al., 1978; Stollery et al., 1980a) or a spin label derivative of iodoacetamide (Boggs et al., 1980c; Stollery et al., 1980c), and studying the interaction of the protein with different lipids using ^{13}C-NMR or ESR spectroscopy. It was possible to use the spin-labeled protein at low concentrations (4 wt %) as a probe for the lipid in the microenvironment of the protein. However, high concentrations (29–36%) of the ^{13}C-labeled protein were necessary in order to observe the signal.

Changes in motion of the spin-labeled protein throughout the phase transition were monitored using three lipids with which the protein interacted differently,

DPPG, DMPA, and DMPE (Boggs et al., 1980b). The spin-labeled methionines were in a polar environment but were sensitive to the phase transition for all three lipids on heating scans. This probably reflects the motion of the lipid polar head groups during the phase transition. The phase transition of DPPG, as sensed by the labeled methionines, was much broader and at a lower temperature on cooling than on heating, in agreement with the DSC results described earlier. This indicated that the hydrophobic interaction of other regions of the protein perturbed the lipid in the vicinity of the methionines so that they no longer detected a distinct phase transition. The phase transition in DMPE was seen distinctly on both heating and cooling, in agreement with the conclusion that little hydrophobic interaction occurred with this lipid. In DMPA the phase transition was broader than in DMPE but not quite as broad as in DPPG; it was similar on heating and cooling, in agreement with the DSC results, indicating that the hydrophobic interaction with this lipid was not reversed on cooling and that the lipid in the vicinity of the methionines was perturbed as with DPPG. The motion of the methionine region of the ^{13}C-enriched basic protein also increased above the phase transition, although the effect was not studied in detail (Stollery et al., 1980a).

The motion of the spin-labeled methionines was restricted to a varying extent by different lipids (Stollery et al., 1980b). The restriction in motion may have been due to perturbation and restriction of the motion of the lipid in the vicinity of the methionine spin labels due to hydrophobic interaction of other portions of the protein as well as to a deeper location within the polar head group region of the methionine spin label. If the lipids were compared at the same temperature, their motional restricting effects on the protein decreased in the order PG > PA > PS > cardiolipin $\geq$ PE. Since the fluidity of these lipids varied, the motion of the spin-labeled protein was also measured at temperatures where all the lipids possessed equal order parameters as measured with a fatty acid spin label. The motional restricting effect decreased in the order PG $\geq$ PA > cerebroside sulfate $\geq$ PS > cardiolipin > PE. This resembles the order of effect of the protein on fatty acid motion and phase transition temperature of the lipids, except for PS and cerebroside sulfate. The results with these two lipids are similar to each other and indicate much less hydrophobic interaction than with either PA or PG.

These studies suggested that the conformation of the protein and/or exposure of its antigenic determinants on the surface of the bilayer might also vary with lipid composition. This was tested by measuring antibody-induced precipitation of lipid vesicles containing [^{125}I]basic protein bound to the outside (Boggs et al., 1981b). Precipitation increased in the order PA < PS < PG < cerebroside sulfate < PE $\leq$ protein in solution, in fairly good agreement with the biophysical studies indicating decreasing degree of hydrophobic interactions in the order PA > PG > PS > cerebroside sulfate > PE. However, more antibody-induced precipitation of basic protein-containing PG vesicles was obtained than expected on the basis of the biophysical studies.

The similar interactions of basic protein with PS, cerebroside sulfate, and cardiolipin found in these studies do not agree with the results of London and colleagues, who have found evidence of substantially greater hydrophobic interaction with cerebroside sulfate and cardiolipin than with PS. Some of this discrepancy may be due to the pH used for the study and to different sources and method of preparation of the lipids. Different techniques also may not be equally sensitive to the interaction of a particular region of the protein with the lipid, or they may be subject to other interpretations.

Differences in the hydrophobic interactions of basic protein with acidic phospholipids must be due primarily to differences in the lipid polar head groups. Although PG and PA are not present in myelin (there may be 1–2% PA in myelin) or in most mammalian membranes, their use allows the polar head group to be varied further and helps one to understand the contribution of the polar head group to interactions with protein. It has been suggested that PE, PS, PA, and cerebroside sulfate are capable of intermolecular hydrogen bonding through the head groups in the case of PE, PS, and PA and through the sphingosine amide and fatty acid hydroxyls for cerebroside sulfate (Papahadjopoulos & Miller, 1967; Hitchcock et al., 1974; Jacobson & Papahadjopoulos, 1975; MacDonald et al., 1976; Pascher, 1976; Eibl & Blume, 1979). (For a recent review see Boggs, 1980.) These interactions increase the phase transition temperature and decrease the fluidity in the liquid crystalline phase of these lipids. The results with basic protein indicated that these interactions also caused resistance of protein-induced expansion of PE, PS, and cerebroside sulfate and restriction of hydrophobic interaction with the protein. PG is probably not capable of intermolecular hydrogen bonding, as is evident from its relatively low phase transition temperature. PA, however, does interact intermolecularly via hydrogen bonding, as demonstrated recently by Eibl and Blume (1979), resulting in a high phase transition temperature for this lipid (Jacobson & Papahadjopoulos, 1975). This hydrogen bonding involves the phosphate moiety, however; thus it is likely that electrostatic interaction with the protein would be more favorable. Upon binding to the protein the intermolecular interactions between adjacent lipids are inhibited and hydrophobic interaction of the protein with the lipid can take place. The protein can bind to PS and cerebroside sulfate without affecting the intermolecular hydrogen bonding interactions because PS has two negative charges per molecule and the negatively charged sulfate of cerebroside sulfate is probably not the moiety involved in hydrogen bonding. Intermolecular interactions between adjacent PE molecules would compete favorably over interaction with the protein, and consequently less protein would bind to PE than to any other acidic lipid (Boggs & Moscarello, 1978c).

PS, PE, and cerebroside sulfate are the major basic protein binding lipids in myelin. Therefore, it must be questioned whether hydrophobic interaction between basic protein and lipids occurs to a significant extent in myelin. To help answer this question, the spin-labeled protein was added in low concentration to a myelin suspension (Stollery et al., 1980b). Myelin restricted the motion of the spin-labeled protein at 37°C more than model systems of any of its major acidic lipids, but not as much as PG and PA. This may be due to the presence of other proteins in myelin, the high degree of order of myelin, or the fact that a mixture of lipids is present which alters the occurrence of intermolecular interactions among the lipids. Consistent with the latter interpretation, London and colleagues found that a mixed acidic lipid extract from myelin was affected more by basic protein than was cerebroside sulfate (Demel et al., 1973; London et al., 1973). In mixtures of lipids intermolecular interactions may be broken up and in some cases new ones formed. Thus the interaction of this protein with a mixture of lipids may be much different from that with pure lipids. This must be investigated further. However, these studies with different pure lipids do show that the type of interaction of basic protein with the lipid bilayer, and probably also the conformation and exposure of antigenic determinants of the protein, can depend on the type of lipid and the intermolecular interactions that occur among the lipid molecules.

4. Preferential Binding to Acidic Lipids

By using DSC in a similar manner to the study with lipophilin, basic protein has been shown to bind acidic lipids preferentially when added to vesicles containing a random mixture of an acidic and a neutral lipid (Boggs et al., 1977b). In the case of basic protein, however, since the protein decreased the phase transition temperature of a pure lipid, it was necessary to use mixtures where the acidic lipid was either the higher melting lipid or the lower melting lipid and to show that the protein shifted the temperature in the direction of the PC component regardless of whether it was the higher or lower melting lipid. From a phase diagram it was estimated that basic protein bound 27–34 molecules of acidic lipid, in agreement with the Mn^{2+} binding inhibition studies. Thus basic protein in myelin would probably be in its own acidic lipid domain as shown in Fig. 8. This domain may contain neutral lipids mixed in. The basic residues in the protein are not adjacent to each other and there would be room for other lipids in the vicinity of the protein, even if every basic residue were bound to an acidic lipid. Thus intermolecular interactions between like lipids may be interrupted.

V. ORGANIZING ROLE OF MYELIN PROTEINS

A. Phase Separation and Asymmetry

Both basic protein and lipophilin have been shown to cause lipid phase separation when added to mixtures of lipids, as discussed above. They should be able to organize the lipids in myelin into domains of varying composition. This may partly account for the asymmetric structure of myelin, although it has not yet been shown that vesicles containing either protein are asymmetric. The only other study of the effect of these proteins on the organization of lipid mixtures is an X-ray diffraction study by Mateu et al. (1973) of basic protein in vesicles of an acidic lipid extract from myelin. They showed that basic protein induced formation of a lamellar phase with a repeat distance of 154 Å, similar to that in central nervous system myelin. Unlike native myelin the two bilayers in the unit were not asymmetric. Rather there were two different but symmetric bilayers that alternated and were held together by the protein. It was concluded that one bilayer was cerebroside sulfate in the gel phase and the other contained the other acidic lipids in the mixture in the liquid crystalline phase. These lipids do not mix well at room temperature due to the high phase transition temperature of cerebroside sulfate, and they may have separated into two phases on their own but were then bound together in an organized fashion by the protein. This is supported by the observation that 20% cholesterol (which abolishes the phase transition), or an increase in temperature, resulted in a lamellar phase with repeat distance of 80 Å, indicating that mixing of the lipids had occurred. Nevertheless it is significant that the protein bound the two phases together in an alternating manner, whereas other basic proteins such as cytochrome c and poly-lysine did not (Gulik-Krzywicki, 1969). Furthermore it was reported that the N- and C-terminal fragments of basic protein produced by BNPS-skatole also did not bind two different bilayers together; the largest repeat distance obtained with the fragments was 80 Å (Mateu et al., 1973). Cytochrome c and polylysine as well as the fragments of basic protein can cause aggregation of lipid vesicles; thus their inability to organize the lipid into two phases suggests that basic protein must have a more specific organizing effect. Further studies would be desirable and might help to clarify whether basic protein has a unique organizing effect not shared by other proteins.

B. Forces Responsible for the Multilayered Structure of Myelin

From the weight percentages of acidic lipids (O'Brien & Sampson, 1965) and protein in myelin (Rumsby & Crang, 1977) it can be calculated that there are about 110 μmol of PS plus cerebroside sulfate, 3.6 μmol of basic protein, and 3.7 μmol of proteolipid per gram of myelin. Since at pH 7.4 there are 31 positively charged residues per mol of basic protein and 14 per mol of lipophilin, the number of positively charged residues from both proteins totals 163 μmol per gram of myelin. There is also 180 μmol of PE per gram of myelin, some of which can also bind to the protein. Thus, there appears to be enough acidic lipid to bind to both basic protein and proteolipid to neutralize the surface charge of myelin. There are strong repulsive forces between charged membranes, and neutralization of the acidic and basic groups by interaction between lipid and protein may help to facilitate asso-ciation of the myelin layers. However, it has been pointed out that even neutral membranes repel each other (B. E. Chapman et al., 1977) at close apposition, and it has been shown that the extracellular surfaces of myelin separate in distilled water due to electrostatic repulsion (Rand et al., 1979). Ions help to prevent this repulsion, but there must be some other mechanism to maintain the multilayered structure, particularly at the cytoplasmic surfaces which separate only slightly in distilled water.

As mentioned earlier, basic protein can aggregate lipid vesicles indicating an interlamellar interaction. By freeze-fracture electron microscopy basic protein has been shown to fuse small unilamellar vesicles of PE into multilayered vesicles (Stollery & Vail, 1977). X-Ray diffraction of PG vesicles prepared by dialysis from 2-chloroethanol with and without basic protein showed that in the absence of protein, the lipid is present in the form of ill-defined multilayers. Basic protein causes these multilayers to become well defined, until at 30% concentration of basic protein, multilayer repeat lines are present (Brady et al., 1981).

It has often been suggested that basic protein is responsible for maintaining the multilayered structure of myelin, particularly for the association of the cytoplasmic surfaces, although it has not been explained why this protein may be better suited for this function than other basic proteins. The reason may lie in its hypothesized hairpin structure, in which the two strands could associate either with the same bilayer or, equally well, with adjacent bilayers as shown in Fig. 8. Rumsby and Crang (1977) have suggested that the N-terminal half interacts hydrophobically with one bilayer surface and the C-terminal half interacts electrostatically with the adjacent bilayer surface. However, the hydrophobic and positively charged amino acids are distributed in equal proportion in both halves of the molecule. There are 17 positively charged residues in the N-terminal portion before the tri-proline sequence and 14 positively charged residues in the C-terminal portion. As mentioned earlier, both the N-terminal and C-terminal fragments obtained by cleavage at the tryptophan can interact hydrophobically and electrostatically to a similar extent (Boggs et al., 1981a). Furthermore, since the two surfaces being joined in myelin are similar (both are the intracellular surfaces of the membrane), the protein would be expected to interact in a similar way with each surface. Interaction of each half of the protein with adjacent bilayers would allow bridging of the bilayers as shown in Fig. 8, which would be an effective and nonrandom mechanism for holding the layers together. Rumsby and Crang (1977) have estimated that if basic protein is located only at the intracellular surface and joining the two opposing surfaces, and if it has dimensions of 15×150 Å, it would occupy 54% of the total cytoplasmic surface area. This estimate is based on many approximations, however.

Basic protein has been found to dimerize above pH 6 both in solution (Smith, 1980; Golds & Braun, 1978a) and in myelin (Golds & Braun, 1978b), although the dimer is in equilibrium with the monomer and only one species with a molecular weight similar to that of the monomer is seen by sedimentation equilibrium studies (Smith, 1980). It has been proposed by Golds and Braun (1978b) and Smith (1977b; 1980) that myelin basic protein links two bilayers by dimerizing. Cross-linking studies have led to the proposal that the C-terminal end of one monomer is adjacent to the N-terminal end of the other monomer, and vice versa (Golds & Braun, 1978b). It has been suggested that dimerization occurs through hydrophobic interactions, but it is possible that the mechanism could also be electrostatic interactions between negatively and positively charged amino acids. If so, this would utilize a maximum of 11 positively charged residues, which would leave 20 available for binding to acidic lipids. The sequence of basic protein is highly conserved (Carnegie & Dunkley, 1975) and in particular, all the basic (Lys, Arg) and acidic residues are either conserved or replaced with equivalent residues (Fig. 2), suggesting their importance either for interaction with acidic lipids or intra- or intermolecular protein interactions.

A study by Smith and MacDonald (1979) showed that basic protein bound micelles of dodecyl sulfate together in a ratio of one molecule of basic protein per two micelles and that the protein was a monomer except at low detergent concentrations, where it dimerized. Thus one molecule of basic protein would be capable of joining two surfaces together. On the basis of presently available evidence, the mechanism for association of the intracellular surfaces in myelin is not clear.

It has also been suggested that basic protein interacts hydrophobically with proteolipid protein to form a 1 : 1 molar complex that helps stabilize the interaction between the two intracellular surfaces (Rumsby & Crang, 1977). Although such an interaction is possible, there is as yet no evidence in support of this hypothesis. As mentioned earlier, however, there is freeze-fracture evidence suggesting that under some conditions interlamellar interactions may occur between proteolipid molecules at both the intracellular and extracellular surfaces (Pinto da Silva & Miller, 1975). This may be partly through hydrophobic interactions of the protein with adjacent bilayers, possibly facilitated by the fatty acids covalently bound to the protein.

Although the swelling experiments with myelin in hypotonic salt solutions indicated that the interactions between the extracellular surfaces are much weaker than those between the intracellular surfaces, some sort of attraction between the extracellular surfaces must occur in order for the oligodendroglial plasma membrane to wrap around the axon (Fig. 1). Indeed interlamellar association between the extracellular surfaces should be more important for maintenance of the multilamellar structure of myelin than association at the intracellular surfaces. A recent X-ray diffraction study of peripheral nervous system myelin (Padron et al., 1979) suggested that divalent cations may be involved in stabilizing interactions between the extracellular surfaces of adjacent bilayers, presumably through interlamellar bridging of acidic lipids on opposing surfaces, such as occurs with vesicles of acidic lipids (Papahadjopoulos et al., 1976b). Myelin was swollen in distilled water and rehydrated with isotonic salt solution with or without divalent cations. The membrane pairs reassociated in either case, but reswelling with distilled water readily occurred for myelin lacking divalent cations and occurred only partially and very slowly for myelin containing divalent cations.

VI. CONCLUDING REMARKS

The structural and functional role of myelin proteins is still not established, although their location and conformation in the bilayer and effects on lipid organization are becoming better understood through studies of reconstituted model membranes containing the proteins. Such studies have provided the opportunity to apply various physical and chemical techniques that have given detailed information at the molecular level. In this chapter, we have described work from our and other laboratories on the properties of the two major proteins from myelin both in the purified soluble state and after reconstitution into vesicles composed of purified phospholipids.

The proteolipid protein is of the hydrophobic intrinsic type that probably spans the bilayer and has sites exposed to the aqueous phase on both sides of the bilayer. It is 75% α-helical in the lipid bilayer and may be organized as an oligomer of 2–4 monomers. It restricts the fatty acid motion of a "boundary lipid" population and can cause lipid phase separation by associating preferentially with acidic lipids, probably through electrostatic interactions.

The basic protein is of the extrinsic type, but portions of it can also interact hydrophobically with the lipid acyl chains. It can also bind acidic lipids preferentially to its 31 positively charged amino acid residues. Both proteins have an organizing effect on lipid mixtures and thus may contribute to the asymmetric structure of myelin. In myelin they would be expected to be localized in acidic lipid domains rather than in areas characteristic of the lipid mixture as a whole. This may confer unique properties and conformations to these proteins. Additional effects of lipid specificity are observed with the basic protein, for which the degree of hydrophobic interaction with the bilayer varies with different lipids, depending on the ability of the lipids to interact with adjacent molecules via intermolecular hydrogen bonds.

Both proteins have been shown to produce experimental allergic encephalomyelitis, which implicates their involvement in various pathologic states. The observed species variation in encephalitogenic sites of basic protein may be due to differences in conformation of basic protein in the myelin of different species. Thus it is possible that different regions of the protein are exposed on the surface of myelin for recognition by lymphocytes. This might be due either to differences in lipid composition or to posttranslational modifications of the protein, since the sequence is highly conserved.

Cellular response against basic protein may also occur in multiple sclerosis. Alterations in lipid or protein composition of myelin, posttranslational modifications of the protein, or factors that alter the acidic lipids available to basic protein (such as complex formation by divalent cations) could result in a change in the degree of hydrophobic interaction and conformation of basic protein leading to recognition by the immune system as a foreign antigen.

Basic protein may be involved in the association of the intracellular surfaces of myelin either through interlamellar bridging by a monomer of this protein or through dimerization. The mechanism of association of the extracellular surfaces is even less well understood but may occur through divalent cation bridging. A role for the proteolipid protein in interlamellar association cannot be ruled out.

Continuing studies on the interaction of purified myelin proteins with chemically defined lipid membranes of varying composition are likely to contribute additional information on their structural and functional role in myelin membranes. In addition, they provide a valuable opportunity for studying the general principles involved in lipid-protein interactions as they apply to biological membranes of various cells.

ACKNOWLEDGMENTS

We thank J. G. Stollery and Dr. W. J. Vail for permission to publish their freeze-fracture electron micrographs and Mrs. I. R. Clement for her technical work in obtaining other freeze-fracture electron micrographs.

REFERENCES

Adams, C. W. M., Bayliss, O. B., Hallpike, J. F., & Turner, D. R. (1971). Histochemistry of Myelin.XII. Anionic Staining of Myelin Basic Proteins for Histology, Electrophoresis and Electron Microscopy. *J. Neurochem.* **18,** 389–394.

Anthony, J. S., & Moscarello, M. A. (1971). A Conformation Change Induced in the Basic Encephalitogen by Lipids. *Biochim. Biophys. Acta* **243,** 429–433.

Baldwin, G. S., & Carnegie, P. R. (1971). Isolation and Partial Characterization of Methylated Arginines from the Encephalitogenic Basic Protein. *Biochem. J.* **123,** 69–74.

Bergstrand, H. (1972). Localization of Antigenic Determinants on Bovine Encephalitogenic Protein; Studies in Rabbits with the Blood Leukocyte Transformation Test. *Eur. J. Immunol.* **2,** 266–269.

Bieri, V. G., & Wallach, D. F. H. (1975). Variations of Lipid-Protein Interactions in Erythrocyte Ghosts as a Function of Temperature and pH in Physiological and Non-Physiological Ranges. *Biochim. Biophys. Acta* **406,** 415–423.

Blaurock, A. E. (1971). Structure of the Nerve Myelin Membrane: Proof of the Low Resolution Profile. *J. Mol. Biol.* **56,** 35–52.

Boggs, J. M. (1980). Intermolecular Hydrogen Bonding Between Lipids—Effects on Organization and Function of Lipids in Membranes. *Can. J. Biochem.* **58,** 755–770.

Boggs, J. M., Vail, W. J., & Moscarello, M. A. (1976). Preparation and Properties of Vesicles of a Purified Hydrophobic Myelin Protein and Phospholipid: A Spin Label Study. *Biochim. Biophys. Acta* **448,** 517–530.

Boggs, J. M., Wood, D. D., Moscarello, M. A., & Papahadjopoulos, D. (1977a). Lipid Phase Separation Induced by a Hydrophobic Protein in Phosphatidylserine-Phosphatidylcholine Vesicles. *Biochemistry* **16,** 2325–2329.

Boggs, J. M., Moscarello, M. A., & Papahadjopoulos, D. (1977b). Phase Separation of Acidic and Neutral Phospholipids Induced by Human Myelin Basic Protein. *Biochemistry* **16,** 5420–5426.

Boggs, J. M., & Moscarello, M. A. (1978a). Structural Organization of the Human Myelin Membrane. *Biochim. Biophys. Acta* **515,** 1–21.

Boggs, J. M., & Moscarello, M. A. (1978b). Dependence of Boundary Lipid on Fatty Acid Chain Length in Phosphatidylcholine Vesicles Containing a Hydrophobic Protein from Myelin Proteolipid. *Biochemistry* **17,** 5734–5739.

Boggs, J. M., & Moscarello, M. A. (1978c). Effect of Basic Protein from Human CNS Myelin on Lipid Bilayer Structure. *J. Membrane Biol.* **39,** 75–96.

Boggs, J. M., Clement, I. R., & Moscarello, M. A. (1980a). Similar Effect of Proteolipid Apoproteins from Human Myelin (Lipophilin) and Bovine White Matter on the Lipid Phase Transition. *Biochim. Biophys. Acta* **601,** 134–151.

Boggs, J. M., Stollery, J. G., & Moscarello, M. A. (1980b). Effect of lipid environment on the motion of a spin-label covalently bound to myelin basic protein. *Biochemistry* **19,** 1226–1233.

Boggs, J. M., Wood, D. D., & Moscarello, M. A. (1981a). Participation of N-Terminal and C-Terminal Portions of Human Myelin Basic Protein in Hydrophobic and Electrostatic Interactions with Lipid. *Biochemistry* **20,** 1065–1073.

Boggs, J. M., Clement, I. R., Moscarello, M. A., Eylar, E. H., & Hashim, G. (1981b). Antibody Precipitation of Lipid Vesicles Containing Myelin Proteins: Dependence on Lipid Composition. *J. Immunol.* **126,** 1207–1211.

Boggs, J. M., Stamp, D., & Moscarello, M. A. (1981c). Interaction of Myelin Basic Protein with Dipalmitoylphosphatidylglycerol: Dependence on the Lipid Phase and Investigation of a Metastable State. *Biochemistry* **20,** 6066–6072.

Boggs, J. M., Stamp, D., & Moscarello, M. A. (1982). Effect of pH and Fatty Acid Chain Length on the Interaction of Myelin Basic Protein with Phosphatidylglycerol. *Biochemistry* (in press).

Boggs, J. M., & Moscarello, M. A. (1982). Interdigitation of Fatty Acid Chains of Dipalmitoylphosphatidylglycerol Due to Intercalation of Myelin Basic Protein. *Biophys. J.* (in press).

Brady, G. W., Birnbaum, P. S., Moscarello, M. A., & Papahadjopoulos, D. (1979). The Model Membrane System: Egg Lecithin + Myelin Protein ($N - 2$), Effect of Solvent Density Variation on the X-Ray Scattering. *Biophys. J.* **26,** 49–60.

Brady, G. W., Murthy, N. S., Fein, D. B., Wood, D. D., & Moscarello, M. A. (1981). The Effect of Basic Myelin Protein on Multilayer Membrane Formation. *Biophys. J.* **34,** 345–350.

Braun, P. E. (1977). Molecular Architecture of Myelin. In *Myelin* (P. Morell, Ed.), pp. 91–115, Plenum Press, New York.

Braun, P. E., & Radin, N. S. (1969). Interactions of Lipids with a Membrane Structural Protein. *Biochemistry* **8,** 4310–4318.

Brostoff, S., Reuter, W., Hichens, M., & Eylar, E. H. (1974). Specific Cleavage of the A_1 Protein from Myelin with Cathepsin D. *J. Biol. Chem.* **249,** 559–567.

Brotherus, J. R., Jost, P. C., Griffith, O. H., Keana, J. F. W., & Hokin, L. E. (1980). Charge Selectivity at the Lipid-Protein Interface of Membranous Na,K-ATPase. *Proc. Natl. Acad. Sci. U.S.A.* **77,** 272–276.

Burnett, P., & Eylar, E. H. (1971). Allergic Encephalomyelitis. Oxidation and Cleavage of the Single Tryptophan Residue of the A_1 Protein from Bovine and Human Myelin. *J. Biol. Chem.* **246,** 3425–3430.

Cabantchik, Z. I., & Rothstein, A. (1974). Membrane Proteins Related to Anion Permeability of Human Red Blood Cells. II. Effects of Proteolytic Enzymes on Disulfonic Stilbene Sites of Surface Proteins. *J. Membrane Biol.* **15,** 227–248.

Carnegie, P. (1971). Properties, Structure and Possible Neuroceptor Role of the Encephalitogenic Protein of Human Brain. *Nature (London)* **229,** 25–28.

Carnegie, P. R., Bencina, B., & Lamoureux, G. (1967). Experimental Allergic Encephalomyelitis. Isolation of the Basic Proteins and Polypeptides from Central Nervous Tissue. *Biochem. J.* **105,** 559–568.

Carnegie, P. R., Kemp, B. E., Dunkley, P. R., & Murray, A. W. (1973). Phosphorylation of Myelin Basic Protein by Adenosine 3′,5′-Cyclic Monophosphate-Dependent Protein Kinase. *Biochem. J.* **135,** 569–572.

Carnegie, P. R., & Dunkley, P. R. (1975). Basic Proteins of Central and Peripheral Nervous System Myelin. In *Advances in Neurochemistry* (B. S. Agranoff & M. H. Aprison, Eds.), Vol. 1, pp. 95–126, Plenum Press, New York.

Caspar, D. L. D., & Kirshner, D. A. (1971). Myelin Membrane Structure at 10 Å Resolution. *Nature (London) New Biol.* **231,** 46–52.

Chan, D. S., & Lees, M. B. (1974). Gel Electrophoresis Studies of Bovine Brain White Matter Proteolipid and Myelin Proteins. *Biochemistry* **13,** 2704–2712.

Chan, D. S., & Lees, M. B. (1978). Tryptic Peptides from Bovine White Matter Proteolipids. *J. Neurochem.* **30,** 983–990.

Chao, I. P., & Einstein, E. R. (1970). Physical Properties of the Bovine Encephalitogenic Protein: Molecular Weight and Conformation. *J. Neurochem.* **17,** 1121–1132.

Chapman, B. E., & Moore, W. J. (1976). Conformation of Myelin Basic Protein in Aqueous Solution from Nuclear Magnetic Resonance Spectroscopy. *Biochem. Biophys. Res. Commun.* **73,** 758–765.

Chapman, B. E., Littlemore, L. T., & Moore, W. J. (1977). Conformation of Myelin Basic Protein and its Role in Myelin Formation. In *Myelination and Demyelination* (J. Palo, Ed.), pp. 207–220, Plenum Press, New York.

Chapman, B. E., & Moore, S. J. (1978). NMR Studies on Myelin Basic Protein. I. ^{13}C Spectra in Aqueous Solutions. *Aust. J. Chem.* **31,** 2367–2385.

Chapman, D., Cornell, B. A., Eliasy, A. W., & Perry, A. (1977). Interactions of Helical Polypeptide Segments Which Span the Hydrocarbon Region of Lipid Bilayers. Studies of the Gramicidin A-Lipid-Water System. *J. Mol. Biol.* **113**, 517–538.

Chapman, D., Gomez-Fernandez, J. C., & Goni, F. M. (1979). Intrinsic Protein-Lipid Interactions. Physical and Biochemical Evidence. *FEBS Lett.* **98**, 211–223.

Chen, R. F. (1967). Removal of Fatty Acids from Serum Albumin by Charcoal Treatment. *J. Biol. Chem.* **242**, 173–181.

Chou, F. C.-h., Chou, J. C-h., Shapira, R., & Kibler, R. F. (1976). Basis of Microheterogeneity of Myelin Basic Protein. *J. Biol. Chem.* **251**, 2671–2679.

Cockle, S. A., Epand, R. M., & Moscarello, M. A. (1978a). Resistance of Lipophilin, a Hydrophobic Myelin Protein, to Denaturation by Urea and Guanidinium Salts. *J. Biol. Chem.* **253**, 8019–8026.

Cockle, S. A., Epand, R. M., Boggs, J. M., & Moscarello, M. A. (1978b). Circular Dichroism Studies on Lipid-Protein Complexes of a Hydrophobic Myelin Protein. *Biochemistry* **17**, 624–629.

Cockle, S. A., Epand, R. M., & Moscarello, M. A. (1978c). The Intrinsic Fluorescence of a Hydrophobic Myelin Protein and Some Complexes with Phospholipids. *Biochemistry* **17**, 630–636.

Cockle, S. A., Stollery, J., Epand, R. M., & Moscarello, M. A. (1980). Nature of Cysteinyl Residues and Covalent Attachment of Fatty Acids in Lipophilin from Human Myelin. *J. Biol. Chem.* **255**, 9182–9188.

Crang, A. J., & Rumsby, M. G. (1977). The Labelling of Lipid and Protein Components in Isolated Central Nervous System Myelin with Dansyl Chloride. *Biochem. Soc. Trans.* **5**, 110–112.

Curatolo, W., Sakura, J. D., Small, D. M., & Shipley, G. G. (1977). Protein-Lipid Interactions: Recombinants of the Proteolipid Apoprotein of Myelin with Dimyristoyl Lecithin. *Biochemistry* **16**, 2313–2319.

Curatolo, W., Verma, S. P., Sakura, J. D., Small, D. M., Shipley, G. G., & Wallach, D. F. H. (1978). Structural Effects of Myelin Proteolipid Apoprotein on Phospholipids: A Raman Spectroscopic Study. *Biochemistry* **17**, 1802–1807.

Davison, A. W., & Peters, A. (1970). Biochemistry of the Myelin Sheath. In *Myelination* (A. N. Davison & A. Peters, Eds.), pp. 80–161, Thomas, Springfield, IL.

Deber, C. M., Moscarello, M. A., & Wood, D. D. (1978). Conformational Studies on ^{13}C-Enriched Human and Bovine Myelin Basic Protein in Solution and Incorporated into Liposomes. *Biochemistry* **17**, 898–903.

Deibler, G. E., & Martenson, R. E. (1973). Chromatographic Fractionation of Myelin Basic Protein. *J. Biol. Chem.* **248**, 2392–2396.

De Kruijff, B., van Dijck, P. W. M., Demel, R. A., Schuijff, A., Brants, F., & van Deenen, L. L. M. (1974.) Non-Random Distribution of Cholesterol in Phosphatidylcholine Bilayers. *Biochim. Biophys. Acta* **356**, 1–7.

Demel, R. A., London, Y., Geurts van Kessel, W. S. M., Vossenberg, F. G. A., & van Deenen, L. L. M. (1973). The Specific Interaction of Myelin Basic Protein with Lipids at the Air-Water Interface. *Biochim. Biophys. Acta* **311**, 507–519.

Eibl, H., & Blume, A. (1979). The Influence of Charge on Phosphatidic Acid Bilayer Membranes. *Biochim. Biophys. Acta* **553**, 476–488.

Epand, R. M., Moscarello, M. A., Zierenberg, B., & Vail, W. J. (1974). The Folded Conformation of the Encephalitogenic Protein of the Human Brain. *Biochemistry* **13**, 1264–1267.

Eylar, E. H. (1972). The Structure and Immunological Properties of Basic Proteins of Myelin. *Biochemistry* **14**, 3049–3056.

Eylar, E. H., Salk, J., Beveridge, G., & Brown, L. (1969). Experimental Allergic Encephalomyelitis. An Encephalitogenic Basic Protein from Bovine Myelin. *Arch. Biochem. Biophys.* **132**, 34–38.

Eylar, E. H., & Thompson, M. (1969). Allergic Encephalomyelitis: The Physico-Chemical Properties of the Basic Protein Encephalitogen from Bovine Spinal Cord. *Arch. Biochem. Biophys.* **129,** 469–479.

Eylar, E. H., Caccam, J., Jackson, J., Westall, F., & Robinson, A. B. (1970). Experimental Allergic Encephalomyelitis: Synthesis of Disease Inducing Site of the Basic Protein. *Science* **168,** 1220.

Eylar, E. H., Brostoff, S., Hashim, G., Caccam, J., & Burnet, P. (1971a). Basic A₁ Protein of the Myelin Membrane. *J. Biol. Chem.* **246,** 5770–5784.

Eylar, E. H., Jackson, J., Bennet, C., Kniskern, P., & Brostoff, S. (1971b). Allergic Encephalomyelitis. An Encephalitogenic Peptide Derived from the Basic Protein of Myelin. *J. Biol. Chem.* **246,** 3418–3424.

Feinstein, M. B., & Felsenfeld, H. (1975). Reactions of Fluorescent Probes with Normal and Chemically Modified Myelin Basic Protein and Proteolipid. Comparisons with Myelin. *Biochemistry* **14,** 3049–3056.

Folch, J., & Lees, M. (1951). Proteolipids, a New Type of Tissue Lipoproteins. *J. Biol. Chem.* **191,** 807–817.

Folch-Pi, J., & Stoffyn, P. J. (1972). Proteolipids from Membrane Systems. *Ann. N.Y. Acad. Sci.* **195,** 86–107.

Gagnon, J. (1976). Studies of a Hydrophobic Myelin Protein, Ph.D. thesis, Department of Biochemistry, University of Toronto, Toronto, Ont.

Gagnon, J., Finch, P. R., Wood, D. D., & Moscarello, M. A. (1971). Isolation of a Highly Purified Myelin Protein. *Biochemistry* **10,** 4756–4763.

Golds, E. E., & Braun, P. E. (1976). Organization of Membrane Protein in the Intact Myelin Sheath. *J. Biol. Chem.* **251,** 4729–4735.

Golds, E., & Braun, P. E. (1978a). Crosslinking Studies on the Conformation and Dimerization of Myelin Basic Protein in Solution. *J. Biol. Chem.* **253,** 8171–8177.

Golds, E., & Braun, P. E. (1978b). Protein Associations and Basic Protein Conformation in the Myelin Membrane. *J. Biol. Chem.* **253,** 8162–8170.

Gould, R. M., & London, Y. (1972). Specific Interaction of Central Nervous System Myelin Basic Protein with Lipids. Effects of Basic Protein on Glucose Leakage from Liposomes. *Biochim. Biophys. Acta.* **290,** 200–218.

Gulik-Krzywicki, T., Schechter, E., Luzzati, V., & Faure, M. (1969). Interactions of Proteins and Lipids: Structure and Polymorphism of Protein-Lipid-Water Phases. *Nature (London)* **223,** 116–1121.

Hagopian, A., & Eylar, E. H. (1969). Glycoprotein Biosynthesis: The Purification and Characterization of a Polypeptide *N*-Acetylgalactosaminyl Transferase from Bovine Submaxillary Glands. *Arch. Biochem. Biophys.* **129,** 515–524.

Hantke, K., & Braun, V. (1973). Covalent Binding of Lipid to Protein, Diglyceride and Amide-Linked Fatty-Acid at the *N*-Terminal End of the Murein-Lipoprotein of the *Escherichia coli* Outer Membrane. *Eur. J. Biochem.* **34,** 284–296.

Hashim, G. (1977). Experimental Allergic Encephalomyelitis in Lewis Rats: Chemical Synthesis of Disease Inducing Determinants. *Science* **196,** 1219–1221.

Hashim, G. (1978). Myelin Basic Protein: Structure, Function and Antigenic Determinants. *Immunol. Rev.* **39,** 61–107.

Hashim, G., Carvalho, E. F. & Sharpe, R. D. (1978). Definition and Synthesis of the Essential Amino Acid Sequence for Experimental Allergic Encephalomyelitis in Lewis Rats. *J. Immunol.* **121,** 665–670.

Hashim, G., Sharpe, R., & Carvalho, E. (1979). Experimental Allergic Encephalomyelitis: Sequestered Encephalitogenic Determinant in Bovine Myelin Basic Protein. *J. Neurochem.* **32,** 73–77.

Hashim, G., Wood, D. D., & Moscarello, M. A. (1980). Myelin-Lipophilin Induced Experimental Allergic Encephalomyelitis. *Neurochem. Res.* **5**, 1137–1145.

Herndon, R. M., Rauch, H. C., & Einstein, E. R. (1973). Immunoelectron Microscopic Localization of the Encephalitogenic Basic Protein in Myelin. *Immunol. Commun.* **2**, 163–172.

Hinz, H. J., & Sturtevant, J. M. (1972). Calorimetric Studies of Dilute Aqueous Suspensions of Bilayers Formed from Synthetic L-α-Lecithins. *J. Biol. Chem.* **247**, 6071–6075.

Hitchcock, P. B., Mason, R., Thomas, K. M., & Shipley, G. G. (1974). Structural Chemistry of 1,2-Dilauroyl-DL-Phosphatidylethanolamine: Molecular Conformation and Intermolecular Packing of Phospholipids. *Proc. Natl. Acad. Sci. U.S.A.* **71**, 3036–3040.

Jackson, J., Brostoff, S., Lampert, P., & Eylar, E. H. (1972). Allergic Encephalitogenic Monkeys Induced with A_1 Protein. *Neurobiology* **2**, 83–88.

Jacobson, K., & Paphadjopoulos, D. (1975). Phase Transitions and Phase Separations in Phospholipid Membranes Induced by Changes in Temperature, pH, and Concentration of Bivalent Cations. *Biochemistry* **14**, 152–161.

Jollés, J., Nussbaum, J. L., Schoentgen, S., Mandel, P., & Jolles, P. (1977). Structural Data Concerning the Major Rat Brain Myelin Proteolipid P7 Apoprotein. *FEBS Lett.* **74**, 190–194.

Jollés, J., Schoentgen, F., Jollés, P., Vacher, M. Nicot, C., & Alfsen, A. (1979). Structural Studies of the Apoprotein of the Folch-Pi Bovine Brain Myelin Proteolipid: Characterization of the CNBr-Fragments and of a Long C-Terminal Sequence. *Biochem. Biophys. Res. Commun.* **87**, 619–626.

Jones, A. J. S., & Rumsby, M. G. (1975). The Intrinsic Fluorescence Characteristics of the Myelin Basic Protein. *J. Neurochem.* **25**, 565–572.

Jones, A. J. S., & Rumsby, M. G. (1977). Localization of Sites for Ionic Interaction with Lipid in the C-Terminal Third of the Bovine Myelin Basic Protein. *Biochem. J.* **167**, 583–591.

Jost, P. C., Griffith, O. H., Capaldi, R. A., & Vanderkooi, G. (1973a). Evidence for Boundary Lipid in Membranes. *Proc. Natl. Acad. Sci. U.S.A.* **70**, 480–484.

Jost, P. C., Griffith, O. H., Capaldi, R. A., & Vanderkooi, G. (1973b). Identification and Extent of Fluid Bilayer Regions in Membranous Cytochrome Oxidase. *Biochim. Biophys. Acta.* **311**, 141–152.

Jost, P. C., & Griffith, O. H. (1980). The Lipid-Protein Interface in Biological Membranes. *Ann. N.Y. Acad. Sci.* **348**, 391–405.

Katona, E., Wood, D. D., Neumann, A. W., & Moscarello, M. A. (1978). The Application of Surface Tension Measurements to the Study of Protein-Lipid Interactions Between a Hydrophobic Myelin Protein (Lipophilin) and Lipids. *J. Colloid Interface Sci.* **67**, 108–117.

Keniry, M. A., & Smith, R. (1979). Circular Dichroic Analysis of the Secondary Structure of Myelin Basic Protein and Derived Peptides Bound to Detergents and to Lipid Vesicles. *Biochim. Biophys. Acta* **578**, 381–391.

Kibler, R. F., & Shapira, R. (1968). Isolation and Properties of an Encephalitogenic Protein from Bovine, Rabbit and Human Central Nervous Tissue. *J. Biol. Chem.* **243**, 281.

Kies, M. W. (1965). Chemical Studies on an Encephalitogenic Protein from Guinea Pig Brain. *Ann. N.Y. Acad. Sci.* **122**, 161–170.

Kies, M. S., Thompson, B. E. & Alvord, E. C. (1965). The Relationship of Myelin Proteins to Experimental Allergic Encephalomyelitis. *Ann. N.Y. Acad. Sci.* **122**, 148–160.

Krigbaum, W. R., & Hsu, T. S. (1975). Molecular Conformation of Bovine A_1 Basic Protein, a Coiling Macromolecule in Aqueous Solution. *Biochemistry* **14**, 2542–2546.

Lamoureux, G., Thibeault, G., Richer, G., & Bernard, C. (1972). Induction de l'encéphalite allergique expérimentale avec des encéphalitogènes humaines de synthèse. *Union Med. Can.* **101**, 674–680.

Lavialle, F., & Levin, I. W. (1980). Raman Spectroscopic Study of the Interactions of Dimyristoyl- and 1-palmitoyl-2-oleoyl-phosphatidylcholine Liposomes with Myelin Proteolipid Apoprotein. *Biochemistry* **19**, 6044-6050.

Lees, M. B., Sakura, J. D., Sapirstein, V. S., & Curatolo, W. (1978).   Structure and Function of Proteolipids in Myelin and Non-Myelin Membranes. *Biochim. Biophys. Acta* **559**, 209–230.

Liebes, L. F., Zand, R., & Phillips, W. D. (1975).   Solution Behavior, Circular Dichroism and 220 MHz PMR Studies of the Bovine Myelin Basic Protein. *Biochim. Biophys. Acta*. **405**, 27–39.

Liebes, L. F., Zand, R., & Phillips, W. D. (1976).   The Solution Behavior of the Bovine Myelin Basic Protein in the Presence of Anionic Ligands. Binding Behavior with the Red Component of Trypsan Blue and Sodium Dodecyl Sulfate. *Biochim. Biophys. Acta* **427**, 392–409.

Littlemore, L. A. T. (1978).   NMR Studies on Myelin Basic Protein. II. ^{1}H-NMR Studies of the Protein and Constituent Peptides in Aqueous Solutions. *Aust. J. Chem.* **31**, 2387–2398.

London, Y., & Vossenberg, F. G. A. (1973).   Specific Interactions of Central Nervous System Myelin Basic Protein with Lipids. Specific Regions of the Protein Sequence Protected from the Proteolytic Action of Trypsin. *Biochim. Biophys. Acta*. **307**, 478–490.

London, Y., Demel, R. A., Guerts van Kessel, W. S. M., Vossenberg, F. G. A., & van Deenen, L. L. M. (1973).   The Protection of A₁ Myelin Basic Protein Against the Action of Proteolytic Enzymes After Interaction of the Proteins with Lipids at the Air-Water Interface. *Biochim. Biophys. Acta* **311**, 520–530.

London, Y., Demel, R. A., Guerts van Kessel, W. S. M., Zahler, P., & van Deenen, L. L. M. (1974).   The Interaction of the "Folch-Lees" Protein with Lipids at the Air-Water Interface. *Biochim. Biophys. Acta* **332**, 69–84.

Lowden, J. A., Moscarello, M. A., & Morecki, R. (1966).   The Isolation and Characterization of an Acid-Soluble Protein from Myelin. *Can. J. Biochem.* **44**, 567–577.

MacDonald, R. C., Simon, S. A., & Baer, E. (1976).   Ionic Influence on the Phase Transition of Dipalmitoylphoshatidylserine. *Biochemistry* **15**, 885–891.

MacLennan, D. H., Yip, C. C., Iles, G. H., & Seeman, P. (1973).   Isolation of Sarcoplasmic Reticulum Proteins. *Cold Spring Harbor Symp. Quant. Biol.* **33**, 469–477.

Martenson, R. E., Deibler, G. E., & Kies, M. W. (1969).   Microheterogeneity of Guinea Pig Myelin Basic Protein. *J. Biol. Chem.* **244**, 4261–4267.

Martenson, R. E., Nomura, K., Levine, S., & Sowinski, R. (1977).   Experimental Allergic Encephalomyelitis in the Lewis Rat: Further Delineation of Active Sites in Guinea Pig and Bovine Myelin Basic Proteins. *J. Immunol.* **118**, 1280–1285.

Mateu, L., Luzzatti, V., London, Y., Gould, R. M., Vossenberg, F. G. A., & Olive, J. (1973).   X-Ray Diffraction and Electron Microscopic Study of the Interactions of Myelin Components. The Structure of a Lamellar Phase with a 150–180 Å Repeat Distance Containing Basic Proteins and Acidic Lipids. *J. Mol. Biol.* **75**, 697–709.

McIntosh, T. J., & Robertson, J. D. (1976).   Observations on the Effect of Hypotonic Solutions on the Myelin Sheath in the Central Nervous System. *J. Mol. Biol.* **100**, 213–217.

Melchior, V., Hollingshead, C. J., & Casper, D. L. D. (1979).   Divalent Cations Cooperatively Stabilize Close Membrane Contacts in Myelin. *Biochim. Biophys. Acta* **554**, 204–226.

Minassian-Saraga, L. T., Albrecht, G., Nicot, C., Nguyen Le, T., & Alfsen, A. (1973).   Electrostatic Interactions Between Phosphatidylinositol and the Apoprotein of Folch-Lees Proteolipid in Mixed Spread Films. *J. Colloid Interface Sci.* **44**, 542–545.

Miyamoto, E., & Kakiuchi, I. (1974).   *In Vitro* and *In Vivo* Phsophorylation of Myelin Basic Protein by Exogenous and Endogenous Adenosine-5'-Monophosphate-Dependent Protein Kinases in Brain. *J. Biol. Chem.* **249**, 2769–2777.

Moscarello, M. A. (1976).   Chemical and Physical Properties of Myelin Proteins. In *Current Topics in Membranes and Transport* (F. Bronner & A. Kleinzeller, Eds.), Vol. 8, pp. 1–28, Academic Press, New York.

Moscarello, M. A., Gagnon, J., Wood, D. D., Anthony, J., & Epand, R. (1973).   Conformational Flexibility of a Myelin Protein. *Biochemistry* **12**, 3402–3406.

Moscarello, M. A., Katona, E., Neumann, A. W., & Epand, R. M. (1974). The Ordered Structure of the Encephalitogenic Protein from Normal Human Myelin. *Biophys. Chem.* **2,** 290–295.

Nakao, A., Davis, W. J., & Einstein, E. R. (1966). Basic Protein from the Acidic Extract of Spinal Cord. I. Isolation and Characterization. *Biochim. Biophys. Acta.* **130,** 163–170.

Neumann, A. W., Moscarello, M. A., & Epand, R. M. (1973). The Application of Surface Tension Measurements to the Study of Conformational Transitions in Aqueous Solutions of Poly-L-Lysine. *Biopolymers* **12,** 1945–1957.

Nicot, C., Nuyen, Le T., Vacher Lepetre, M., & Alfsen, A. (1973). Study of Folch-Pi Apoprotein. I. Isolation of Two Components, Aggregation During Delipidation. *Biochim. Biophys. Acta* **322,** 109–123.

Norton, W. T., & Poduslo, S. E. (1973). Myelination in Rat Brain: Changes in Myelin Composition During Brain Maturation. *J. Neurochem.* **21,** 759–773.

Nussbaum, J. L., Rouayrenc, J. F., Mandel, P., Jollés, J., & Jollés, P. (1974a). Isolation of Terminal Sequence Determination of the Major Rat Brain Myelin Proteolipid P7 Apoprotein. *Biochem. Biophys. Res. Commun.* **57,** 1240–1247.

Nussbaum, J. L., Rouayrenc, J. F., Jollés, J., Jollés, P., & Mandel, P. (1974b). Amino Acid Analysis and *N*-Terminal Sequence Determination of P7 Proteolipid Apoprotein from Human Myelin. *FEBS Lett.* **45,** 295–298.

O'Brien, J. S., & Sampson, E. L. (1965). Lipid Composition of the Normal Human Brain: Grey Matter, White Matter and Myelin. *J. Lipid Res.* **6,** 537–544.

Padron, R., Mateu, L., & Kirschner, D. A. (1979). X-Ray Diffraction Study of the Kinetics of Myelin Lattice Swelling. Effect of Divalent Cations. *Biophys. J.* **28,** 231–240.

Palmer, F. B., & Dawson, R. M. C. (1969). The Isolation and Properties of Experimental Allergic Encephalitogenic Protein. *Biochem. J.* **111,** 629–636.

Papahadjopoulos, D., & Miller, N. (1967). Phospholipid Model Membranes. I. Structural Characteristics of Hydrated Liquid Crystals. *Biochim. Biophys. Acta* **135,** 624–638.

Papahadjopoulos, D., Cowden, M., & Kimelberg, H. (1973). Role of Cholesterol in Membranes. Effects on Phospholipid-Protein Interactions, Membrane, Permeability and Enzymatic Activity. *Biochim. Biophys. Acta* **330,** 8–26.

Papahadjopoulos, D., Vail, W. J., & Moscarello, M. A. (1975a). Interaction of a Purified Hydrophobic Protein from Myelin with Phospholipid Membranes: Studies on Ultrastructure, Phase Transitions and Permeability. *J. Membrane Biol.* **22,** 143–164.

Papahadjopoulos, D., Moscarello, M., Eylar, E. H., & Isac, T. (1975b). Effects of Proteins on Thermotropic Phase Transitions of Phospholipid Membranes. *Biochim. Biophys. Acta* **401,** 317–335.

Papahadjopoulos, D., Hui, S., Vail, W. J., & Poste, G. (1976a). Studies on Membrane Fusion. 1. Interactions of Pure Phospholipid Membranes and the Effect of Myristic Acid, Lysolecithin, Proteins and Dimethylsulfoxide. *Biochim. Biophys. Acta* **448,** 245–264.

Papahadjopoulos, D., Vail, W. J., Pangborn, W. A., & Poste, G. (1976b). Studies on Membrane Fusion. II. Induction of Fusion in Pure Phospholipid Membranes by Calcium Ions and Other Divalent Metals. *Biochim. Biophys. Acta* **448,** 265–283.

Pascher, I. (1976). Molecular Arrangements in Sphingolipids. Conformation and Hydrogen Bonding of Ceramide and Their Implication on Membrane Stability and Permeability. *Biochim. Biophys. Acta* **455,** 433–451.

Peters, A., & Vaughn, J. E. (1970). Morphology and Development of the Myelin Sheath. In *Myelination* (A. N. Davison & A. Peters, Eds.), Thomas, Springfield, IL.

Pinto da Silva, P., & Miller, R. G. (1975). Membrane Particles on Fracture Faces of Frozen Myelin. *Proc. Nat. Acad. Sci. U.S.A.* **72,** 4046–4050.

Poduslo, J. F., & Braun, P. E. (1975). Topographical Arrangement of Membrane Proteins in the Intact Myelin Sheath. *J. Biol. Chem.* **250,** 1099–1105.

Rand, R. P., Papahadjopoulos, D., & Moscarello, M. A. (1976). On the Interaction of Lecithins Both with a Purified Hydrophobic Protein from Myelin and with Fatty Acids and Lysolecithin. Paper presented at 20th Annual Meeting, Biophysical Society.

Rand, R. P., Fuller, N. L., & Lis, L. J. (1979). Myelin Swelling and Measurement of Forces Between Myelin Membranes. *Nature (London)* **279,** 258–260.

Rice, D. M., Meadows, M. D., Scheinman, A. O., Goni, F. M., Gomez, J. C., Moscarello, M. A., Chapman, D., & Oldfield, E. (1979). Protein-Lipid Interaction. An NMR Study of SR. ATPase, Lipophilin and Proteolipid Apoprotein-Lecithin Systems and a Comparison with the Effects of Cholesterol. *Biochemistry* **18,** 5893–5903.

Roboz-Einstein, E., Robertson, D., DiCaprio, J., & Moore, W. (1962). The Isolation from Bovine Spinal Cord of a Homogeneous Protein with Encephalitogenic Activity. *J. Neurochem.* **9,** 353–361.

Rumsby, M. G. (1978). Organization and Structure in Central-Nerve Myelin. *Biochem. Soc. Trans.* **6,** 448–462.

Rumsby, M. G., & Crang, A. J. (1977). The Myelin Sheath—A Structural Examination, *Cell Surf. Rev.* **4,** 247–362.

Schmitt, F. O., Bear, R. S., & Palmer, K. J. (1941). X-Ray Diffraction Studies on the Structure of the Nerve Myelin Sheath. *J. Cell Comp. Physiol.* **18,** 31–42.

Schnapp, B., & Mugnaini, E. (1975). The Myelin Sheath: Electron Microscopic Studies with Thin Sections and Freeze Fracture. In *Proceedings of the Golgi Centennial Symposium* (M. Santini, Ed.), p. 209, Raven Press, New York.

Schnapp, B., & Mugnaini, E. (1976). Freeze Fracture Properties of Central Myelin in the Bullfrog. *Neuroscience* **1,** 459–467.

Sherman, G., & Folch-Pi, J. (1970). Rotary Dispersion and Circular Dichroism of Brain Proteolipid Protein. *J. Neurochem.* **17,** 597–605.

Singer, S. J. (1971). The Molecular Organization of Biological Membranes. In *Structure and Function of Biological Membranes* (L. I. Rothfield, Ed.), pp. 145–222, Academic Press, New York.

Singer, S. J. (1974). The Molecular Organization of Membranes. *Annu. Rev. Biochem.* **43,** 805–833.

Smith, R. (1977a). The Secondary Structure of Myelin Basic Protein Extracted by Deoxycholate. *Biochim. Biophys. Acta* **491,** 581–590.

Smith, R. (1977b). Non-Covalent Cross-Linking of Lipid Bilayers by Myelin Basic Protein. A Possible Role in Myelin Formation. *Biochim. Biophys. Acta* **470,** 170–184.

Smith, R. (1980). Sedimentation Analysis of the Self-Association of Bovine Myelin Basic Protein. *Biochemistry* **19,** 1826–1831.

Smith, R., & McDonald, B. J. (1979). Association of Myelin Basic Protein with Detergent Micelles. *Biochim. Biophys. Acta* **554,** 133–147.

Smyth, D., & Utsumi, S. (1967). Structure at the Hinge Region in Rabbit Immunoglobulin G. *Nature (London)* **216,** 332–335.

Spitler, L., von Muller, C., Fudenberg, H., & Eylar, E. H. (1972). Experimental Allergic Encephalitis. Dissociation of Cellular Immunity to Brain Protein and Disease Production. *J. Exp. Med.* **136,** 156–174.

Steck, A. J., & Appel, S. H. (1974). Phosphorylation of Myelin Basic Protein. *J. Biol. Chem.* **249,** 5416–5420.

Steck, A. J., Siegrist, H. P., Zahler, P., & Herschkowitz, N. N. (1976). Lipid-Protein Interactions with Native and Modified Myelin Basic Protein. *Biochim. Biophys. Acta* **455,** 343–352.

Stoffyn, P., & Folch-Pi, J. (1971). On the Type of Linkage Binding Fatty Acids Present in Brain White Matter Proteolipid Apoprotein. *Biochem. Biophys. Res. Commun.* **44,** 157–161.

Stollery, J. G., & Vail, W. J. (1977). Interactions of Divalent Cations or Basic Proteins with Phosphatidylethanolamine Vesicles. *Biochim. Biophys. Acta* **471,** 372–390.

Stollery, J. G., Boggs, J. M., Moscarello, M. A., & Deber, C. M. (1980a). Direct Observation by ^{13}C NMR of Membrane Bound Human Myelin Basic Protein. *Biochemistry* **19**, 2391–2396.

Stollery, J. G., Boggs, J. M., & Moscarello, M. A. (1980b). The Variable Interaction of Human Myelin Basic Protein with Different Acidic Lipids. *Biochemistry* **19**, 1219–1225.

Sturgess, J. M., Moscarello, M. A., & Schachter, H. (1978). The Structure and Biosynthesis of Membrane Glycoproteins. In *Current Topics in Membranes and Transport* (F. Bronner & A. Kleinzeller, Eds.), Academic Press, New York.

Ting-Beall, H. P., Lees, M. B., & Robertson, J. D. (1979). Interactions of Folch-Lees Proteolipid Apoprotein with Planar Lipid Bilayers. *J. Membrane Biol.* **51**, 33–46.

Tomasi, L. G., & Korngurth, S. E. (1967). Purification and Partial Characterization of a Basic Protein from Pig Brain. *J. Biol. Chem.* **242**, 4933–4938.

Uda, Y., & Nakazawa, Y. (1973). Proteolipid of Bovine Brain White Matter. Relationship Between Phospholipid and Protein. *J. Biochem.* **74**, 545–549.

Usher, J. R., Epand, R. M., & Papahadjopoulos, D. (1978). The Effect of Free Fatty Acids on the Thermotropic Phase Transition of Dimyristoylphosphatidylcholine. *Chem. Phys. Lipids.* **22**, 245–253.

Vail, W. J., Papahadjopoulos, D., & Moscarello, M. A. (1974). Interaction of a Hydrophobic Protein with Liposomes. Evidence for Particles Seen in Freeze Fracture as Being Proteins. *Biochim. Biophys. Acta* **345**, 463–467.

Van Dijck, P. W. M., Verivergaert, P. H. J. T., Verkleij, A. J., van Deenen, L. L. M., & de Gier, J. (1975). Influence of Ca^{2+} and Mg^{2+} on the Thermotropic Behaviour and Permeability Properties of Liposomes Prepared from Dimyristoylphosphatidylglycerol and Mixtures of Dimyristoylphosphatidylglycerol and Dimyristoylphosphatidylcholine. *Biochim. Biophys. Acta* **406**, 465–478.

Verkleij, A. J., de Kruijff, B., Verivergaert, P. H. J. T., Tocanne, J. F., & van Deenen, L. L. M. (1974). The Influence of pH, Ca^{2+} and Protein on the Thermotropic Behaviour of the Negatively Charged Phospholipid, Phosphatidylglycerol. *Biochim. Biophys. Acta* **339**, 432–437.

Verma, S. P., Wallach, D. F. H., & Sakura, J. D. (1980). Raman Analysis of the Thermotropic Behavior of Lecithin–Fatty Acid Systems and of Their Interaction with Proteolipid Apoprotein. *Biochemistry* **19**, 547–579.

Waksman, B. H., Porter, H., Lees, M. B., Adams, R. D., & Folch-Pi, J. (1954). A Study of the Chemical Nature of Components of Bovine White Matter Effective in Producing Allergic Encephalomyelitis in the Rabbit. *J. Exp. Med.* **100**, 451–471.

Warren, G. B., Houslay, M. D., Metcalfe, J. C., & Birdsall, N. J. M. (1975). Cholesterol Is Excluded from the Phospholipid Annulus Surrounding an Active Calcium Transport Protein. *Nature (London)* **255**, 684–687.

Westall, F. C., Robinson, A. B., Caccam, J., Jackson, J., & Eylar, E. H. (1971). Essential Chemical Requirements for Induction of Allergic Encephalomyelitis. *Nature (London)* **229**, 22–24.

Whitaker, J. N., Chou, C-h. Jen, Chou, Frank G.-H. & Kibler, R. F. (1977). Molecular Internalization of a Region of Myelin Basic Protein, *J. Exp. Med.* **146**, 317–331.

Wolfgram, F. (1965). Macromolecular Constituents of Myelin. *Ann. N.Y. Acad. Sci.* **122**, 104–115.

Wood, D. D., Gagnon, J., Finch, P. R., & Moscarello, M. A. (1971). The Isolation of an Electrophoretically Homogeneous Protein from Myelin. *Am. Assoc. Neurochem. Trans.* **2**, 117.

Wood, D. D., Epand, R. M., & Moscarello, M. A. (1977). Localization of the Basic Protein and Lipophilin in the Myelin Membrane. *Biochim. Biophys. Acta* **467**, 120–129.

Wood, D. D., & Moscarello, M. A. (1977). The Degree of Penetration of Lipophilin into Phosphatidylcholine Vesicles. *Am. Soc. Neurochem. Trans.* **8**, 78.

Wood, D. D., Boggs, J. M., & Moscarello, M. A. (1980). The Transmembrane Orientation of Lipophilin in Phosphatidylcholine Vesicles. *Neurochem. Res.* **5**, 745–756.

Worthington, C. R., & Blaurock, A. E. (1969). A Structural Analysis of Nerve Myelin. *Biophys. J.* **9**, 970–990.

Two

Spin Labeling and Lipid-Protein Interactions in Membranes

DEREK MARSH AND ANTHONY WATTS

Max-Planck-Institut für biophysikalische Chemie, Göttingen, Federal Republic of Germany

ABBREVIATIONS

ASL, androstanol spin label
DMPC, dimyristoylphosphatidylcholine
doxyl, 4,4-dimethyl-oxazolidine-*N*-oxyl

ESR, electron spin resonance
(Na^+,K^+)-ATPase, sodium-potassium adenosinetriphosphatase
NMR, nuclear magnetic resonance
proxyl, 2,2-dialkyl-5,5-dimethylpyrrolidine-N-oxyl
ROS, rod outer segment

I. INTRODUCTION

A. The Lipid-Protein Interface

Biological membranes are composed of lipid and protein in comparable quantities by weight. A substantial part of the membrane lipid is known to be arranged as a phospholipid bilayer, forming the membrane matrix and the basic permeability barrier. Of the membrane proteins, the integral membrane proteins are those that penetrate and mostly, if not exclusively, span the membrane. It is these proteins that are responsible for the specific transport properties and many of the enzymatic functions of the membrane. In many cases these activities have been found to be sensitive to the physical state of the membrane lipids (see, e.g., Melchior & Steim, 1976). Lipid-protein interactions are therefore likely to play an important role in the structure and function of biological membranes, both from the point of view of sealing the proteins into the membrane, thus maintaining the structural integrity of the membrane permeability barrier, and of controlling the conformational stability of the protein, interfacing it with the fluid environment of the bulk bilayer lipids.

A central role is thus assumed by the protein-lipid interface and this chapter is devoted to the study of this interface using the electron spin resonance (ESR) spectra of lipid spin labels. In many natural membranes and reconstituted lipid-protein systems a spin label population is observed that is not present in the spectra of bilayers of the lipids alone. The rotational mobility of this population differs significantly from that of fluid bilayer lipids. The relative amount of this second, more motionally restricted, lipid component seems to correlate roughly with protein size, and in reconstituted or progressively delipidated systems there is found to be a fixed amount of this spectral component relative to protein, independent of the total lipid to protein ratio. It was these properties that led Jost et al. (1973a), who originally observed this effect with cytochrome oxidase, to attribute the motionally restricted spectral component to lipid at the protein-lipid interface. Since the amount of immobilized lipid appeared to correlate roughly with the perimeter of the protein as observed by electron microscopy, it was assigned to a first shell or *boundary layer* of lipid surrounding the protein. The motional characteristics of this boundary lipid are that it is relatively immobilized and is in slow exchange with the fluid lipids on the spin label ESR time scale, since two distinct spin label components are resolved. Since ESR spectroscopy corresponds to rather fast motions, this means that typical rotational correlation times for the immobilized lipid exceed approximately 10^{-8} s, and exchange rates are less than about 10^8 s^{-1}. Thus although the motional characteristics of the boundary lipid are quite distinct from those of fluid bilayer lipid, it is still possible that the boundary lipid has appreciable rotational motion and exchanges relatively rapidly with the fluid bilayer lipid: the exchange frequency within fluid lipid bilayers is about 10^7 s^{-1} (Scandella et al., 1972; Träuble & Sackmann, 1972). In deuterium nuclear magnetic resonance (NMR) for instance, only a single lipid component is observed, indicating that the exchange between the boundary and the rest of the lipid is fast on the deuterium NMR time scale, that is, an exchange rate faster than 10^4 s^{-1}, see Chapter 3.

Spin label ESR studies thus provide a convenient time window that is sufficiently fast that even if the exchange rate between the shells of lipid surrounding the protein is as rapid as in unperturbed bilayers, spectra from the individual lipid shells can be resolved, provided the mobility difference between them is large enough. In this way it is possible to look directly at the properties of the lipid molecules at the lipid-protein interface, which is then defined as boundary lipid and in no way implies a rigid shell of specialized tightly bound lipid.

Several reviews have already appeared on spin label studies of lipid-protein interactions and boundary lipids in particular. These include Griffith and Jost (1978, 1979) and Jost and Griffith (1978a, 1980). For reviews covering most aspects of spin label ESR studies in membranes see Berliner (1976) and Marsh (1981); and for an introduction to the application of ESR methods to biological systems see Knowles et al. (1976).

B. Integral Protein Structure

It is clear from the previous discussion that any detailed interpretation of the data from the spin labels at the lipid-protein interface requires a knowledge of the structure of the intramembranous sections of the integral membrane proteins. Whereas the phospholipid component of biological membranes is rather well characterized both chemically and structurally, the protein component is less so. The full amino acid sequence is known for only relatively few integral proteins. Glycophorin A, the major glycoprotein of the erythrocyte membrane, is a good example. The complete amino acid sequence is given in Fig. 1. The protein is relatively small and the major feature of its primary structure is the nonuniform distribution of polar and apolar residues throughout the sequence. In particular, there is a hydrophobic stretch of 23 completely nonpolar residues in the center of the sequence that is identified as the intramembranous section of the protein. As is indicated in Fig. 2, this nonpolar sequence is sufficient to span the 35 Å hydrophobic region of the lipid bilayer if the residues are arranged in an α-helical conformation.

The sequence of glycophorin is thus characterized by an amphiphilic distribution of amino acid residues matching the polarity profile of the membrane. The N-terminal region is hydrophilic and contains all the carbohydrate residues, in particular sialic acid. This then passes via the hydrophobic central region to the hydrophilic C-terminal region, which is rich in proline residues. Hence for this small protein the polypeptide chain passes once through the membrane. Evidence for an α-helical structure of the central, nonpolar region comes from circular dichroism (CD) measurements on the hydrophobic tryptic peptide in nonpolar solvents (Schulte

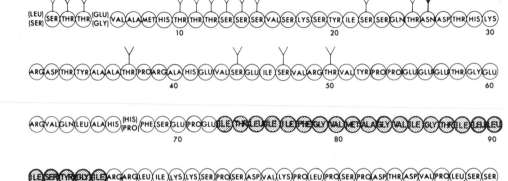

Figure 1. Complete amino acid sequence of human glycophorin A. The shaded residues indicate the intramembranous hydrophobic domain, and the sticks indicate glycosylated residues. Adapted from Tomita et al. (1978).

& Marchesi, 1979). The intramembranous section may be a single α-helix or an oligomer of α-helices. The α-helical net in Fig. 2 shows that the helix is divided into domains, one face having bulky, very hydrophobic residues and the opposite face having essentially neutral, nonbulky residues, at least in its central portion. This suggests that the helices may well tend to key together in some oligomeric structure, with the hydrophobic residues exposed to the lipid phase.

Few complete structures are known for integral membrane proteins. The best characterized is that of bacteriorhodopsin, the retinal-containing proton pump from the halophilic bacterium *Halobacteria halobium*. Electron diffraction studies to a resolution of 7 Å (Henderson & Unwin, 1975) have revealed a structure that consists essentially of seven α-helices, 35–40 Å in length, spanning the membrane and oriented essentially perpendicular to the membrane surface (see Fig. 3). These seven α-helices account for approximately 70–80% of the total molecular weight (26,000), indicating that the protein does not project very far from either surface of the membrane, in agreement with its predominantly apolar amino acid composition. A probable model based on the complete amino acid sequence (Ovchinnikov et al., 1979) places the charged residues in the helices along surfaces that face one another toward the center of the molecule, whereas the faces of the helices that are directed outward toward the lipid phase consist solely of hydrophobic, uncharged residues (Engelman et al., 1980; see Fig. 4). Thus in this particular instance, the structural and chemical nature of the protein-lipid interface is reasonably well defined. For other integral membrane proteins this is unfortunately not the case.

A more complicated situation arises in the case of larger, multi-subunit proteins of which cytochrome oxidase, the terminal member of the mitochondrial electron transport chain, may be taken as an example. Cytochrome oxidase is composed of 7–10 polypeptides with molecular weights varying from approximately 40,000 to 5,000. The three larger polypeptides are synthesized on mitochondrial ribosomes and have a rather hydrophobic amino acid composition (see Table 1). Two of these polypeptides are particularly hydrophobic, containing a high proportion of leucine. Most of the remaining, smaller polypeptides are synthesized on cytoplasmic ribosomes and have a less predominantly hydrophobic amino acid composition. It seems probable that the differential polypeptide composition of the larger protein may determine the disposition of the protein in the membrane in much the same way as does the amphiphilic distribution of amino acid residues in the primary sequence

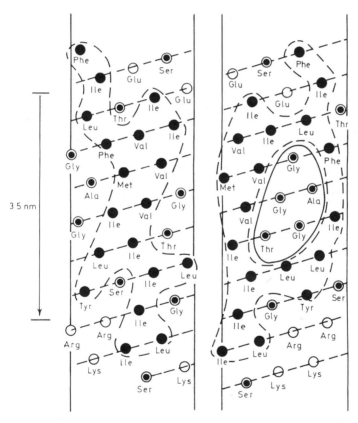

Figure 2. Proposed α-helical net of the nonpolar domain of glycophorin A. Each turn of the helix (3.6 residues) is represented by the diagonal broken lines. *Left:* the hydrophobic residues form a cluster that is outlined by a dashed line; *right:* the central cluster of neutral residues is outlined with a solid line; the center of the projection has been rotated slightly from that on the left. From Segrest (1977).

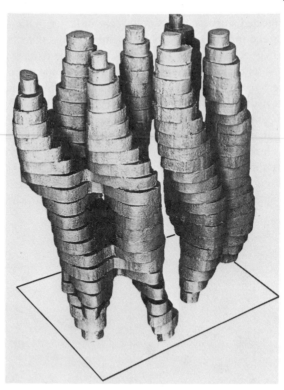

Figure 3. Three-dimensional model of the bacteriorhodopsin molecule in the purple membrane as obtained by electron microscopy–electron diffraction of unstained specimens. The model is viewed roughly parallel to the plane of the membrane. The top and bottom of the model correspond to the parts of the molecule in contact with the aqueous phase, the rest being in contact with lipid. The most strongly tilted α-helices are in the foreground. From Henderson and Unwin (1975).

of smaller proteins such as glycophorin A. Surface labeling, antibody binding, and photolabeling with hydrophobic probes [see, e.g., Cerletti & Schatz (1979) and Bisson et al. (1979)] are providing information regarding the disposition of the cytochrome oxidase components and clearly, since the complete molecule spans the membrane and has a nonsymmetrical distribution, some polypeptides (most probably the predominantly hydrophobic ones) will contribute more to the intramembranous section of the protein than others. Electron microscopy and electron diffraction results on the shape of this protein are discussed in subsequent sections.

The preceding discussion summarizes most of the important known features of the structure of the intramembranous section of integral membrane proteins. For many proteins the information presently available is rather fragmentary. For an interpretation of the spin label results, the most important single quantity is the perimeter of the intramembranous section of the protein, since it is this that defines the protein-lipid interface. Information on this comes from electron diffraction studies as discussed for bacteriorhodopsin; from normal freeze-fracture and negative

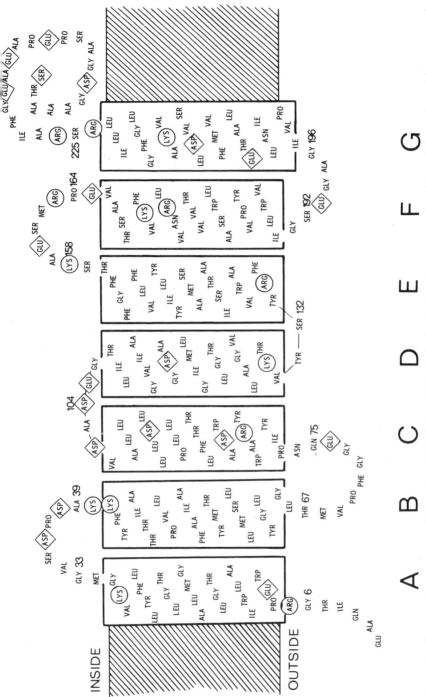

Figure 4. Suggested arrangement of the polypeptide chain of bacteriorhodopsin, based on structural studies and the amino acid sequence of Ovchinnikov et al. (1979). The seven α-helices are labeled *A–G*, starting from the amino terminus. The hatch marks indicate the approximate locations of the lipid hydrocarbon regions; ○ and ◇ indicate positively and negatively charged residues, respectively. From Engelman et al. (1980).

TABLE 1 Amino Acid Composition of the Subunits of Bovine Heart Cytochrome Oxidase

Amino Acid	Subunit Number							Totals
	I	II	III	IV	V	VI	VII	
Lysine	10	5	13	8	7	6	4	53
Histidine	12	6	3	4	3	1	2	31
Arginine	8	6	5	6	7	6	4	42
Aspartic acid	24	16	12	11	13	12	7	95
Threonine	24	13	6	8	5	5	4	65
Serine	19	23	10	6	4	5	3	70
Glutamic acid	16	22	15	15	12	10	8	98
Proline	18	13	7	8	7	4	4	61
Glycine	30	8	8	11	6	5	6	74
Alanine	30	9	11	9	8	7	5	79
Half-cystine	1	2	0.3	~3	1	3	0.24	~10
Valine	24	13	11	7	8	5	5	73
Methionine	15	15	2	2	1	1	1	37
Isoleucine	22	21	6	7	7	4	3	70
Leucine	47	36	12	9	11	3	7	125
Tyrosine	12	12	6	3	4	4	2	43
Phenylalanine	25	7	6	2	3	4	2	49
Tryptophan	5	6	5	3	2	2	1	24
Total residues	342	233	138	122	109	87	68	1,099 (~1,110)
MW (× 115)	39,330	26,795	15,870	13,524		10,005	7,800	~125,000
MW (sequence)					12,436			

From Yasunobu et al. (1979).

TABLE 2 Molecular Weights and Dimensional Information of Various Integral Membrane Proteins

Protein	Molecular Weight (daltons)	Number of Subunits	Shape[a]	Size[b]	Method and References[c]
Bacteriorhodopsin	26,000	1	Seven-helix prism	20×40 Å	EM (1)
Cytochrome c oxidase	200,000	7–10	"Y"	2×20 Å	EM (2, 3)
Ca^{2+}-ATPase	115,000 ± 7,000	1	Elongated (ellipsoid)	21–25 Å (210–150 Å)	Sedimentation (4)
(Na^+,K^+)-ATPase (renal)	280,000	2×2	Dimer (cylinder)	2×50 Å	EM (5)
Band III	95,000	1	—	80 Å (dimer)	EM (6)
Acetylcholine receptor (*Torpedo*)	250,000	4	Rosette	75–80 Å	EM (7, 8)
Rhodopsin (bovine)	37,000	1	(Ellipsoid or cylinder)	34 Å (125 Å) or 31 Å (97 Å)	X-Ray scattering (9)
Gap junction "connexon"	150,000–180,000	6×1	Cylinder	50–60 Å diam. 75 Å	EM (10)
Cytochrome b_5	16,000 $(5,000)^d$	1	(Helix)	10 Å	Sequence (11, 12)
Glycophorin A	30,000 $(3,700)^e$	1	(Helix)	10 Å	Sequence (13)

[a]Assumed shapes, that is, those not observed by electron microscopy (EM), are given in parentheses. The dimensions under "Size" refer to these shapes.
[b]Cross-sectional dimensions: diameter or (breadth × width). Longitudinal dimensions, where available, are given in parentheses.
[c]References: (1) Henderson & Unwin (1975); (2) Henderson et al. (1977); (3) Fuller et al. (1979); (4) Le Maire et al. (1976b); (5) Deguchi et al. (1977); (6) Barret et al. (1977); (7) Reynolds & Karlin (1978); (8) Zingsheim et al. (1980); (9) Sardet et al. (1976); Osborne et al. (1978); (10) Unwin & Zampighi (1980); (11) Ozols & Gerard (1977); (12) Fleming et al. (1978); (13) Tomita et al. (1978).
[d]Hydrophobic segment.
[e]Monomer.

contrast electron microscopy; from primary sequence data as discussed for glycophorin A; and from sedimentation analysis in detergent solutions as discussed in Chapter 5. Clearly the precision and degree of resolution of such experiments varies enormously. The available data for various membrane proteins of interest are summarized in Table 2.

II. BILAYER LIPID IN MEMBRANES

The lipid-protein composition of a variety of biological membranes is given in Table 3. It can be seen that depending on the function and source, a fairly wide range of lipid to protein ratios is found in natural membranes.

The protein determination measures peripheral as well as integral proteins, and among the integral proteins there is a wide variation in the fraction of the protein extending beyond the bilayer. Therefore, it is not possible in a heterogeneous membrane to make a direct correlation between the lipid to protein ratio and the hydrophobic protein surface area in the membrane. Nevertheless, it is to be expected that membranes with high lipid to protein ratios will display lipid properties that are dominated by the lipid-lipid interactions in the membrane, whereas those with lower lipid to protein ratios will show stronger effects from lipid-protein interactions.

In practice it is found that the ESR spectra of lipid spin labels in membranes with high lipid to protein ratios are very similar to those of the same labels in aqueous bilayer dispersions of the extracted membrane lipids. This is because the bilayer line shapes are narrower than the protein-associated spectral line shape to be discussed later; hence the bilayer line shape dominates the spectrum. The spectra display features of anisotropic motion and flexibility gradient (McConnell & McFarland, 1972) that are highly characteristic of these labels in bilayer structures and that can be used as a fingerprint to identify bilayer regions in membranes. These lipid bilayer properties are considered in this section, using the chromaffin granule membrane as an example. This forms the basis for subsequent discussion of membranes with lower lipid to protein ratios and reconstituted systems in which the lipid to protein ratio can be varied. Here the effects of the protein on the bilayer can be studied as a function of protein content and specific effects of the protein-lipid interface, if any, might be observed.

The chemical structures of a number of lipid spin labels that can be used to study the lipid chain motion in model or natural membranes are shown in Fig. 5. The nitroxide reporter group for each of these labels is covalently attached to one methylene group of the lipid molecule, which, when introduced into the lipid bilayer, orients in the same way as the endogenous membrane lipids. Details of the synthesis of these labels are given in the Appendix. The $(n + 2)$ doxyl fatty acid spin labels, $I(m,n)$, have been most commonly used as probes of membrane structure, but further synthesis to produce phospholipid labels of the type $II(m,n)$, improves the chemical similarity between host and probe. Labels are normally used at a concentration of one or less per 100 host phospholipid molecules, in order to minimize spin-spin interactions.

TABLE 3 Lipid-Protein Weight Ratios of Various Biological Membranes

Membrane	Ratio of Phospholipid to Protein (wt/wt)	Ratio of Cholesterol to Protein (wt/wt)	Ratio of Total Lipid to Protein (wt/wt)	References[a]
Myelin (human CNS)	1.52	1.0	4.0	(1)
Chromaffin granule (bovine adrenal medulla)	1.46	0.47	1.93	(2)
Erythrocyte ghost (human)	0.94	0.67	1.61	(3)
Rod outer segment (bovine)	1.08	0.04	1.35	(4)
Gap junction (mouse liver)	0.89	0.16	1.08	(5)
Enriched (Na$^+$,K$^+$)-ATPase (rabbit kidney outer medulla)	1.33	0.27	1.50	(6, 7)
Kidney microsomes (rabbit)	0.73	0.30	1.09	(8)
Duodenal brush border (mouse)	0.21	0.15	0.63	(9)
Enriched acetylcholine receptor (*Torpedo marmorata*)	0.58	0.24	0.83	(10)
Sarcoplasmic reticulum (rabbit skeletal muscle)	0.44	0.04	0.57	(11)
Rhodopseudomonas sphaeroides	0.33	0	0.33	(12)
Inner mitochondrial membrane (beef heart)	0.30	0.02	0.32	(13)
Purple membrane (*Halobacterium halobium*)	0.27	0	0.30	(14)

[a]References: (1) O'Brien & Sampson (1965); (2) Winkler sermiller (1967); (4) Daeman (1973); (5) Casper et al. (1977); (6) De Pont et al. (1978); (7) De Pont personal communication; (8) Jørgenson (1975); (9) Billington & Nayudu (1978); (10) Marsh et al. (1981); (11) Fiehn & Hasselbach (1970); (12) Birrell et al. (1968); and (14) Kushwaha et al. (1975).

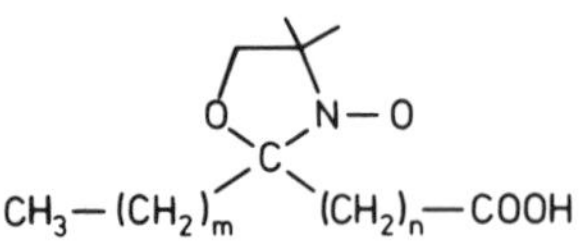

Figure 5. Lipid spin labels commonly used in studying lipid chain fluidity and molecular motion in biological membranes. $I(m,n)$, positional isomers of $(n + 2)$-doxyl stearic acids [$(n + 2)$-(4',4'-dimethyloxazolidinyl-N-oxy)stearic acid]; $II(m,n)$, positional isomers of $(n + 2)$-doxyl phosphatidylcholine [β-$(n + 2)$-(4',4'-dimethyloxazolidinyl-N-oxy)stearoyl-γ-acyl-α-phosphatidylcholine]; III, doxyl cholestane (3-spiro-[2'-(N-oxyl-4',4'-dimethyloxazolidine)] cholestane). From Fretten et al. (1980).

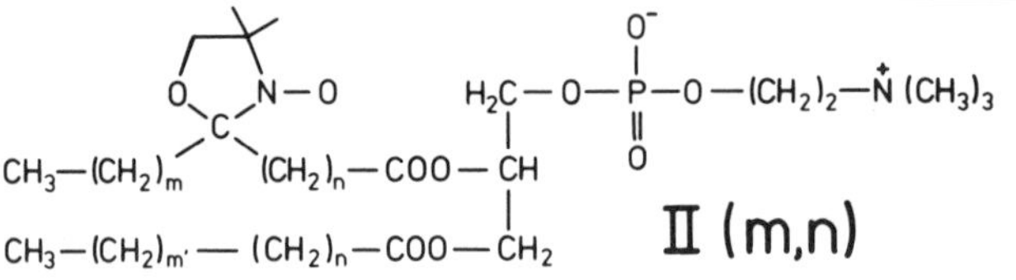

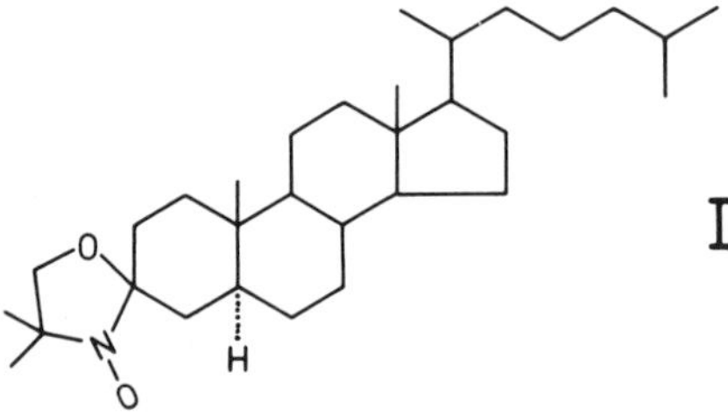

Typical ESR spectra of various positional isomers of the $I(m,n)$ stearic acid spin labels in chromaffin granule membranes are given in Fig. 6. The chromaffin granule has a membrane with a high lipid-protein ratio, which is consistent with its function as an exocytotic storage vesicle.

The spectra in Fig. 6 closely resemble those of the same labels in aqueous bilayer dispersions of the extracted membrane lipids throughout the whole temperature range studied. Figure 6 shows that the extent of spectral anisotropy, measured by the difference between the splittings of the outer peaks $2A_\parallel$ and of the inner peaks $2A_\perp'$ decreases as the label is stepped down the chain toward the center of the bilayer. Qualitatively this decrease in spectral anisotropy with increasing n represents an increase in angular amplitude of motion of the lipid chain segments on proceeding toward the terminal methyl end of the chain.

The overall motion of the lipid chain results from rotations about individual C—C bonds. If the lipid molecule is effectively anchored at the membrane polar-apolar

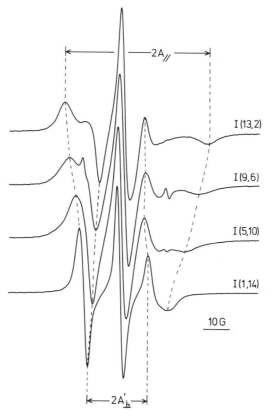

Figure 6. ESR spectra of the stearic acid spin labels *I(m,n)* in chromaffin granule membranes at 30°C. (See Fretten et al., 1980).

interface on the time scale of C—C bond rotations, then the number of possible *trans-gauche* isomerisms that can occur increases on proceeding toward the center of the bilayer. Intermolecular interactions between adjacent chains within the bilayer will modify this profile of rotational isomerism, giving rise to an increase in the intrinsic probability of *gauche* conformations toward the ends of the chains, but basically this is the origin of the flexibility gradient. Although different in detail from the results obtained by ²H-NMR (Seelig & Seelig, 1974), the spin label flexibility gradient (McConnell & McFarland, 1972; Hubbell & McConnell, 1971; Jost et al., 1971) is diagnostic of a lipid bilayer type of structure, and its existence in the membrane spectra indicates that a substantial part of the lipid in the chromaffin granule membrane is present in a bilayer form.*

---

*The spin label flexibility gradient (McConnell & McFarland, 1972) does not necessarily distinguish between bilayers and other lyotropic, liquid cystalline phases. For this, additional information is required from electron microscopy or X-ray diffraction to show that the extracted lipids form *lamellar* structures in water. The spin label results can then be used to identify bilayer regions in membranes.

Quantitatively the decrease in spectral anisotropy and the flexibility gradient are best described by an order parameter, S. In the limit of fast motional averaging ($10^{-11} \leq \tau_R \leq 3.10^{-9}$ s) this is related to the time-averaged amplitude of angular motion and is given by

$$S = \frac{1}{2}(3 < \cos^2\theta > -1) \tag{1}$$

where θ is the instantaneous angular orientation of the chain segment relative to the bilayer normal. [See Griffith & Jost (1976), Gaffney (1976), and Marsh (1981) for details of order parameter measurements.] Experimentally the order parameter is given by the ratio of the spectral anisotropy in the bilayer ($A_\parallel - A_\perp$) to the maximum obtainable in a rigidly oriented system (defined by A_{zz}, A_{xx} and A_{yy}, the principal values of the spin label hyperfine tensor):

$$S = \frac{A_\parallel - A_\perp}{A_{zz} - \frac{1}{2}(A_{xx} + A_{yy})} \cdot \frac{a_0'}{a_0} \tag{2}$$

where $A_\parallel$ and $A_\perp$ are the hyperfine splittings corresponding to the magnetic field oriented parallel and perpendicular to the plane of the membrane, respectively, as would be obtained in an oriented sample, for example. In the case of a random, unoriented membrane dispersion, $A_\parallel$ is given by the maximum hyperfine splitting in the pseudo-powder pattern, as defined in Fig. 6. $A_\perp$ is related to, but not identical to, the minimum hyperfine splitting, $A_\perp'$, defined in Fig. 6. Line shape simulations for unoriented dispersions have indicated the necessary correction (Gaffney, 1976). The isotropic hyperfine splitting constant in the bilayer is given by $a_0 = \frac{1}{3}(A_\parallel + 2A_\perp)$, whereas the isotropic hyperfine splitting constant corresponding to the single crystal environment in which A_{xx}, A_{yy}, A_{zz}, are measured, is $a_0' = \frac{1}{3}(A_{xx} + A_{yy} + A_{zz})$. For motion in the slow correlation time regime of spin label ESR ($3.10^{-9} \leq \tau_R \leq 10^{-7}$ s), the effective hyperfine splittings depend not only on the amplitude but also on the rate of motion, as discussed in Section III.C. Thus in the case of slow motion, Eq. 2 will give an overestimate of the order parameter.

The order parameter thus gives a quantitative method of comparing the lipid chain motion in lipid bilayers with that in the lipid bilayer component of membranes. Such a comparison is shown in Fig. 7, in which the order parameters for six positional isomers of the $I(m,n)$ stearic acid spin labels are plotted as a function of temperature for both chromaffin granule membranes and bilayers of their extracted membrane lipids. On the whole, the absolute values of the order parameters and their temperature dependences are very similar for both the membrane and the extracted lipids, indicating that the bilayer properties of the membrane lipids are relatively unperturbed by the presence of the membrane protein.

This result is also in agreement with the analysis of the perturbation of the extra-boundary lipid shells in reconstituted lipid-protein complexes with low protein content (see Section III.E, below). For membranes with lower lipid to protein ratios, there is in general an apparent increase in the order parameters of the membranes relative to the lipids. A critical examination of the spectra indicates that in many cases this apparent increase in order parameter may be artifactual for a number of

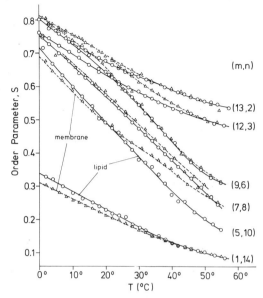

Figure 7. Temperature dependence of the effective order parameters (*S*) of the stearic acid spin labels *I(m,n)* in chromaffin granule membranes (△) and in bilayers of the extracted membrane lipids (○). From Fretten et al. (1980).

reasons. First, in spectra from many membrane preparations, there is probably a second, unresolved, more immobilized lipid component in addition to the fluid bilayer component. This corresponds to the boundary lipid component discussed in Sections III and IV, and for spin label isomers for which the two components are not well resolved, this will tend to bias the fluid lipid spectra toward higher apparent order parameters. Second, the order parameter as given by Eq. 2 is a well-defined quantity only when there exists limited amplitude motion that is rapid on the ESR time scale. In many cases, particularly when the label is near the polar head group and at lower temperatures, the motion is not sufficiently rapid and an apparent increase in order parameter reflects a decrease in the rate as well as the amplitude of motion.

For the present discussion, however, the important result is that the flexibility gradient, that is, the decrease in *S* with *n*, is characteristic of the behavior of these spin labels in pure lipid bilayers. Also, these characteristics are not appreciably perturbed by the presence of the integral membrane proteins away from the first boundary shell, as discussed later.

The fluidity of membranes is expressed not only in the segmental motion of the lipid chains, but also in the rotation of the entire lipid molecule about its long molecular axis. This motion is readily studied using rigid sterol spin labels, such as the cholestane label III (Fig. 5), which does not undergo segmental motion but is free to rotate about its long axis. The nitroxide axis with the largest hyperfine splitting is in this case directed at right angles to the long molecular axis and motional averaging for this label reduces the maximum outer hyperfine splitting,

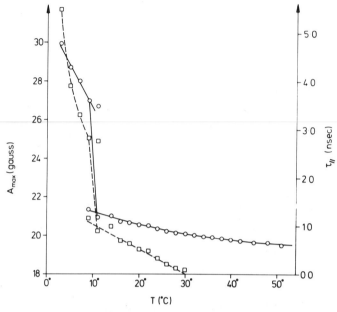

Figure 8. Temperature dependence of the outer hyperfine splitting A_{max} of the cholestane spin label III in bilayers of chromaffin granule membrane lipids (○) and of the correlation time $\tau_{\parallel}$ for rotation about the long molecular axis of the spin label (□) deduced from the A_{max} splittings. From Fretten et al. (1980).

A_{max}, from 32 gauss to ½ (32 + 6) = 19 gauss. This label is therefore diagnostic for rotation about the long axes of the molecules in the membrane. Figure 8 shows the outer splittings of the cholestane label III as a function of temperature in bilayers of the lipid extracted from the chromaffin granule membrane. The rotational correlation times calculated from the outer splitting are also given, and they clearly indicate that the discontinuity at 10°C in the outer splitting measured in Fig. 8 corresponds to a cooperative increase in the rate of long axis motional averaging (Fretten et al., 1980). Similar motion effects have been investigated in synthetic lipid bilayers (Marsh, 1980; Watts & Marsh, 1980; Polnaszek et al., 1981).

The spectral analysis of lipid spin labels in bilayers has been treated in a number of reviews. More extensive coverage is given by Smith and Butler (1976), Griffith and Jost (1976), Knowles et al. (1976), Marsh (1981), and Marsh and Watts (1981).

III. BOUNDARY REGION IN RECONSTITUTED SYSTEMS

The interaction of lipids with a membrane protein, in particular the stoichiometry of the interaction, is best determined in reconstituted or recombined systems, since it is then possible to vary the lipid to protein ratio in the samples. In the absence of any concentration-dependent protein aggregation, the amount of lipid at the protein interface should remain constant (per protein), independent of the total lipid to protein ratio. Lipid to protein titrations should thus be able to test whether there is a fixed stoichiometry in the interaction, and also to determine the range of interaction beyond the first stoichiometric shell. Spin labels with the nitroxide group attached close to the end of the lipid chain have normally been used for such titrations, since these labels will give rise to a narrow, extensively motional-averaged spectrum in fluid lipid bilayers (see Fig. 6). Thus a large range of averaged spectral anisotropy is available to detect any immobilization induced by the protein. This section discusses the properties of the motionally restricted boundary lipid observed with such labels. Extensive use is made of ESR difference spectroscopy, to separate the two components, and reconstituted cytochrome oxidase is used as the principal example.

A. Stoichiometry

Figure 9 shows the ESR spectra of a phospholipid spin label in cytochrome oxidase–lipid complexes at various lipid to protein ratios. The complexes used for Fig. 9 are lipid-substituted preparations in which all the endogenous lipid of yeast cytochrome oxidase has been replaced by the synthetic phospholipid, dimyristoyl-phosphatidylcholine (DMPC), using cholate-mediated exchange in the presence of a large excess of the substituting lipid. Samples with different lipid to protein ratios were prepared either by adding extra lipid in the presence of cholate, or by precipitating with ammonium sulfate in the presence of varying concentrations of cholate, before dialysis (Knowles et al., 1979).

The spectra of Fig. 9 (with the exception of the bottom spectrum) consist of two components. One component is a narrow, three-line spectrum similar to that found in bilayers of the lipid alone. The other component is a broad spectrum similar to that found at very low lipid to protein ratios, and is characteristic of spin labels that are immobilized on the ESR time scale. The proportion of the broad spectral component increases systematically with increasing protein content of the complexes and arises directly from interaction of the lipid with the protein, since it is not present in the spectrum of bilayers of the pure lipid (bottom spectrum of Fig. 9). The partial resolution of the two components makes it possible to perform spectral subtractions to obtain the spectra of the separate components and to determine the relative amounts of spin label intensity contributing to them. Difference spectra have been obtained (see Section III.F) by subtracting a suitable immobilized component from the spectra of Fig. 9, until a satisfactory single-component, fluid endpoint spectrum is achieved. The relative proportions of the fluid and immobilized components, n_f^*/n_b^*, obtained by double integration are given by the lipid to protein titration in Fig. 10. The titration can be expressed as $n_f^*/n_b^* = n_t/n_b - 1$, where n_t is the lipid to protein ratio and it is assumed that the lipid spin label reflects the distribution of the unlabeled host lipid. The data in Fig. 10 for DMPC-substituted cytochrome oxidase are linear for all lipid to protein ratios greater than $n_t = 100$, with intercept -1 on the n_f^*/n_b^* axis. The intercept on the n_t axis shows that a constant number, $n_b = 55 \pm 5$, of immobilized lipid molecules is associated with

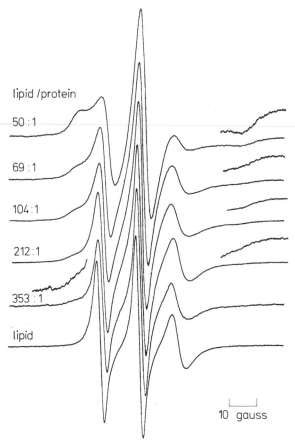

Figure 9. ESR spectra of 14-doxyl phosphatidylcholine spin label in yeast cytochrome oxidase–dimyristoylphosphatidylcholine (DMPC) complexes of different lipid to protein ratios prepared by cholate-mediated exchange at 32°C. From Knowles et al. (1979).

each 200,000 dalton protein, independent of the total amount of lipid in the complex. In this sense the immobilized component can be defined as the first shell of motionally restricted lipid associated with the protein.

The agreement of the data for the phosphatidylcholine spin label in DMPC-substituted cytochrome oxidase with the formulation given above justifies the assumption made in this equation, that the spin label directly reflects the distribution of the host lipid. An alternative treatment of the lipid to protein titration (Griffith & Jost, 1978; Brotherus et al., 1981) is to model the association of the immobilized spin label with the protein as an exchange equilibrium between labeled lipids L^* and unlabeled lipids L, occupying n_1 independent sites on the protein, P:

$$L^* + L_{n_1-m} L_m^* P \overset{K_r}{\rightleftharpoons} L + L_{n_1-m-1} L_{m+1}^* P$$

where K_r is the relative binding constant of the labeled lipid compared with the unlabeled lipid (see Chapter 6 and Brotherus et al., 1981). It can then be shown that the titration has the form:

$$\frac{n_f^*}{n_b^*} = \frac{n_t}{(n_1 K_r)} - \frac{1}{K_r} \tag{3}$$

where the ratio n_f^*/n_b^* refers to the spin label and is obtained from double integration of the component spectra, and n_t is the unlabeled lipid to protein ratio and is obtained by chemical analysis of the sample. From Fig. 10, $K_r \simeq 1$ for the phosphatidylcholine spin label, again implying no selectivity between the spin-labeled phosphatidylcholine and the unlabeled lipid (DMPC).

The earlier data of Jost et al. (1973a) for aqueous acetone-extracted cytochrome oxidase are also plotted in Fig. 10. These points lie very close to the straight line for the DMPC-substituted cytochrome oxidase, and give rise to only a slightly smaller number of immobilized lipids per 200,000 dalton protein: $n_b = 45–50$, and an effective binding constant only slightly smaller than unity with $K_r \sim 0.95$. There are, however, two interesting differences between this and the data for the DMPC-substituted complexes. The first concerns the behavior at low lipid to protein ratios.

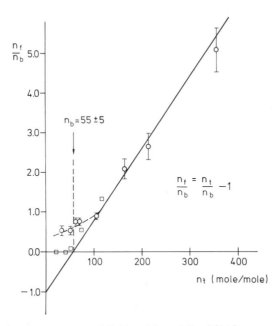

Figure 10. Lipid-protein titrations of fluid and immobilized lipid components deduced from double integrated intensities of the spin label ESR spectra: n_b, the immobilized part of the total lipid to protein ratio n_t, which must be subtracted from the spectra of the complexes to yield the fluid component n_f; $\bigcirc$, yeast cytochrome oxidase–DMPC complexes (data from Knowles et al., 1979); $\square$, aqueous acetone extracted bovine cytochrome oxidase (data from Jost et al., 1973b).

Figure 10 shows that departures from linearity occur for our lipid-substituted complexes at lipid to protein ratios less than 100 : 1, at which the quantity of immobilized lipid exceeds that of the fluid lipid. This can possibly be interpreted as an effect of the cholate that is used to mediate the lipid substitution. Apparently random protein-protein contacts occur when removing the detergent, giving rise to a heterogeneous sample, which is not the case with aqueous acetone extraction. The second distinction is that the lipid-substituted preparations reach lipid to protein ratios much higher than those found in the inner mitochondrial membrane, even approaching those of the more relatively lipid-rich membranes of Table 3. Extrapolation to the natural endogenous level from higher lipid to protein ratios in the case of the lipid-substituted preparations, and from lower lipid to protein ratios in the case of the lipid-extracted preparations, leads to essentially the same result: a fixed stoichiometry of the immobil..ed lipid with respect to protein, independent of the total lipid to protein ratio. This is an important result, since a criticism sometimes leveled against the immobilized or boundary lipid concept is that it is observed *only* at unnaturally low lipid to protein ratios at which there may be considerable protein aggregation. Clearly this is not the case.

To attempt some more precise structural definition of the protein-immobilized lipid component it is necessary to compare the observed values of n_b with the size of the intramembranous protein perimeter. This was first done by Jost et al. (1973a) for cytochrome oxidase and led to the boundary layer concept. From the electron microscopy of negatively stained, two dimensional lattice preparations of cytochrome oxidase [see, e.g., Henderson et al. (1977) and references therein] it was estimated that the protein cross section corresponded roughly to a rectangle of 52 × 60 Å. This has a perimeter that could accommodate about 50 lipid chains. Allowing for two chains per phospholipid and the two sides of the bilayer, this corresponds to 50 phospholipid molecules per cytochrome oxidase monomer. This is in good agreement with the number of immobilized lipids per 200,000 dalton cytochrome oxidase, suggesting that the first shell might, to a first approximation, correspond to a complete boundary layer encircling the protein. However, this is a rather crude estimate for the perimeter of the protein, as higher resolution studies of the cytochrome oxidase structure have later shown (Henderson et al., 1977; Fuller et al., 1979). A further complication regards the molecular weight of cytochrome oxidase, which on the basis of sedimentation and other data, was taken to be about 200,000 [see, e.g., Capaldi & Briggs (1976)]. However, present evidence suggests that the monomer consists of a two heme and two copper complex which is the smallest unit capable of accepting four electrons for reaction with O_2. This leads to a monomer molecular weight of 140,000 (Hartzell et al., 1978). The minimum apparent molecular weight of the seven subunits is 125,000–130,000. Thus it appears that a value of 200,000 might be rather high for the monomer. A lower molecular weight would, of course, result in a linear scaling down of the number of binding sites reported here. On the other hand, if cytochrome oxidase exists as a dimer, then n represents half the number of binding sites per dimer.

Results of electron diffraction studies on lattice preparations of cytochrome oxidase are shown in Fig. 11. The monomer resembles a lopsided "Y", the intra-

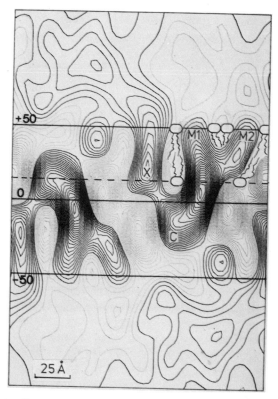

Figure 11. Projection down the a-axis of the three-dimensional map of monomeric, deoxycholate-derived two-dimensional crystals of cytochrome oxidase. Heavy lines, negative stain-excluding regions; M_1 and M_2, membrane-bound domains that face the mitochondrial matrix; C, cytochrome c-binding domain, which is on the cytoplasmic face of the inner mitochondrial membrane (X is thought to be artifactual). From Fuller et al. (1979). We have superimposed our estimate of the approximate location of the lipid bilayer, thickness 35 Å, and the first-shell lipids.

membranous portion consisting of two domains whose centers are approximately 30–35 Å apart. Projections within the plane of the membrane are given in Fig. 12 for two different crystal forms. The agreement between the two is rather good. The approximate intramembranous perimeter is indicated on these projections. It appears from Fig. 11 that there will probably be two different contour levels with different perimeters corresponding to the two halves of the bilayer. The mean number of lipid chains per monomer is 34 for the perimeters in Fig. 12A and 39 for Fig. 12B. This similarity suggests that it is not very critical whether it is assumed that the lipid is or is not accommodated between the forks of the "Y" shape. Corrected to a molecular weight of 200,000, these figures give a mean value for the number of first shell lipids of 51 molecules. An alternative estimate is to assume two circular domains centered at a separation of ~35 Å, and that either the two domains are in contact or there is just enough space for two layers of lipid between them. The

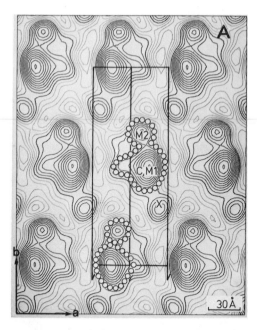

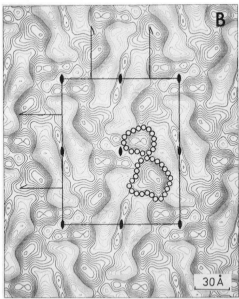

Figure 12. Projection down the *c*-axis of the three-dimensional map of two-dimensional cyto-chrome oxidase crystals. (*A*) Negatively stained monomeric, deoxycholate-derived crystals, as in Fig. 11, to a resolution of 25 Å. (From Fuller et al., 1979). (*B*) Unstained, dimeric Triton-derived crystals to a resolution of about 12.5 Å. Heavy lines indicate protein electron density. (From Henderson et al., 1977). An interpretation of the approximate intramembranous perimeter is indicated by the enclosing shell of circles, which are drawn to scale for lipid chains. In projection (*A*) two possibilities are indicated, corresponding to the two opposing monolayers.

boundary layer estimate in either case is about 40 lipids per monomer, which on correcting for molecular weight yields ~57 lipids per 200,000 dalton protein, in surprisingly good agreement with the number of first-shell lipids in cytochrome oxidase. Thus in spite of the rather complicated intramembranous shape of cytochrome oxidase, it appears that the immobilized first-shell lipid may form a continuous boundary around it.

Coexistence of fluid and immobilized lipid similar to that observed with cytochrome oxidase has also been seen with the Ca^{2+}-ATPase of sarcoplasmic reticulum, the integral membrane transport enzyme responsible for active reaccumulation of Ca^{2+} released as the trigger for muscular contraction. Again the immobilized lipid component has been observed both with lipid-substituted Ca^{2+}-ATPase (Hesketh et al., 1976) and with progressively delipidated-relipidated samples of intact sarcoplasmic reticulum membranes (Nakamura & Ohnishi, 1975; Griffith & Jost, 1978). A lipid to protein titration for sarcoplasmic reticulum membranes is given in Fig. 13. Approximately 80% of the total protein in the intact sarcoplasmic reticulum is the Ca^{2+}-ATPase pump protein. The titration again conforms to the linear form of Eq. 3 and implies that approximately $n_b \approx 17$ lipids per protein are immobilized independent of total lipid to protein ratio (i.e., $K_r \approx 1$). Other experiments on sarcoplasmic reticulum have given an estimate of $n_b \approx 20$ lipids per ATPase (Nakamura & Ohnishi, 1975), in reasonable agreement with the titration in Fig. 13.

Only a limited amount of dimensional information is available on the Ca^{2+}-ATPase. Le Maire et al. (1976b) have obtained a frictional ratio (f/f_{min}) of 1.4–1.6 from sedimentation studies on the detergent-solubilized delipidated monomer. This

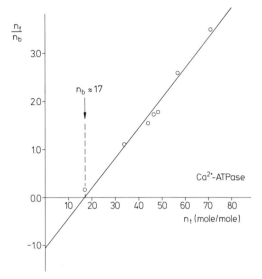

Figure 13. Lipid-protein titration of sarcoplasmic reticulum Ca^{2+}-ATPase preparations using the 16-doxyl stearic acid spin label. Nomenclature as in Fig. 10. Data replotted from Jost & Griffith (1978a).

indicates a rather asymmetric, elongated shape, with an axial ratio of 6–10 : 1 for the equivalent prolate ellipsoid. Calculations of the number of first-shell lipids made on the basis of this elongated ellipsoid (see Table 2) are given in Table 4. Quite good correspondence is obtained with the measured numbers of motionally restricted lipids measured with lipid spin labels. The extremely elongated ellipsoidal shape (cf. Table 2) cannot be taken too seriously, but the results do suggest that the intramembranous section may be relatively small compared with the rest of the protein.

An additional complication that arises in the estimate of the first-shell occupancy regards the state of aggregation of the protein in the membrane. There is some evidence that the Ca^{2+}-ATPase exists as an oligomer, possibly a tetramer, in the sarcoplasmic reticulum membrane (Murphy, 1976; Le Maire et al., 1976a). If there are direct protein-protein contacts at the monomer interfaces, the number of boundary shell molecules per monomer would be reduced accordingly. This is illustrated in Table 4, from which it is clear that with an intramembrane radius of 20 Å, the calculated boundary layer for a tetramer is consistent with the observed values for the number of immobilized lipids per monomer. Recent X-ray scattering data (Le Maire et al., 1981) have suggested that the effective intramembrane radius of the solubilized monomer is in fact close to 20 Å.

A somewhat similar analysis of the lipid environment of the Ca^{2+}-ATPase has been made based on activity measurements as a function of lipid to protein ratio in lipid-substituted complexes (Warren et al., 1974; Hesketh et al., 1976). In this way an annulus of lipid surrounding the protein was defined by the number of lipid molecules (~30 lipids per ATPase) required to maintain full activity of the protein (see Fig. 14). Beyond this value the activity is independent of lipid to protein ratio. This estimate is somewhat larger than the structural first shell from the boundary layer titration of Fig. 13, determined using spin labels. The latter corresponds more closely to the lipid to protein ratio at which all activity is lost. Thus it may be that rather more lipid than the structurally defined boundary layer is required to maintain full enzymatic activity, though the stoichiometries of the two are clearly very closely related. Perhaps not too much should be read into these differences, since the

TABLE 4 Calculated First-Shell Occupancies n_1 per Monomer for Ca^{2+}-ATPase Oligomers, as a Function of Intramembrane Radius r_1

r_1 (Å)	Monomer	Dimer	Tetramer
10.7[a]	17	13	9
12.7[a]	19	15	10
20.0[b]	29	23	16

[a]Major axis of assumed prolate ellipsoid. These two values correspond to the range of values deduced from the sedimentation data of le Maire et al. (1976b).
[b]Radius of assumed intramembranous section, based on the double-cylinder model suggested from the X-ray scattering measurements of le Maire et al. (1981).

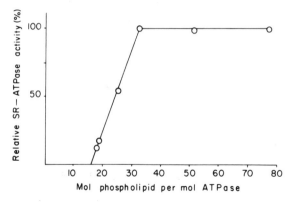

Figure 14. Specific activity of sarcoplasmic reticulum (SR) Ca^{2+}-ATPase at 37°C as a function of lipid to protein ratio. Adapted from Hesketh et al. (1976).

experiments were performed on different systems, and at low lipid to protein ratios the preparations might be heterogeneous. With cytochrome oxidase, for example, a reasonably good correlation was obtained between enzyme activity and amount of fluid bilayer lipid, the stoichiometry of inactivation agreeing fairly well with that of the immobilized shell lipids (Knowles et al., 1979).

B. Selectivity

In the preceding section it was shown by lipid to protein titration that there was no selectivity between the labeled and unlabeled phosphatidylcholine molecules with regard to the motionally restricted lipid sites in cytochrome oxidase–DMPC complexes. An interesting question arises as to whether there is any phospholipid head group specificity in the competition for first shell lipid sites in cytochrome oxidase. This can be checked using phospholipid spin labels with different head groups, but the same positionally spin-labeled chains (such as the labels shown in Fig. 15). Differences in the relative proportions of the fluid and immobilized ESR spectral components, for complexes of the same lipid to protein ratio, would then indicate preference for specific phospholipid head groups. Normalization relative to spin-labeled phosphatidylcholine would also give any preferential association relative to unlabeled phosphatidylcholine.

This approach has been used to obtain the spectra in Fig. 16 of the 14-doxyl phospholipid spin labels with various head groups, all in cytochrome oxidase–DMPC complexes of identical lipid to protein ratio. Qualitatively, it can be seen that the cardiolipin (diphosphatidylglycerol) spin label spectrum displays a considerably larger immobilized component than does the phosphatidylcholine spin label in this complex. The phosphatidic acid spin label also displays a somewhat larger immobilized component, whereas the spectra of the phosphatidylserine, phosphatidylglycerol, and phosphatidylethanolamine labels are all very similar to those

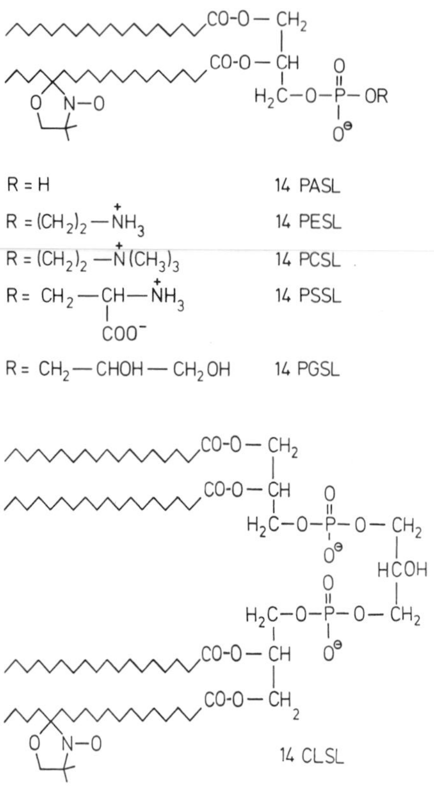

CO-O—CH₂
CO-O—CH
H₂C—O—P—OR

R = H   14 PASL

R = (CH₂)₂—NH₃⁺   14 PESL

R = (CH₂)₂—N(CH₃)₃⁺   14 PCSL

R = CH₂—CH—NH₃⁺   14 PSSL
          |
         COO⁻

R = CH₂—CHOH—CH₂OH   14 PGSL

14 CLSL

Figure 15. Phospholipid spin labels of different head groups, labeled with the doxyl moiety on the C-14 atom of the β-chain [β-14-(4′,4′-dimethyloxazolidine-N-oxyl)stearoyl-γ-acyl-α-phospholipid]: 14-CLSL, cardiolipin; 14-PASL, phosphatidic acid; 14-PGSL, phosphatidylglycerol; 14-PSSL, phosphatidylserine; 14-PESL, phosphatidylethanolamine; 14-PCSL, phosphatidylcholine.

of the phosphatidylcholine label. In addition, the fluid lipid spectral line widths are very similar for all the different labels.

The selectivity for cardiolipin and phosphatidic acid is seen very clearly when pairwise subtractions are performed between the spectra of the different labels [see Brotherus et al. (1980) for a detailed discussion of this method of subtraction]. Subtracting the phosphatidylcholine spectrum from the cardiolipin (or phosphatidic acid) spectrum gives a difference spectrum corresponding to an immobilized component, whereas subtracting the cardiolipin (or phosphatidic acid) spectrum from the phosphatidylcholine spectrum gives a difference spectrum corresponding to the fluid component. Similar subtractions between the other labels and the phosphatidylcholine spectrum yield essentially a flat baseline.

The quantitative endpoints of the pairwise subtractions can be used to determine the relative amounts of fluid and immobilized components in the two spectra being subtracted. This method takes advantage of the fact that both spectra refer to exactly the same lipid to protein ratio complex. Thus the line shapes of the two components are more likely to match between any two spectra than if the spectrum of the pure lipid or another complex were used as a reference spectrum. From intersubtractions with the cardiolipin label it is found that $\alpha^{CL} = 0.82$ and $\alpha^{PC} = 0.46$, and with

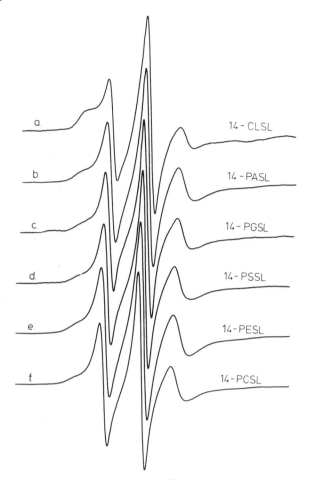

a.   14 - CLSL

b.   14 - PASL

c.   14 - PGSL

d.   14 - PSSL

e.   14 - PESL

f.   14 - PCSL

Figure 16. ESR spectra of different 14-doxyl phospholipid head group spin labels (see Fig. 15) added exogenously to separate aliquots taken from the same cytochrome oxidase–DMPC complex; lipid to protein ratio, 95 : 1; temperature, 32°C. From Knowles et al. (1981).

the phosphatidic acid label that $\alpha^{PA} = 0.63$ and $\alpha^{PC} = 0.47$, where α is the fraction of immobilized lipid. Thus a consistent value between the two sets of subtractions is obtained for the phosphatidylcholine reference spectrum.

The selectivity of the first-shell lipids for cardiolipin and phosphatidic acid can be interpreted either as a larger effective binding constant for these phospholipids or as an increased number of sites. For the first case, assuming that the effective number of sites is the same for both labels, the ratio of the two effective binding constants, K_r^A/K_r^B, can be obtained from the two measurements on complexes with the same lipid to protein ratio, n_t. From Eq. 3 of the preceding section, one then simply gets:

$$\frac{(n_f^*/n_b^*)^A}{(n_f^*/n_b^*)^B} = \frac{K_r^B}{K_r^A} \tag{4}$$

where the ratios on the left-hand side of the equation are obtained from the inter-subtractions. The difference in the effective free energy of association of the two lipids with the protein is then given by:

$$\Delta G_A^\circ - \Delta G_B^\circ = -RT \cdot \ln \frac{K_r^A}{K_r^B} \tag{5}$$

For the second case, assuming that the relative binding constants are the same, the observed selectivity would reflect a difference in the number of available binding sites accessible to the labeled lipids.

The data on the head group specificity in cytochrome oxidase–DMPC complexes (Fig. 16) is analyzed according to both models in Table 5. Thus, on the basis of the first possibility discussed above, cardiolipin has a more than fivefold greater effective affinity for the first-shell immobilized lipid sites than phosphatidylcholine, and phosphatidic acid has an approximately twofold greater affinity. The values for the increase in effective free energy of association (Table 5) are relatively modest and do not imply a highly specific form of interaction; for instance, the electrostatic interaction energy between two elementary electric charges 5 Å apart in a medium of dielectric constant $\varepsilon = 80$ (water) is about 830 cal/mol. An interpretation of the selectivity in terms of different numbers of sites seems less probable. It is unlikely, for example, that there would be extra sites for which the cardiolipin label could compete with the unlabeled phosphatidylcholine, but that were nonetheless totally inaccessible to labeled phosphatidylcholine, especially since the number of these extra sites (~ 35) is so large. Similarly it is less likely that there is such a large number of specific sites for cardiolipin. A hybrid situation is, of course, possible with some specific sites and others nonspecific but for which cardiolipin has a greater affinity (see pp. 227–231).

Titrations on a different cytochrome oxidase–phosphatidylcholine system (Cable & Powell, 1980) have also indicated a specificity for cardiolipin in terms of an

TABLE 5 Head Group Selectivity for the Immobilized Lipids in Cytochrome Oxidase–Dimyristoyl Phosphatidylcholine Complexes; Relative Association Constants K_r or First-Shell Occupancies n_1

L[a]	K_r^L/K_r^{PC}	$\Delta G_L^\circ - \Delta G_{PC}^\circ$ (cal/mol)	n_1^L	n_1^{PC}
CL*	5.4	1020	77	43
PA*	1.9	390	59	44
PE*, PS*, PG*	1.0	0	$n_1^L = n_1^{PC}$	

From Knowles et al. (1981).

[a]Asterisk indicates that the phospholipid is a spin label.

increased affinity rather than a greater number of sites. Preliminary experiments involving reconstitution of cytochrome oxidase in which unlabeled cardiolipin is mixed with phosphatidylcholine as the reconstituting host lipid have also revealed evidence for a specificity for cardiolipin (Jost & Griffith, 1978a). This is an alternative method for investigating selectivity in the boundary layer shell. By comparison with a sample of equal lipid to protein ratio reconstituted solely with phosphatidylcholine, it was found that the presence of 26% cardiolipin in the reconstitution reduced the amount of phosphatidylcholine spin label in the immobilized component by ~13%. This shows that the unlabeled cardiolipin competes more effectively with the spin-labeled phosphatidylcholine for boundary layer sites than does the unlabeled phosphatidylcholine. Clearly there are at least some sites that are not absolutely specific for cardiolipin but for which cardiolipin has a preferential specificity.

Thus the results with different head group spin labels and different lipids suggest that the immobilized lipid sites on cytochrome oxidase do not have a high degree of specificity. Phosphatidylcholine, phosphatidylserine, phosphatidylglycerol, and phosphatidylethanolamine were all found to be equivalent. This is consistent with a picture of a first shell of lipid whose chains occupy relatively nonspecific positions on the hydrophobic face of the protein. Head group specificities were seen, however, with certain phospholipids, namely, cardiolipin and phosphatidic acid. This indicates some degree of specific interaction of the phospholipid head group with the protein, further strengthening the assertion that the immobilized lipid is situated in sites that are in direct contact with the protein, rather than simply being trapped nonspecifically between protein aggregates. Selectivity is discussed further below for rhodopsin (Section IV.A) and the (Na^+, K^+)-ATPase (Section IV.D).

C. Degree of Immobilization

An important characterization of the properties of the first-shell boundary lipids is the degree to which their hydrocarbon chains are motionally restricted. For this it is necessary to obtain difference spectra corresponding to the more immobilized component in the spectra of the protein-lipid complexes. Such spectra for the cytochrome oxidase–DMPC system are given in Fig. 17. All lie close to the limit of motional sensitivity of conventional ESR spectroscopy, and it is for this reason that they were said to correspond to a lipid population that is immobilized on the ESR time scale. For the relatively simple case of *isotropic* Brownian motion, an estimate of the spin label correlation time τ_R giving rise to this type of spectrum is given by $\tau_R = a(1 - A_{max}/A_{zz})^b$, where $a = 5.4 \times 10^{-10}$ and $b = -1.36$. A similar empirical equation also gives the correlation time from the spectral line widths (Freed, 1976). Applying these equations, the observed reduction of splittings and increase in line widths when compared to the no-motion limits, gives values of τ_R in the 10^{-8} s time regime for the spectra of Fig. 17 (Knowles et al., 1979). In the case of anisotropic motions, of limited amplitude, it is clear that such estimates made using an *isotropic* formulation will give an approximate lower limit for the possible rate of motion. Clearly, regardless of the approach, the protein-associated

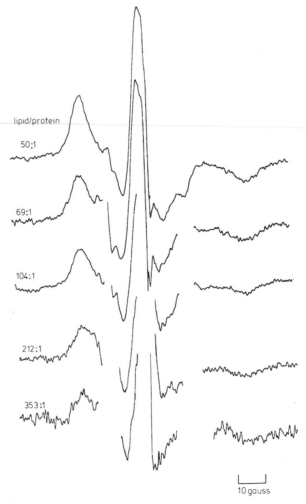

Figure 17. ESR difference spectra of the immobilized lipid component in cytochrome oxidase–DMPC complexes labeled with 14-doxyl phosphatidylcholine spin label, obtained by subtracting the fluid component from the spectra of Fig. 9. From Knowles et al. (1979).

lipid is *not* completely immobilized on the protein surface even on the short ESR time scale, but undergoes significant motion.

From an analysis of the difference spectra of Fig. 17, where the line widths and maximum hyperfine splittings can be fairly readily measured, the label correlation times all have approximately the same value, $\tau_R \sim 50$ ns, independent of the lipid to protein ratio (Knowles et al., 1979). This implies that the motion of the immobilized component is dominated by the protein rather than by the lipid. Thus, it appears that the first-shell lipid chains have rotational rates $\precsim 50$ times slower than the fluid bilayer lipids, independent of the amount of bilayer lipid present.

However, since the spectra are approaching the motional limit, this must be considered to be a relatively insensitive estimate.

D. Degree of Order

ESR experiments with oriented samples are required to determine the degree of order of the lipid chains giving rise to the motionally restricted lipid component. This is because the immobilized component occupies the full spectral anisotropy of the nitroxide group, independent of the amplitude of motion. In this motional limit the order parameter, defined previously in Section II.A, Eq. 1, refers to an orientational distribution of (apparently) static molecular tilt angles, rather than a time-averaged amplitude of angular motion.

The spectra of 5-doxyl stearic acid in (partially) oriented samples of cytochrome oxidase of various lipid to protein ratios are given in Fig. 18. From the discussion of the bilayer properties in Section II.A, it is clear that although this label will not give such good resolution of the immobilized component, it will have greater anisotropy, hence it will be better able to detect the orientation of the bilayer component. The samples with high lipid to protein ratios in Fig. 18 show a considerable degree of anisotropy between the spectra taken with the magnetic field oriented parallel and perpendicular to the plane of the substrate. At lower lipid to protein ratios, however, the spectra consist only of the immobilized lipid component, which showed no anisotropy with respect to the orientation of the magnetic field. This suggests either that the immobilized lipid is disordered on the surface of the protein, or that the samples with low lipid to protein ratio are macroscopically disordered on the glass substrate. Spectral subtractions performed on the spectra

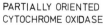

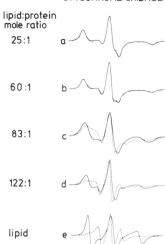

PARTIALLY ORIENTED
CYTOCHROME OXIDASE

lipid:protein
mole ratio

25:1 a

60:1 b

83:1 c

122:1 d

lipid e

Figure 18. ESR spectra of the 5-doxylstearic acid label in cytochrome oxidase–lipid preparations of various lipid to protein ratios, oriented on a glass slide. The spectra were recorded with the magnetic field parallel (solid line) and perpendicular (dashed line) to the normal to the slide. Samples a–d were recorded at 100% relative humidity and room temperature; sample e was recorded at 93% relative humidity. Adapted from Jost et al. (1973c).

of the samples at higher lipid to protein ratio indicate that the former is most probably the case (Jost et al., 1973c).

Deuterium NMR spectra also suggest that the degree of order of the first-shell lipid chains are less than those of the other bilayer lipids (see Chapter 3). Both methods seem to agree that there is some disordering, and certainly no ordering, of the first-shell lipid chains in cytochrome oxidase.

E. Exchange Rate

The exchange rate between the first-shell lipids and those of the rest of the bilayer determines whether the associated integral membrane protein rotates and translates within the plane of the membrane together with its first-shell lipids, or on the contrary, whether the first-shell exchange might provide a molecular mechanism for such rotational and translational motion. For integral membrane enzymes and transport proteins it is of considerable interest to know whether the first shell is essentially at exchange equilibrium during the catalytic or transport cycle, or whether the exchange rate is such that it could participate directly in the enzymatic or translocation steps.

The first-shell boundary lipids do exchange with the bulk bilayer lipids—they are not tightly bound—since lipid spin labels that are added exogenously are incorporated into the complexes and display the immobilized component (see, e.g., Fig. 16). An example of this exchange was given by Jost et al. (1977) in the fusion of spin-labeled phosphatidylcholine vesicles with unlabeled cytochrome oxidase vesicles. Lateral diffusion after fusion leads to dilution of the spin label in the unlabeled cytochrome oxidase lipids, and the spectrum changes from the single exchange-narrowed line arising from label-label interactions to the normal, three-line spectrum. The endpoint of the fusion is a spectrum indistinguishable from spectra in which the spin label has been homogeneously mixed with the lipid in cholate during reconstitution. Thus, the boundary layer shell exchanges and is at equilibrium with the bulk bilayer lipids. The kinetics of label incorporation are most probably limited by the fusion process. Thus only a lower limit can be put on the exchange rate between boundary and bilayer. This is $\nu_{ex} > 1$ h^{-1}, that is, the first time at which the incubation mixture was sampled.

Since the immobilized and fluid lipids are clearly resolved as separate components in the ESR spectrum, the two populations must be in slow exchange on the ESR time scale. This puts an upper limit $\nu_{ex} < (A_{zz} - a_0)/h \approx 5.10^7$ s^{-1} on the exchange rate between the first boundary layer shell and the subsequent fluid bilayer shells (see also Chapter 3, Section V). The correlation times for the motionally restricted lipid (see Section III.C) also put an upper limit on the exchange rate (i.e., $\tau_{ex} > 50$ ns) if there is no other independent motion of the lipids relative to the protein. However, this probably refers to τ_k, the rotational correlation time for segmental motion of the lipid chains. From considerations of lipid lateral diffusion rates in terms of τ_k, it has been argued that the exchange between the boundary and fluid lipid can be at least 10 times slower than the lipid-lipid exchange in unperturbed bilayers; that is, $\nu_{ex} \approx 10^5$–10^6 s^{-1} (Knowles et al., 1979).

Differential exchange rates of different lipid types are implied by the results on lipid head group selectivity given in Section III.B. If in the exchange equilibrium given in Section III.A

$$L^* + L_{n_1 - m} \overset{*}{L}_m P \underset{k_{\text{off}}}{\overset{k_{\text{on}}}{\rightleftharpoons}} L + L_{n_1 - m - 1} \overset{*}{L}_{m+1} P$$

it is assumed that the on-rate constant is the same for all species (i.e., diffusion controlled), then a relative equilibrium association constant, $K_r = k_{\text{on}}/k_{\text{off}} > 1$, implies a slower off-rate constant. Thus the data of Table 5 indicate that the average exchange rates $\nu_{\text{ex}} = k_{\text{off}}$, are five times and twice as slow for cardiolipin and phosphatidic acid, respectively, as for phosphatidylcholine.

In contrast to the spin label ESR situation, the first-shell and subsequent shell lipids appear to be in fast exchange on the deuterium NMR time scale. Deuterium NMR experiments on cytochrome oxidase complexes with specifically chain-deuterated phosphatidylcholines resolve only a single component spectrum, implying that the bulk bilayer and first-shell boundary lipids are in fast exchange on this time scale (Seelig & Seelig, 1978; Kang et al., 1979; Rice et al., 1979). This puts a lower limit on the exchange rate, $\nu_{\text{ex}} \gtrsim 10^4$ s^{-1}, determined by the ^{2}H quadrupole splitting (see Chapter 3). Thus the first boundary layer shell of lipids appears to be in slow exchange on the spin label ESR time scale and in fast exchange on the deuterium NMR time scale, with the remaining bilayer lipids. This result holds for several different lipid-protein systems and implies exchange rates in the range $10^4 < \nu_{\text{ex}} < 10^7$ s^{-1}.

F. Subsequent Shells

It is unlikely that the perturbations of the lipid chain motion by the protein would extend only to the first boundary shell, although they might be very rapidly attenuated. Information on perturbation of the subsequent lipid shells comes from the changes in the spectra of the fluid bilayer component with lipid to protein ratio.

Difference spectra corresponding to the fluid component in the spectra of cytochrome oxidase–DMPC complexes of varying lipid to protein ratios are given in Fig. 19. [For a detailed discussion of the techniques of spectral subtraction, see Jost & Griffith (1978b).] These difference spectra show a systematic change with lipid to protein ratio corresponding to the perturbation by the protein of the lipids beyond the first boundary shell. These changes are quantitated in Fig. 20 in terms of the empirical parameter $(A_\parallel - A_\perp)_{\text{eff}}$, defined in Fig. 19, which is determined by both line splittings and line widths. The changes in $(A_\parallel - A_\perp)_{\text{eff}}$ with lipid to protein ratio are given in Fig. 20a (broken line). It is not known whether the difference spectra in Fig. 19 really consist of a single component, corresponding to fast exchange between the more distant shells, or whether they consist of a superposition of unresolved components from shells in slow exchange.

The data for the more distant shells in cytochrome oxidase–DMPC complexes given in Fig. 20 are consistent with either fast or slow exchange but cannot distin-

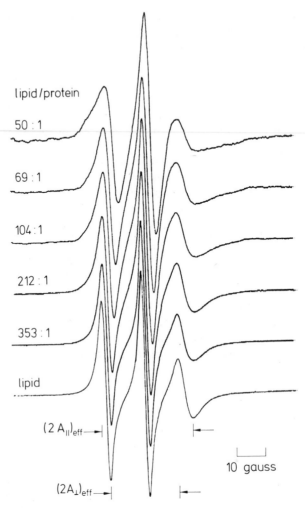

lipid/protein

50 : 1

69 : 1

104 : 1

212 : 1

353 : 1

lipid

$(2 A_{\parallel})_{eff}$

$(2A_{\perp})_{eff}$

10 gauss

Figure 19.   ESR difference spectra of the 14-doxyl phosphatidylcholine spin label in the fluid lipid bilayer component of cytochrome oxidase–DMPC complexes at 32°C; obtained by subtracting the immobilized component from the spectra of the complexes in Fig. 9. From Knowles et al. (1979).

guish between them (Knowles et al., 1979). It is clear that in either case the results indicate that the influence of the protein extends beyond the first boundary layer shell. At least two further shells are perturbed and possibly up to three further shells are weakly perturbed, depending whether the shells are in fast exchange on the ESR time scale. Regarding the nature of the perturbation of the mobility of the subsequent lipid shells by the protein, it is not possible from the spectra of Fig. 19 to determine whether the observed perturbation is accompanied by a change in the degree of order (or angular amplitude of motion) of the lipid chains or by a change in their rate of motion. The fluid components of Fig. 19 are in the spectral regime in which it is very difficult to distinguish between changes in the rate of motion and in order (Schreier et al., 1978).

Somewhat similar results have been obtained with the Ca^{2+}-ATPase from sarcoplasmic reticulum (see Fig. 21). In this case the data were analyzed only in terms of slow exchange between the subsequent shells. Again perturbation is detected beyond the first boundary shell, and this perturbation is much less than that of the boundary layer and falls off very quickly away from the protein.

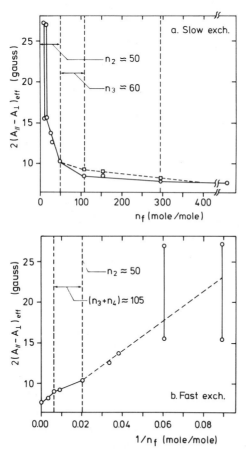

Figure 20. Variation of the empirical line width and splitting parameter $\Delta = (A_{\parallel} - A_{\perp})_{eff}$ of the 14-PCSL label in the fluid lipid component spectra of Fig. 19, as a function of lipid to protein ratio in the cytochrome oxidase–DMPC complex. (a) Variation with n_f both without ($\square$) and with ($\bigcirc$) correction for the underlying shells by subtraction, assuming slow exchange. (b) Variation with $1/n_f$ assuming fast exchange between the extra-boundary shells: $\Delta \sim (1/n_f) \cdot \Sigma n_i \cdot \Delta_i$. This would require an exchange rate exceeding $(\Delta_2 - \Delta_\infty)g\beta$ (i.e., $> 8 \times 10^6 \ s^{-1}$). From Knowles et al. (1979).

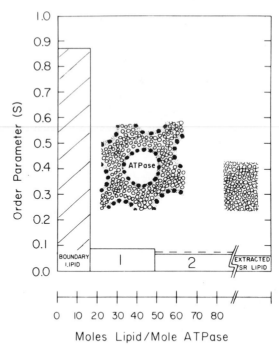

Figure 21. Variation of the empirical line width and splitting parameter $(A_\parallel - A_\perp)_{eff}$ of the 16-doxylstearic acid spin label as a function of lipid to protein ratio in sarcoplasmic reticulum Ca^{2+}-ATPase samples at 24°C. The spectra of the various lipid shells were obtained by subtraction from the spectra of different lipid to protein ratios, assuming slow exchange. From Jost and Griffith (1978a).

IV. IMMOBILIZED LIPID IN NATURAL MEMBRANES

A number of natural membranes have been studied using spin labels of the type shown in Figs. 5 and 15. In some cases, especially those membranes with low lipid to protein ratios, the spectra do not appear to be homogeneous, but rather consist of two components—one corresponding to fluid bilayer and the other to motionally restricted lipid. This section discusses specific examples that have been characterized. Other examples where immobilized lipid has been observed, although little or no reference has been made to it, include: plasma membranes from *E. coli* with labels *I*(12,3), *I*(5,10) and *I*(1,14) (Sackmann et al., 1973) and also with label *I*(3,10), which was biosynthetically incorporated into phosphatidylglycerol (Takeuchi et al., 1978); yeast cells with label *I*(5,10) (Nakamura & Ohnishi, 1972); and erythrocytes with the *I*(5,10) label (Shiga et al., 1977).

In contrast to the examples given in the preceding sections, in a natural membrane the lipid to protein ratio cannot usually be manipulated over a wide range. However, one great advantage of natural systems is that they do not rely on the efficiency of reconstitution techniques. Membrane fragments can be prepared by normal subcellular fractionation techniques that do not involve the use of organic solvents or detergents, or other manipulations that are likely to disrupt lipid-protein interactions.

A. Rhodopsin

The retinal rod outer segment (ROS) disc membrane is in many ways an ideal membrane in which to study lipid-protein interactions. It contains predominantly a single protein species, rhodopsin, the visual receptor protein, which accounts for 80–90% of the total membrane protein. The lipid to protein ratio in the membrane is fairly low (see Table 3), which on a mol/mol basis corresponds to approximately 65–70 lipids per rhodopsin molecule. Rhodopsin is a transmembrane integral glycoprotein having a molecular weight of 37,000; it is responsible for the primary step in visual transduction leading from light absorption to nerve excitation. Lipid-protein interactions could be involved both in stabilizing the structure of rhodopsin during the various stages of the photolytic and regenerative cycles and in regulating the conformational changes leading to the second transmitter response.

The ESR spectra of the stearic acid label $I(3,12)$ and the androstanol spin label (ASL) in bovine ROS membranes are given in Fig. 22. For both labels the spectra from the membranes contain an immobilized component that is not present in the spectra from the bilayers of the extracted membrane lipids. Digital subtraction of the lipid spectra from the corresponding membrane spectra yields a component that is immobilized on the conventional spin label ESR time scale for both labels (see Fig. 22). Since the axis of the maximum hyperfine splitting is differently oriented for the two labels (see Section II), it is clear that the motion of the lipids in this second membrane component are inhibited both with respect to angular motions of the chains and to rotation around the long molecular axis.

Spectral subtractions were performed to determine the proportion of the motionally restricted lipid component as a function of temperature. Above 15°C an immobilized component spectrum was used for the subtraction, rather than a fluid bilayer spectrum, as was used in Fig. 22, because of the difficulty in exactly matching the line shape of the very sharp fluid component at the higher temperatures. Two similar, but differently derived, simulated spectra, whose outer splitting was adjusted to allow for the temperature dependence, were used for the immobilized component (Watts et al., 1979); both gave essentially the same quantitative results. The percentages of the doubly-integrated intensity of the immobilized components for the stearic acid and androstanol labels are given in Fig. 23. From this it is seen that the percentage of immobilized spin label remains essentially constant with temperature up to 37°C, at a value of 37–39%. Similar experiments have been performed with phospholipid spin labels of different head groups (see Fig. 15), and results of the quantitation of the immobilized component are given in Table 6. Except for the phosphatidylserine spin label, which shows a slight specificity, the immobilized component corresponds to ~37% of the total spectral intensity for all spin labels.

If the lipid motion is restricted by direct contact with the hydrophobic surface of the protein, the fraction of immobilized lipid would be expected to be relatively

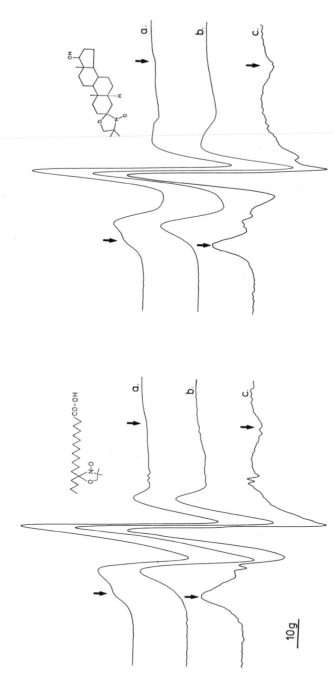

Figure 22. ESR spectra of lipid spin labels in bovine rod outer segment disc membranes. *Left:* line shape of stearic acid spin label *I*(3,12) in ROS membranes at 3°C (a), aqueous dispersion of extracted membrane lipids at 3°C (b), and difference spectrum (a minus b) to yield the immobilized component in the membrane spectrum (c); *right:* androstanol spin label (ASL) in ROS membranes at 3°C (a), aqueous dispersion of extracted membrane lipids at 5°C (b), and difference spectrum (a minus b) to give the immobilized component in the membrane spectrum (c). From Watts et al. (1979).

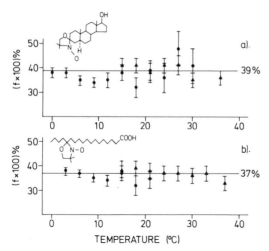

Figure 23. Percentage of integrated spectral intensity in the immobilized component of ROS membranes being probed by (a) the androstanol spin label (ASL) and (b) the stearic acid spin label $I(3,12)$ from 0 to 37°C. Quantitation by two methods of subtraction: ●, fluid spectrum subtraction to yield the immobilized spectrum; ▲, immobilized spectrum subtraction to yield fluid spectrum. From Watts et al. (1979).

TABLE 6   Percentage ($f \times 100$) of Immobilized Lipid in ROS Membranes at Different Temperatures, Measured by Spectral Subtraction.

| Lipid[b] | Temperature[a] | | | |
	3°C	15°C	24°C	37°C
14-SASL	37 ± 5 37 ± 3	37 ± 4 37 ± 4	37 ± 4	37 ± 4
ASL	44 ± 4 38 ± 3	41 ± 4 37 ± 4	41 ± 4 37 ± 6	38 ± 3
14-PCSL	42 ± 4 39 ± 4	38 ± 3	37 ± 3	37 ± 3
14-PESL	38 ± 5	37 ± 4	37 ± 4	38 ± 3
14-PSSL	45 ± 4 45 ± 5	43 ± 4	44 ± 3	44 ± 3
14-PASL	36 ± 5 30 ± 6	35 ± 5	33 ± 4	32 ± 4
14-PGSL	37 ± 3 38 ± 3	37 ± 3	37 ± 3	36 ± 3

From Watts et al. (1979).

[a]First line, from subtraction of an immobilized spectrum from the membrane spectrum; second line, from subtraction of a lipid spectrum from the membrane spectrum.

[b]14 refers to the position of the nitroxide moiety (see Fig. 15).

insensitive to temperature, as is observed, and it would be directly correlated with the size of rhodopsin. Taking the molecular weight of rhodopsin as 37,000 (Table 2) and the lipid to protein ratio in bovine disc membranes as 1.35 wt/wt (Table 3), it can be calculated that the value of $f = 0.37$ in Table 6 corresponds to $n_b \simeq 24$ immobilized lipids per protein. Information on the overall shape and size of rhodopsin has come from X-ray scattering and neutron scattering measurements of the radius of gyration of the protein. Three possible combinations of shapes and dimensions consistent with the observed radius of gyration are given in Fig. 24. Calculated values for the numbers of lipid molecules in the first three shells surrounding rhodopsin based on these shapes are given in Table 7. According to these estimates, the number of immobilized lipids, $n_b = 24$, would be sufficient to form a single boundary layer in direct contact with the hydrophobic surface of the protein. Thus an interpretation of the motionally restricted lipid is possible in this natural membrane system that is similar to that deduced from the lipid-protein titrations of the delipidated and reconstituted systems of Section III. Some caution is necessary, however, in such interpretations, since the detailed shape of rhodopsin is not yet known, although a cylindrical shape has been suggested by Albert and Litman (1978) from proteolysis of rhodopsin and analysis of the fragments by circular dichroism.

Some estimate of the degree of motion of the motionally restricted lipid component in ROS membranes and its temperature dependence can be obtained from the difference spectra, as was done with cytochrome oxidase in Section III.C. The temperature dependence of the outer line splittings A_{max}, and outer line widths ΔH, deduced from the immobilized component difference spectra for 14-doxyl stearic acid such as in Fig. 22c (left side) are given in Fig. 25. At the higher temperature

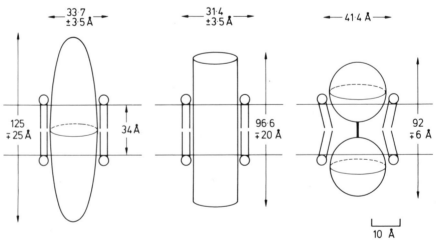

Figure 24. Dimensions of the rhodopsin molecule consistent with the radius of gyration measured by X-ray scattering (Sardet et al., 1976) for an ellipsoid, a cylinder, and a dumbbell. The putative location of the lipid bilayer is indicated.

TABLE 7   Calculated Radii and Numbers of Molecules in the Lipid Shells Surrounding Rhodopsin [a]

Shell, i	Ellipsoid		Cylinder		Dumbbell	
	r_i (Å)	n_i	r_i (Å)	n_i	r_i (Å)	n_i
1	18.6 ± 1.4	24 ± 2	18.1 ± 1.8	24 ± 2	21.4 ± 1.3	28 ± 2
2	23.4 ± 1.4	31 ± 2	22.9 ± 1.8	30 ± 2	26.2 ± 1.3	34 ± 2
3	28.2 ± 1.4	37 ± 2	27.7 ± 1.8	36 ± 2	31.0 ± 1.3	41 ± 2

From Watts et al. (1979).

[a]Calculated from X-ray scattering measurements of Sardet et al. (1976). Radii are calculated to the center of a lipid chain (diameter = 4.8 Å) at a height ± 17 Å from the center of the membrane, assuming the shapes indicated—see Fig. 24.

the quality of the difference spectra is not as good as in Fig. 22c because of the more rapid increase in relative amplitude of the fluid component. However, resolution is sufficiently good in the outer wings for measurements to be made. The results in Fig. 25 clearly show a decrease in outer splitting and increase in outer line widths. These spectral changes are characteristic of increasing motion and thus correspond to an increase in mobility of the immobilized component with increasing temperature (cf. Section III. C). Thus the immobilized lipid component in ROS membranes is not totally rigid, but displays some temperature-dependent molecular motion, although not as great as that of the fluid lipid component.

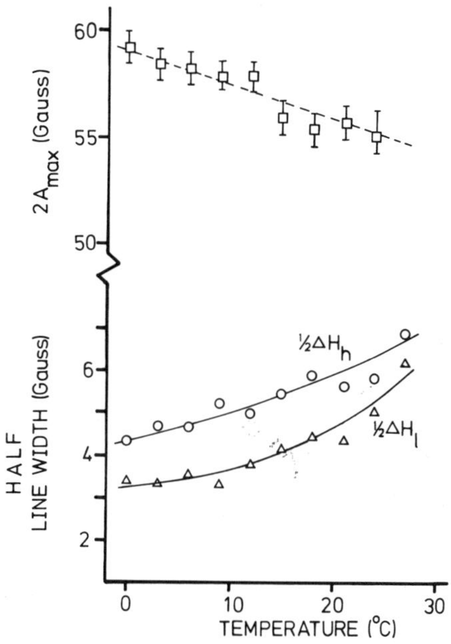

Figure 25. Temperature dependence of the outer splitting A_{zz} (□) and the low field, ΔH_l (△) and high field ΔH_h (○) outer line widths in the immobilized component difference spectra of the stearic acid spin label $I(3,12)$ (cf. Fig. 22c, left-hand side) in bovine ROS membranes. From Watts et al. (1982).

TABLE 8   Effective Rotational Correlation Times[a] for Rhodopsin, the Immobilized Lipid Component, and the Fluid Lipid Component in Bovine Rod Outer Segment Membranes Deduced from Spin-Labeling Studies of Maleimide-Labeled Rhodopsin and 14-Doxylstearic Acid.

ROS Membrane Component	Temperature		
	3–4°C	20°C	30°C
Rhodopsin (rotation axes unknown)[b]			
From H″/H[c]	65 μs[d]	20 μs (20 μs)[d]	15 μs[e]
From L″/L[c]	30 μs[d]	20 μs (18 μs)[d]	30 μs[e]
Immobilized lipid			
From outer line widths (ΔH_l; ΔH_h)[e]	100 ns; 70 ns	40 ns; 60 ns	~20 ns[f]
From outer splitting ($2A_{zz}$)[e]	15 ns	10 ns	~ 6 ns[f]
Fluid lipid		~ 1–3 ns[g]	

From Watts et al. (1982).

[a]Calculated using isotropic models; these are order-of-magnitude estimates only.

[b]Rhodopsin labeled with maleimide spin label on one or more sites whose locations are not known with respect to the protein or membrane. Most of the label is "immobilized" on the conventional ESR time scale, but label-protein motions may occur. [For details see Kusumi et al. (1978); Baroin et al. (1977); Watts et al. (1982).]

[c]Diagnostic line height ratios measured in saturation transfer ESR spectra.

[d]From Baroin et al. (1977).

[e]From Watts et al. (1982).

[f]Extrapolated from 27°C value.

[g]Approximate value from simulations (Schindler & Seelig, 1973).

Effective (isotropic) rotational correlation times for the immobilized and fluid lipid components as well as for the protein rotation are compared in Table 8. [The effective protein rotational correlation times were obtained from saturation transfer ESR* measurements on the covalently spin-labeled protein component of the membrane (Baroin et al. (1977); Watts et al. (1982).] From this it is seen that the immobilized lipid component has an effective rotational correlation time more than an order of magnitude slower than that of the fluid lipid component but at least two to three orders of magnitude faster than that for the protein rotation. Thus, although the immobilized lipid component has its motion quite significantly restricted by its proximity to the protein, it has considerable motion itself relative to the protein.

Reference to Table 7 suggests that the remaining 63% of the membrane lipid that makes up the fluid component is just sufficient to form a second shell of lipid surrounding the protein. ESR data suggest that there is some perturbation of the lipid in this layer, although this is much attenuated compared with the effects on

*For a discussion of the technique of saturation transfer ESR, see Thomas et al. (1976), Hyde (1978), and Marsh (1981).

the first shell (Watts et al., 1979, 1982). Pontus and Delmelle (1975), using a combination of spin labels, have estimated that only two thirds of the lipids in ROS disc membranes are in a fluid state, with the remaining 34% of nonfluid lipids correlating well with the fraction of immobilized lipid given in Table 7. These authors also observed lipid immobilization in partially delipidated ROS disc membranes; this possibly corresponds to enrichment in the immobilized lipid component observed in Fig. 22, but may also result from protein aggregation and denaturation arising from extensive delipidation. In reconstituted rhodopsin-lipid membranes, Hong and Hubbell (1972) have observed an increase in the effective order parameter of the spin-labeled lipid chains (cf. Section II) with increasing protein content of the membranes. This immobilization of the lipid chains by rhodopsin could be the result both of the perturbation of the more distant lipid shells and of the presence of an unresolved immobilized component. Favre et al. (1979) have studied the lipid environment of rhodopsin using spin-labeled fatty acids covalently attached to hydrophobic sites on the protein. Their results with labels far removed from the point of attachment suggest an environment with fluidity similar to that obtained for the second-shell lipid (or intermediate between first and second shells). Spectra of labels close to the point of attachment, on the other hand, gave spectra that were immobilized on the conventional ESR time scale, suggesting a first-shell location for the upper section of the chain.

Favre et al. (1979) have also pointed out conditions under which they can induce a separate immobilized lipid component: delipidation, prolonged bleaching, and high label concentrations. These conditions do not apply to the preparations whose properties are given in Figs. 22–25 (Watts et al., 1979).

Fatty acid labels covalently bound to rhodopsin (less than 1 mol per mol of protein) reconstituted with various lipids (Davoust et al., 1979, 1980) have been found to give rise to a two-component-like spectrum, apparently consisting of both fluid and immobilized components. The apparent proportion of immobilized component was found to vary with temperature and lipid to protein ratio. (Although it has since been pointed out by Brotherus et al. (1980) that by ignoring the temperature dependence of the immobilized component, such estimates could be in error by 50–100%). At least part of these temperature-dependent effects was attributed to protein aggregation. More recently, the data have been interpreted in terms of an exchange of the label between the first and second lipid shells at a rate intermediate-to-slow on the ESR time scale: $\nu \sim 10^7$ s^{-1} (Devaux et al., 1981; Devaux & Davoust, 1980). Such an exchange is both temperature and presumably concentration dependent, and could also have important implications for the quantitation of the immobilized components with freely diffusable labels, if the rates are within this range. In addition, it has recently been shown that the immobilized component seen by the covalently linked labels under conditions of aggregation induced by delipidation or extensive bleaching is quite different from the motionally restricted component observed by freely diffusable labels under normal conditions as described above (Watts et al., 1981). The aggregation-induced component was found to have a considerably larger hyperfine splitting, with relatively little temperature dependence, in contrast to the freely-diffusable motionally restricted lipid component. This definitively demonstrates that the two components are of quite different origin, and that the results of Figs. 22–25 (Watts et al., 1979) refer to lipid-protein interactions in rod outer segment disc membranes in their natural, unperturbed state.

It has recently been demonstrated (Watts et al., 1981, and see below) that the artifactually induced immobilized lipid has quite different spectral characteristics from that found by Watts et al. (1979) in normal, unbleached membranes.

In conclusion, it appears that in a membrane where no protein aggregation occurs (e.g., the ROS membrane) the amount of motionally restricted lipid is independent of temperature but shows sensitivity in its spectral shape and maximum hyperfine splitting (Watts et al., 1979; Brotherus et al., 1980). In contrast, when aggregation occurs, the amount of immobilized lipid is dependent on the degree of aggregation, and the immobilized spectrum in this case is much less temperature dependent (Davoust et al., 1980; Watts et al., 1981).

B. Acetylcholine Receptor

The electroplax membranes of electric fish contain large amounts of the acetylcholine receptor, an integral membrane protein complex that regulates the cation permeability of the postsynaptic membrane in response to acetylcholine released from the presynaptic nerve terminal. Membranes highly enriched in the receptor, up to 50% of the total protein, can thus be obtained simply by density gradient separation of membrane fragments from the electric organ, without the use of detergents or other means of protein extraction or purification.

ESR spectra of two stearic acid spin label isomers $I(m,n)$ and the steroid label ASL (see Figs. 5 and 15 for structures) in acetylcholine receptor-rich membranes from *Torpedo marmorata* are given in Fig. 26. In each case the spectra from the membranes contain an immobilized spin label component (indicated by the arrows) that is not present in the spectra from aqueous dispersions of the extracted membrane lipids. The spectra were recorded at different temperatures for the three labels and correspond to those at which the immobilized spin label component is best resolved.

A similar immobilized component is also seen in the membranes with phospholipid labels, both phosphatidylcholine and phosphatidylethanolamine, as is indicated in Fig. 27. In this case the spectra are complicated by an underlying spin-spin broadened component corresponding to vesicles of unincorporated phospholipid spin label. The immobilized component of the phospholipid spin label in the membrane is seen more clearly after digital subtraction of the spin-spin broadened component. Since an immobilized component is observed in the spectra of both of the fatty acids, the steroid, and two types of phospholipid spin labels, it is most likely that this component corresponds to an immobilized lipid population within the membrane rather than some specific binding of the label to the protein.

Bienvenue et al. (1977), using spin-labeled acylcholines to probe the immediate environment of the receptor binding site, concluded that the spin labels were immobilized only because of tight binding of the choline moiety to the protein and not because of the lipid. It is possible that the bound acylcholine labels do not sense

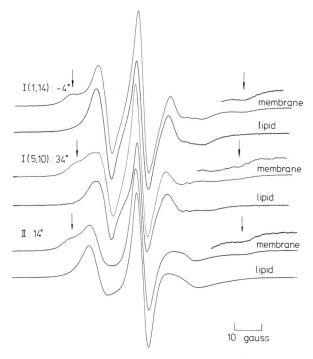

I(1,14) : -4°

membrane

lipid

I(5,10) : 34°

membrane

lipid

II : 14°

membrane

lipid

10 gauss

Figure 26. ESR spectra of lipid spin labels in acetylcholine receptor-rich membranes from *Torpedo marmorata* and aqueous bilayer dispersions of extracted membrane lipids. The upper spectrum of each pair is from the membranes and the lower from the lipid bilayers. *Upper pair of spectra:* stearic acid spin label *I*(1,14) at −4°C; *middle pair of spectra:* stearic acid spin label *I*(5,10) at 34°C; *lower pair of spectra:* androstanol spin label, ASL at 14°C. From Marsh and Barrantes (1978).

exactly the same immobilized sites occupied by the freely diffusable lipid labels of Fig. 26, and 27; since the acetylcholine recognition site located on the 40,000 dalton subunit may protrude well out into the aqueous phase.

The observation of the immobilized lipid component with the phospholipid labels in Fig. 27 is important, since Rousselet et al. (1979b) have reported that an immobilized component is observed with the fatty acid but not with the phosphatidylcholine spin label, implying some specific binding of the fatty acid. Reasons for the discrepancy could lie in the somewhat different preparative procedures used and in the difficulty of incorporation of the phospholipid labels. In particular, Rousselet et al., (1979b) used the phosphatidylcholine exchange protein (see Volume 1, Chapter 4), for which rigid membranes are known to be poorer substrates, to achieve label incorporation. Thus it is possible that the label was preferentially incorporated into more fluid regions of the membrane (and only into the outer monolayer). The results of Figs. 26 and 27 strongly indicate that the effects of lipid-protein interactions in these acetylcholine receptor-rich membranes are to induce immobilization in a substantial portion of the membrane lipid population, although the heterogeneity of the preparations makes it difficult to give a precise molecular interpretation.

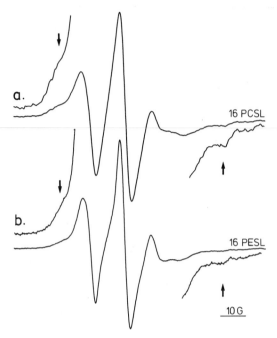

Figure 27. ESR spectra of phospholipid spin labels in acetylcholine receptor-rich membranes from *Torpedo marmorata*. (a) 16-PCSL phosphatidylcholine spin label at −5°C [*II*(1,14)]. (b) 16-PESL phosphatidylethanolamine spin label at −1°C. The spin-spin broadened component from unincorporated spin label vesicles has been digitally subtracted from these spectra. From Marsh et al. (1981).

C. Chromatophores

Chromatophores are homogeneous preparations of photosynthetic membranes from photosynthetic bacteria such as *Rhodopseudomonas spheroides*. More than half the chromatophore proteins is made up of two particular proteins, the reaction center protein (MW 80,000), which contains only a small amount of bacteriochlorophyll, and the light-harvesting protein, which contains the majority of the photopigment. When growing under low light intensity, the bacterium requires more of the light-harvesting protein and although the total lipid to protein ratio remains the same, the proportion of this one protein increases in response to the lower light intensity (Takemoto & Huang Kao, 1977).

The ESR spectra of three different stearic acid spin label isomers *I(m,n)* in chromatophores and in aqueous dispersions of extracted chromatophore lipids have been reported by Birrell et al. (1978). The spectra of the 16-doxylstearic acid isomer [I(1,14)] in chromatophores clearly reveal an immobilized component that is not present in the extracted lipids. Spectral subtraction showed that the immobilized component accounted for 59 ±3% of the total spin label intensity for this label (Birrell et al., 1978). Subtraction endpoints for the other two positional isomers

(7- and 12- positions), were not as clear, but it was concluded that the immobilized component corresponded to 60–75% of the total spin label intensity for these isomers also. Thus it seems most probable that the broad spectral component corresponds to an endogenous lipid population that is relatively immobilized on the ESR tme scale through interaction with the membrane protein. These lipid molecules are motionally restricted along the entire length of their chains by the presence of the proteins.

The content of the light-harvesting protein was varied by changing the light intensity under which the bacteria were grown. The characterization of these different chromatophores is given in Table 9, where the light absorbance at 800 nm is taken as a measure of the amount of light-harvesting protein in the preparation. Clearly, the content of this protein varies considerably (at the expense of the other proteins), while the total lipid to protein ratio remains constant.

As shown in Fig. 28, the percentage of the motionally restricted fatty acid (upper line) increases as the bacteriochlorophyll content increases (i.e., as the light level decreases). To examine the charge effects on these interactions, a positively charged quaternary ammonium group was substituted for the carboxyl group of the spin label (lower line in Fig. 28). The clear difference in the behavior of the two labels strongly suggests that there is some selectivity in the association of the lipid with the protein (cf. Section III. B), although for a multicomponent lipid mixture such as this (23% phosphatidylcholine; 35% phosphatidylethanolamine; 34% phosphatidylglycerol; 4% cardiolipin; 3% phosphatidic acid), it is difficult to distinguish between preferential lipid-lipid and preferential lipid-protein interactions.

Since the amount of spin label immobilized increases with increasing bacteriochlorophyll concentration, it can be concluded that some of the spin labels are interacting directly with the light-harvesting protein. Extrapolation to zero bacteriochlorophyll content in Fig. 28 indicates that roughly 43% of the fatty acid spin labels are motionally restricted by proteins other than the light-harvesting protein (23% for the trimethyl ammonium label). Thus at the highest bacteriochlorophyll content examined (11 nmol/100 μg protein) the fraction of immobilized fatty acid associated specifically with the light-harvesting protein can be calculated from Fig.

TABLE 9 Composition of Chromatophore Material and Percentages of Fatty Acid (16-Doxylstearate) Spin Label Immobilized

Light Intensity During Growth (ft-c)	Specific Bacteriochlorophyll Content[a]	Weight Ratio of Phospholipid to Protein	Light-Harvesting Bacteriochlorophyll (A_{800}/mg of Protein)	16-Doxylstearic acid Immobilized (%)
20	11.0	0.31	6.5	68
600	8.1	0.33	3.6	59
8500	2.1	0.34	0.7	48

From Birrell et al. (1978).
[a]In nmol of bacteriochlorophyll per 100 micrograms of protein.

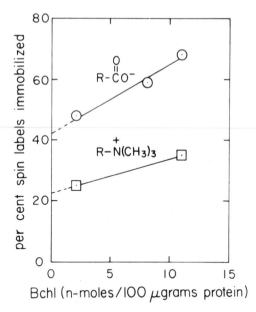

Figure 28. Effect of bacteriochlorophyll (Bchl) content on the amount of spin label immobilized by chromatophore proteins: ⊙, 16-doxylstearic acid [I(1,14)]; ☐, 16-doxyloctadecyltrimethyl-ammonium methanesulfate [l-$N^+(CH_3)_3$(1,14)]. From Birrell et al. (1978).

28 to be at least (68 - 43)/68 = ~37% (neglecting the reduction in binding to other proteins). This indicates that the immobilization of lipids by the light-harvesting protein can be a very significant contribution to the protein-lipid interactions in the chromatophores.

D. Purple Membrane

The purple membrane patches of the halophilic bacterium *Halobacterium halobium* contain a single protein, bacteriorhodopsin, which is arranged as trimers in the membrane in a two-dimensional crystalline array. Bacteriorhodopsin is a retinal-containing protein of molecular weight 26,000, which functions as a light-driven proton pump. As was explained earlier (Section I. B), this is one of the integral membrane proteins whose structure has been best determined. The lipid to protein ratio of the native membrane is very low (see Table 3) and corresponds to only 10 lipid molecules per bacteriorhodopsin monomer. In the two-dimensional hexagonal crystalline structure, three bacteriorhodopsin monomers are arranged around one of the crystallographic three-fold axes in such a way as to suggest that not only protein-protein contacts occur, but also lipid is trapped in the center of the trimer.

Spin-labeling studies using the 16-doxylstearic acid spin label show the presence of two spectral components at 37°C (Chignell & Chignell, 1975; Jost et al., 1978), dominated by the broad spectral component. Quantitative analysis is difficult because the small fraction of bilayer is broadened by the limited size of the lipid pool.

From model building, based on compositional data and the electron diffraction results, it was estimated that all except two to seven lipid chains are in direct contact with the protein (Jost et al., 1978). The models suggested that all the lipid may not exchange with this pool, so the added spin label also would not sample all lipid sites. Thus, the usual method of arriving at the number of sites in the boundary layer from ESR data was considered as probably inappropriate in the native crystalline membrane. The ESR spectra are compatible with models derived from diffraction data in that a very large immobilized component is observed and a few lipid sites have molecular motion that approaches that of a fluid bilayer.

Experiments with oriented membrane stacks, using the 16-doxylstearic acid label, revealed little or no orientation of the immobilized component, which accounted for most of the spin label intensity (Jost et al., 1978). The small degree of anisotropy that was observed was attributed to the limited pools of fluid lipid. Thus it would appear that, in common with the results for cytochrome oxidase given in Section III. D, the chains of the immobilized component are spatially disordered on the protein hydrophobic surface.

In the same study, it was concluded that the degree of association of the 16-doxyl lipid spin labels (R^*) with bacteriorhodopsin was in the order of increasing negative charge: R^*—$COO^- > R^*$—$COOH > R^*$—$N^+(CH_3)_3$. Such studies are of particular relevance in view of the high proportion of negatively-charged lipids (68% phosphatidylglycerophosphate; 17.4% glycolipid sulfate; 7.1% phosphatidylglycerosulfate; 1.3% phosphatidylglycerol) in the membrane, and the preponderance of positively charged amino acid residues in the helix linking regions of the proposed bacteriorhodopsin structure in Fig. 4. Again, as noted in Section IV.C, it is difficult to distinguish between lipid-lipid and lipid-protein interactions.

E. (Na$^+$,K$^+$)-ATPase

The (Na$^+$,K$^+$)-ATPase, a transmembrane protein involved in the active transport of the Na$^+$ and K$^+$ across cell plasma membranes, is normally isolated in enriched membrane fragments with a lipid composition very similar to the original membrane used for its isolation (Jorgensen, 1975; Skou & Nørby, 1979). Spin label studies on the interaction of lipids with the (Na$^+$,K$^+$)-ATPase from the electric organ of *Electrophorus electricus* have been reported by Brotherus et al. (1980). The protein was purified along with some of the native lipids from the microsomal membranes. Four derivatives of the 14-proxyl spin label [$I(3,12)$] were used, with the oxygen of the doxyl ring replaced by carbon to give the more stable proxyl spin label (Keana, 1979). The head group portions of these labels were chosen such that one possessed a residual negative charge, one had a positive charge, and the other two were neutral (an alcohol and a dimethyl phosphate ester).

Pairwise subtractions between the negatively and positively charged labels (see Section III.B) indicated that the spectra consist of the same two components, but in different proportions. Quantitation of the relative amounts from these pairwise subtractions (as described in Section III. B) showed that the fractions were approximately 0.57 and 0.25 for the negatively charged and the positively charged labels, respectively, with an intermediate value of roughly 0.35 for the two neutral labels. Very similar results were obtained by the more conventional single-component subtraction method (see Section III. A), using a partially delipidated (Na^+, K^+)-ATPase sample as the immobilized reference component.

The measured fraction of immobilized spin label as a function of temperature is given in Fig. 29. In common with the results for the ROS membrane discussed in Section IV. A, it is seen that the fraction of lipid immobilized is practically independent of temperature, as might be expected from the boundary layer at the protein interface. The pairwise subtractions showed that the outer splitting of the immobilized component ($2A_{zz}$) varied with temperature from 64.1 ±0.5 gauss at 5°C to 58.9 ±0.5 gauss at 35°C. These data are similar to the results we subsequently measured for the ROS membrane (see Fig. 25), and indicate that the immobilized lipid is not completely rigid, but has some flexibility.

The inset of Fig. 29 demonstrates the effect of subtracting with an immobilized spectrum with constant A_{zz} splitting. Here an apparent decrease in bound fraction is observed, showing that careful subtractions require that the temperature dependence of both spectral components must be taken into account. For labels trapped by protein aggregation, it is possible that this correction may not be necessary (Davoust et al., 1979 and 1980; Watts et al., 1981).

From the values of the fractions of label immobilized by the ATPase it was

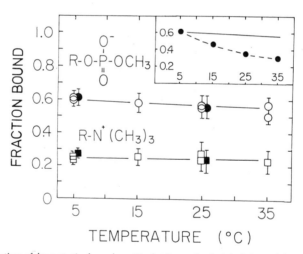

Figure 29. Fraction of the negatively and positively charged spin labels immobilized as a function of temperature in membranous (Na^+,K^+)-ATPase preparations. Open symbols: pairwise (inter-) subtraction method. Closed symbols: single-component subtraction method. Vertical lines: range of systematic error introduced by calculations in both methods when values of endpoints are used that reflect deliberate over- and under-subtractions. All single-component subtractions were corrected to compensate for the presence of 10% bilayer component in the spectrum from the partially delipidated sample. *Inset:* solid circles and dashed line show the artifactual decrease in the immobilized fraction of the methyl phosphate label with increasing temperature caused by using the 5°C reference sample for all single-component subtractions. From Brotherus et al. (1980).

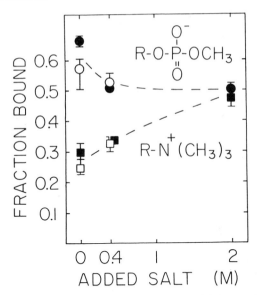

Figure 30. Fraction of negatively and positively charged spin labels immobilized in (Na⁺,K⁺)-ATPase membranes, as a function of salt (LiCl) concentration: circles, label with negatively charged methyl phosphate head group; squares, label with positively charged quaternary amine head group. Two types of pairwise subtraction were used: solid symbols indicate intersubtraction of two spectra of one spin label at different LiCl concentrations; open symbols are points obtained by intersubtraction of spectra from the two oppositely charged labels at the same LiCl concentration. Within experimental error, the salt effect was independent of temperature over the range examined (5–35°C), and the points are the means of two to four pairs of spectra at different temperatures; the vertical bars show the range of values. From Brotherus et al. (1980).

calculated using Eq. 4 that the effective binding constant of the negatively charged label is about four times that for the positively charged label, and 2.5 times larger than for the neutral labels when the lipid background is a mixture of native lipids (Brotherus et al., 1980). These relative affinities correspond to differences in the effective free energy of association of 820 and 540 cal/mol, respectively. These values lie within the range expected for purely electrostatic effects. The effects were shown to be completely screened by increasing the salt concentration (Fig. 30); at high ionic strength in 2 M LiCl, both charged labels showed essentially identical spectra. Thus, both the order of magnitude and the salt dependence suggested a purely electrostatic origin of the observed selectivity. Experiments with the (Na⁺,K⁺)-ATPase in a defined lipid background confirm that these charged head groups have different relative binding constants (Brotherus et al., 1981, and see Chapter 6).

At 2 M LiCl, a limiting value of ~0.48 for the motionally restricted fraction was observed for both charged labels. This corresponds to an effective boundary shell of approximately 61 lipid molecules per functional unit (MW of ~ 314,000). Electron microscopic studies on the (Na⁺,K⁺)-ATPase from rabbit kidney (Deguchi et al., 1977) suggested that the intramembranous section of the enzyme corresponds to two 50 Å diameter cylinders. We have calculated that such a structure could accommodate 57–72 lipid molecules around its perimeter, depending on whether the dimer interface is accessible. Thus the observed number of immobilized lipids would be consistent with a single boundary layer around the protein. In conclusion, it should perhaps be mentioned that earlier spin label studies on the (Na^+, K^+)-ATPase membranes had already revealed an immobilized component (Simpkins & Hokin, 1973; Grisham & Barnett, 1973), although this either was not commented on or was not specifically interpreted as boundary lipid.

V. CONCLUSIONS AND SUMMARY

The spin label studies discussed in Sections III and IV have revealed the existence of a population of relatively immobilized lipid, distinct from the fluid bilayer lipid, in a variety of membranes containing large integral membrane proteins. The immobilized component is structurally part of the bilayer and exchanges rapidly with the fluid part of the bilayer; the restriction of motion arises from the lipid chains making direct contact with the hydrophobic surface of the membrane protein. The amount of immobilized lipid seems to correlate with the shape and size of the protein in a way that suggests that this lipid forms the first shell of lipid surrounding the protein (see Table 10). For instance, for proteins that protrude to equivalent extents from the membrane, the intramembranous perimeter will increase approximately according to the square root of the molecular weight. For cytochrome oxidase, rhodopsin, and the (Na^+, K^+)-ATPase, the numbers of immobilized lipids are in the ratio 1 : 0.48 : 1.22, in good agreement with the ratios of the square roots of their molecular weights, which are 1 : 0.44 : 1.25, respectively. However, the structure of the membrane proteins is not known in sufficient detail or sufficiently detailed ESR experiments have not been performed) to accurately define the intramembranous perimeter.

TABLE 10 Values of the Number of Immobilized Lipids (n_b) from ESR Spin Label Studies, and Calculated Estimates of the Number of First-Shell Lipids (n_1) Interacting with Some Integral Membrane Proteins

Protein	n_b (mol/mol)	n_1 (mol/mol)	References[a]
Cytochrome oxidase	48	47	(1)
	55 ±5		(2)
Rhodopsin	24 ±3	24 ±2	(3)
Ca^{2+}-ATPase	15–20	17–19	(4, 5)
(Na^+, K^+)-ATPase	61 ±4	(64)	(6)

[a]References: (1) Jost et al. (1973a, 1977); (2) Knowles et al. (1979); (3) Watts et al. (1979); (4) Griffith & Jost (1978); (5) Nakamura & Ohnishi (1975); and (6) Brotherus et al. (1980).

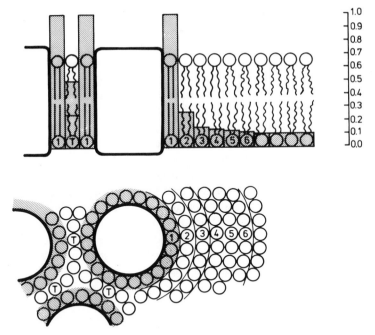

Figure 31. Schematic indication of the degree of chain immobilization in the various shells of lipid surrounding cytochrome oxidase in DMPC complexes. The vertical scale corresponds to values of $(A_{\parallel} - A_{\perp})_{eff}$ normalized with respect to the maximum hyperfine anisotropy of 25 gauss (see Section III.F). Details beyond the second shell depend on whether the lipids are in fast or slow exchange; T is a shared second shell that exists only in regions of high protein packing density (low lipid to protein ratios), where the perturbations from two adjacent protein molecules overlap. From Knowles et al. (1979).

The ESR experiments are consistent with similar measurements using ^{2}H-NMR if it is assumed that the immobilized component is in fast exchange with the fluid component on the ^{2}H-NMR time scale, whereas it is in slow exchange on the ESR time scale. ^{2}H-NMR results are not available on the stoichiometry of the presumed immobilized components, nor is any information available on whether the presumed component, with which the bulk lipid is in fast exchange in the ^{2}H-NMR experiment, corresponds to the first or to subsequent shells of lipid observed by ESR spin labels.

The basic features of the immobilized lipid component as observed by spin label ESR were given in Section III in the discussion of the reconstituted lipid-protein systems. All these features have been observed in one or more of the natural membrane systems discussed in Section IV, indicating that such features are likely to be operative in natural membranes. Clearly, it is also possible that there are proteins with properties different from those discussed here. Additionally, spectral analysis can be difficult because of changing line shapes, and lipid spin labels do introduce some perturbation into the membrane. Considerations of these problems indicate that there is still some uncertainty about the nature of the lipid-protein interface and the role of protein-protein aggregation. However, our present view is that the motionally restricted lipid component: (a) bears a fixed stoichiometry to the protein, independent of the lipid to protein ratio; (b) correlates in number with the intramembranous protein perimeter; (c) displays a limited selectivity for the lipid polar head group; (d) Has appreciable motion compared to that of the protein; (e) Reflects a large degree of spatial disorder compared to lipids in the bulk bilayer phase; and (f) Exchanges with the fluid lipids at a rate that is slow on the ESR time scale and rapid on the ^{2}H-NMR time scale.

The experiments with the reconstituted systems also give some indication of the range dependence of the protein perturbation of the lipid fluidity. This distance dependence is summarized in Fig. 31, which gives a generalized scheme of the lipid interaction with the protein as determined by ESR. The degree of motional perturbation drops off very steeply away from the protein, indicating how effectively the protein integrates into the bilayer structure. The second lipid shell is much less perturbed than the first boundary layer shell, and it may be that the combination of slower motion but greater disorder of the immobilized lipids provides the ideal interface between the protein and the fluid bilayer.

ACKNOWLEDGMENTS

We thank Dr. Richard Henderson for providing the originals of Figs. 3, 4, 11, and 12, Prof. P.F. Devaux for his comments on certain parts of this chapter, and our colleague Dr. Peter F. Knowles, in collaboration with whom all our studies on cytochrome oxidase have been performed. The work described on chromaffin granule membranes was supported in part through grant number Ma/756 to D. Marsh from the Deutsches Forschungsgemeinschaft.

APPENDIX: SPIN LABEL SYNTHESIS

Several complete reviews have recently appeared describing spin label synthesis (Gaffney, 1976; Keana, 1978, 1979). Therefore only the essential details for the synthesis of the labels discussed in this chapter are presented here.

I. FATTY ACID SPIN LABELS, I (m,n)

Overview

Doxyl fatty acid spin labels of the type $I(m, n)$, (see Figs. 5 and 15) have been generally prepared from an $(n + 2)$ keto- fatty acid ester (**8**), the synthesis of which is shown in Scheme A. A standard Grignard reaction with an ester diacid chloride (**4**) and 1-bromo-alkane (**5**) gives the desired keto-acid ester (**8**) in good yields.

Some of the necessary ester diacid chlorides (**4**) are commercially available, and these products and their sources are listed in Table A.1, together with the required starting material for cases requiring the making of the product (**4**). In fact, only for $n = 12$ and 14 is it necessary to perform the complete sequence of Scheme A, beginning at the diacid (**1**). The dimethyl ester (**2**) is made by methylation with HCl gas in methanol and then converted to the half ester with Ba (OH)$_2$ in good yields (Durham *et al.*, 1963). This method is suitable only for $n > 8$ and for methyl esters, but it has the advantage that no purification procedures are required. The acid chloride (**4**) of the half ester is then realized using oxazyl chloride and is used for the Grignard reaction without distillation. Typical yields for $n = 12$ and 14 of the keto-methyl stearates (**8**) from the diacids are 55–65%. Full experimental details for producing the 14-keto-methyl stearate are given below, but the procedure is identical for all isomers.

Condensation of the alcohol 2-amino-2-methyl-1-propanol (**9**) with the keto-stearate ester (**8**) results in an oxazolidine (**10**), as shown in Scheme B (Keana et al., 1967). Oxidation with *m*-chloroperbenzoic acid yields the free-radical nitroxide ester (**11**), which is first purified and then quantitatively hydrolyzed to the fatty acid spin label (**12**), *I(m,n)*.

The yields obtained in our hands for the reaction Scheme B, based on pure *I(m,n)* from starting keto-stearate ester are given in Table A.1. It is clear that some isomers more readily undergo the condensation reaction (e.g., labels with $n \geq 5$), whereas with smaller n many side products appear to be formed, resulting in poor yields. The 3- and 2- isomers have not yet been produced.

A commercially available stearic acid label, $I(5,10)$, can be synthesized from 12-hy-droxystearic acid, a natural product available cheaply commercially (see Table A.1). Oxidation to the keto derivative followed by methylation gives the 12-keto-stearate ester without the need for a Grignard reaction. Complete details for these steps are given by Waggoner et al. (1969).

Stearic acid spin labels, with $m + n = 15$, seem to have been the most commonly used fatty acid labels, although palmitic acid labels have been used by some workers (Hubbell & McConnell, 1971). Schemes A and B can in principle be used for almost any fatty acid isomer and for a variety of chain lengths.

Scheme A. Reaction sequence for synthesizing $(n + 2)$ ketomethyl fatty acids (8) for fatty acid spin labels $I(m,n)$. The first reaction step (i–iv) depends on the available starting material (see Table A.1).

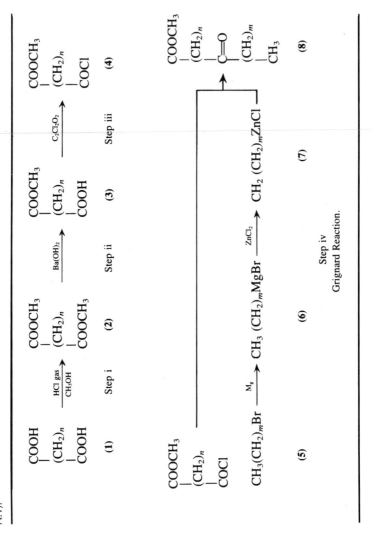

Scheme B. Reaction sequence for condensation of $(n + 2)$-keto—fatty acid ester (**8**)and 2-amino-2-methyl propranol (**9**) to give the fatty acid spin label (**12**) $l(m,n)$.

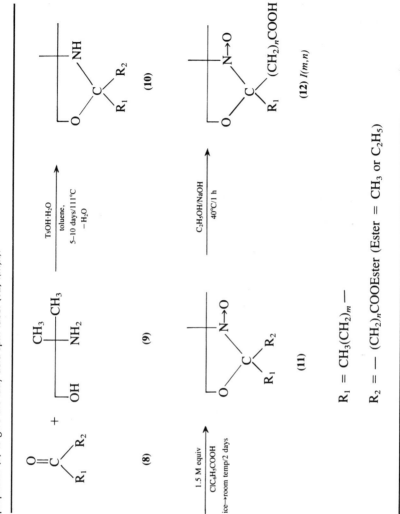

$R_1 = CH_3(CH_2)_m —$

$R_2 = — (CH_2)_n COOEster$ (Ester $= CH_3$ or C_2H_5)

TABLE A.1 Commercially Available Starting Materials for (n + 2)-Keto Stearic Acid Esters (Scheme A) for Stearic Acid Spin Labels I(m,n)

Stearic Acid Spin Label Isomer, $I(m,n)$	First Step in Reaction Scheme A	Sources[a] of $CH_3(CH_2)_mBr$ (1-bromoalkane)	Product used to give $ClOC(CH_2)_nCOOester$; Sources[a] in Brackets	Yield of Pure $I(m,n)$ from Keto-Stearate Ester (%)
(13,2)	iv	[A, F]	$ClOC(CH_2)_2CO_2CH_3$ (Methyl succinyl chloride) [A, F]	8.0
(12,3)[b]	iv	[A, F]	$ClOC(CH_2)_3CO_2CH_3$ (Methyl glutaryl chloride) [A, F]	13.2
(11,4)	iv	[A, F, M]	$ClOC(CH_2)_4CO_2CH_3$ (Adipic acid chloride monomethyl ester) [M]	14.5
(10,5)	iii	[A, F]	$HO_2C(CH_2)_5CO_2CH_3$ (Pimelic acid monomethyl ester) [M]	43.6
(9,6)	iii	[A, F, M]	$HO_2C(CH_2)_6CO_2CH_3$ (Suberic acid monomethyl ester) [M]	34.1
(8,7)	iii	[A, F, M]	$HO_2C(CH_2)_7CO_2CH_3$ (Azelaic acid monomethyl ester) [A, F]	7.4
(7,8)	iii	[A, F, M]	$HO_2C(CH_2)_8CO_2C_2H_5$ (Sebacic acid monoethyl ester) [M]	28.4
(6,9)	ii	[A, F, M]	$CH_3O_2C(CH_2)_9CO_2CH_3$ (Undecanedioic acid dimethyl ester) [F]	11.9
(5,10)[b]	See text		$CH_3(CH_2)_5CO(CH_2)_{10}CO_2H$ (12 Keto stearic acid) [F]	17.4

TABLE A.1   (Continued)

Stearic Acid Spin Label Isomer, $I(m,n)$	First Step in Reaction Scheme A	Sources[a] of $CH_3(CH_2)_mBr$ (1-bromoalkane)	Product used to give $ClOC(CH_2)_nCOOester$; Sources[a] in Brackets	Yield of Pure $I(m,n)$ from Keto-Stearate Ester (%)
(4,11)	ii	[A, F, M]	$CH_3O_2C(CH_2)_{11}CO_2CH_3$ (Tridecanedioic acid dimethyl ester) [E]	—
(3,12)	i	[A, F, M]	$HO_2C(CH_2)_{12}CO_2H$ (Tetradecanedioic acid) [A, M]	15.5
(2,13)	ii	[A, F, M]	$CH_3O_2C(CH_2)_{13}CO_2CH_3$ (Pentadecanedioic acid dimethyl ester) [Ferak]	—
(1,14)[b]	i	[A, F, M]	$HO_2C(CH_2)_{14}CO_2H$ (Hexadecanedioic acid) [A]	17.4

[a]Source abbreviations

[M], E. Merck,Postfach 4119,D-61 Darmstadt 1, West Germany.

[A], Aldrich Chemicals; European Division, Postfach 2004, Poststrasse 56, D-4054 Nettetal 2, Kalderkirchen, West Germany; United States Division, 940 St. Paul Avenue, Milwaukee, WI 53233.

[E], EGA-Chemie, D-7924 Steinheim, West Germany. (EGA and Aldrich have almost identical supplies for tridecanedioic acid dimethyl ester).

[F], Fluka AG, CH-9470 Buchs, Switzerland.

[Ferak], Ferak GmbH, Friedrichsbrunner Str. 3-5, 1 Berlin 47, West Germany. (Pentadecanedioic acid and the monomethyl ester are also available.)

[b]These labels are commercially available from Syva, 3181 Porter Drive, Palo Alto, CA 94304.

Experimental Details for Schemes A and B

The following experimental details have been adapted from published procedures and follow the reaction Schemes A and B for the stearic acid spin label $I(3,12)$, which involves the maximum number of synthetic steps necessary for any label $I(m, n)$ for $m + n = 15$. For other isomers starting at different steps in Scheme A, the procedure is identical in terms of the molar proportions of materials used.

All materials were of highest available quality and were obtained from either Fluka or Merck, except where stated in Table A.1. Thin layer chromatography (TLC) was made on silicic acid self-coated plates (Woelm-Pharma, FRG; Silica Woelm DC) and detection made with conc. H_2SO_4 and charring in hexane-ether (7 : 3 vol/vol) except where stated. Quoted R_f values are for this solvent.

Reaction Scheme A

Step i. Tetradecanedioic Acid Dimethyl Ester (2)

Tetradecanedioic acid (55 g; 0.213 mole) was added to absolute methanol (1200 ml). With stirring, HCl gas was bubbled through the solution until it went clear (about 20 min). TLC showed that no diacid remained (diacid $R_f \sim 0.05$; diester $R_f \sim 0.8$). After rotary evaporation to dryness, the solid was dissolved in diethyl ether and shaken against a 10% Na_2CO_3 solution (2 × 300 ml) until the lower phase was alkaline. The ether phase was then washed against water and dried with Na_2SO_4 and the solvent removed under reduced pressure to give a white solid that was pure on TLC (58.5 g; 96% yield).

Step ii. Tetradecanedioic Acid Monomethyl Ester (3)

Anhydrous barium oxide (47.8 g; 0.31 mol), methanol (625 ml), and water (5.6 ml; 0.31 mol) were mixed in a conical flask fitted with an NaOH drying tube and stirred for 1 h. The filtrate, obtained by filtration under an N_2 atmosphere, was then made up to 600 ml with absolute methanol and the normality determined by titration of 10 ml filtrate with 0.1 N HCl with methyl red. The solution was found to be 0.275 M in 600 ml.

To the melted diester (58.5 g; 0.205 mol) was quickly added benzene (24.4 ml) and freshly made barium hydroxide solution in methanol (212 ml; 0.107 mol), and the resulting mixture was shaken. (Molar ratio of diester to $Ba(OH)_2$, 1 : 0.52). After 17 h the light gray precipitate was washed with ether (100 ml), benzene (100 ml), and again with ether (100 ml) to remove any remaining diester. The solid was then taken up in 10% HCl and shaken against ether with the aqueous phase at pH 1–2. The combined ether extractions were washed with water, dried with Na_2SO_4, and rotary evaporated to dryness to give a white solid (46 g). On TLC the product was more than 98% pure ($R_f \sim 0.4$), with any remaining impurity being diacid.

Step iii. Tetradecanedioic Acid Chloride Monomethyl Ester (4)

Monoester (4) (46 g; 0.17 mol), oxalyl chloride (38.6 g; 0.306 mol), and water-free pyridine (1 ml) were stirred in anhydrous 1,4-dioxane (300 ml) at room temperature. The reaction was checked by TLC by adding absolute methanol to a sample and running against diester. When the reaction was completed as monitored by the absence of monoester (2–4 h), the solution was rotary evaporated to give an oil that was further rotary evaporated twice from toluene (42.0 g) and then stored under N_2 before use in the Grignard reaction.

Step iv. Methyl-14-Ketostearate (8) (Grignard Reaction)

Dried magnesium turnings (4.85 g, 0.2 mol) were warmed in a three-necked flask with a piece of I_2 until it sublimed. On cooling, the magnesium was just covered with anhydrous diethyl ether and 1-bromobutane (5) (27.4 g, 0.2 mol) in anhydrous diethyl ether (200 ml) added slowly enough to maintain effervesence, which was initiated with a drop of CH_3I. Remaining magnesium was reacted by refluxing for 30 min, after which time the reaction volume was brought to 300 ml by adding more ether. By titration of 1 ml of reaction mixture in 25 ml of 0.1 N HCl against 0.1 N NaOH with methyl orange, the Grignard yield was calculated to be 89% (0.177 mol).

Freshly fused zinc chloride (25 g, 0.18 mol) was just covered with ether in a three-necked flask. The Grignard solution (6) (0.177 mol) was added slowly with stirring and then refluxed for 1 h before removing 160 ml of ether by distillation. With vigorous stirring, tetradeca-nedioic acid chloride monomethyl ester (4) (42 g, 0.13 mol) in benzene (100 ml) was added to the reaction mixture over 15 min, followed by 3 h refluxing.

After cooling to room temperature, 2 N HCl (300 ml) was carefully added to the thick reaction product, followed by hot toluene (600 ml). Vigorous shaking dissolved the gel to produce two phases. After separation, the toluene phase was washed twice with warm water (60–70°C), dried over Na_2SO_4, filtered, and rotary evaporated to dryness. Crystallization from n-pentane on dry ice gave pure 14-keto-methyl stearate (8) (38.4 g, 0.123 mol).

Analysis: TLC R_f ~ 0.5; mp (uncorr) 44–46°C ; IR (KBr) keto band at 1718 cm⁻¹ and ester band at 1738 cm⁻¹. Elemental analysis: calculated for $C_{19}H_{36}O_3$, C, 73.07; H, 11.53; 0, 15.38; found: C, 73.18; H, 11.51; 0, 15.50.

Reaction Scheme B

14-(4', 4'-Dimethyloxazolidinyl-N-oxyl) Stearic Acid (12), I (3,12)

Condensation of 2-amino-2-methyl-1-propanol (9) (140 ml; 0.4 mol) (99% pure; Fluka) and the 14-keto-methyl-stearate (8) (22 g; 0.07 mol) was effected in refluxing toluene (500 ml) (or xylene) and catalyzed with p-toluene–sulfonic acid–water (0.1 g) with the removal of water with a Dean-Stark trap. After 5 days, the reaction was checked by TLC. When an appreciable amount of starting keto-stearate ester remained, further alcohol (70 ml; 0.2 mol) was added, and refluxing continued for 5 more days. When little keto-ester was left, the reaction mixture was shaken when cool against saturated $NaHCO_3$ (4 × 350 ml) and saturated NaCl (350 ml), and then dried over Na_2SO_4 , filtered, and rotary evaporated to dryness to give the 14-(4',4'-dimethyloxazolidine)stearic acid methyl ester (10) as a light brown solid.

Oxidation of 10 to the nitroxide (11) was achieved by dissolving 10 in anhydrous diethyl ether (500 ml) and adding dropwise on ice, m-chloroperbenzoic acid (10.5 g; 0.61 mol corresponding to 1.5 mol equiv of 10 as judged by TLC) (90% pract; Fluka, Buchs) dissolved in anhydr. diethyl ether (200 ml). After being allowed to come to room temperature over a period of 2 days with occasional mixing, the yellow solution was washed with saturated $NaHCO_3$ (4 × 350 ml) and water (4 × 350 ml) before drying over Na_2SO_4 and evaporating to dryness to give 29 g of the crude nitroxide ester (11) as a yellow solid.

The product (11) was purified by dry column chromatography (Silica Woelm DCC, activity III/30 mm, particle size 63–200μm; Woelm Pharma GmbH, Postfach 840, D-3440 Eschwege, West Germany). The total product was rotary evaporated from ether together with silica gel (~ 200 g) to dryness and loaded onto four dry columns (100 cm × 5 cm in nylon tubing), which were then eluted with about 1 liter of hexane-ether (7 : 3 vol/vol each). The yellow, UV light-sensitive bands at R_f ~ 0.6 were removed and extracted with refluxing toluene over 18–24 h. Rotary evaporation then gave the pure stearic acid spin label ester (11) as judged by TLC.

The ester group of 11 was hydrolyzed in ethanol (400 ml; 95%) by the addition of solid NaOH (7 g) with stirring at 40°C until the reaction was complete (~ 1 h) as seen by TLC by the disappearance of the starting product. The acid spin label (12) was poured from any remaining NaOH, acidified with 1 N HCl (pH 2–3), and extracted with $CHCl_3$. Rotary evaporation twice from toluene gave the pure stearic acid spin label (5), I(3,12) [4.2 g; 0.011

mol; 15.5% yield from 14-keto-methyl stearate (**8**)]. *Analysis:* TLC in hexane-ether 7 : 3 vol/vol.) $R_f \sim 0.6$; in $CHCl_3$-CH_3OH- 33% NH_4OH (100 : 15 : 1 vol/vol) $R_f \sim 0.5$ (in both cases $I(m,n)$ runs slightly slower than unlabeled stearic acid); IR (KBr) acid band at 1705 cm^{-1}. Elemental anal. calc. for $C_{23}H_{44}NO_4$ (for Me-ester), C, 69.34; H, 11.05; 0, 16.08; N, 3.51; found: C, 69.40; H, 11.05; 0, 16.09; N, 3.40.

II. PHOSPHOLIPID SPIN LABELS

Overview

The phosphatidylcholine spin label PCSL (see Fig. 15), is effectively synthesized from lysolecithin (1-acyl-2-hydroxy-phosphatidylcholine). The condensation of the appropriate fatty acid spin label imidazole is achieved under fairly mild conditions with relatively little acyl chain migration, to give the 1-acyl-2-(acyl spin label) phosphatidylcholine as shown in Scheme C (Boss et al., 1975). Yields of the reaction depend on the position of the spin label group, smaller values of n giving lower yields, but between 15 and 35% of phosphatidyl choline spin label can be obtained based on the starting lysolecithin. Most of the unused fatty acid label is recovered upon purification, if the reaction is quenched with water. Recovered lysolecithin should not be used for further synthesis with different fatty acid isomers, since it may be contaminated with spin label, either PCSL or 1-(fatty acid spin label)-2-hydroxy-3-phosphatidylcholine. Any 1, 3-phosphatidylcholine can be separated as a slightly faster running component on an efficient silicic acid column. Exact experimental details of the synthesis of Scheme C are presented below.

A convenient method of producing phospholipid spin labels with different head groups is shown in Scheme D. Phospholipase D-mediated head group exchange of the PCSL is a fairly mild procedure, unlike some total synthetic pathways, which may result in appreciable reduction. Some free-radical reduction does take place, however (10–20%), especially with charged lipids, and perhaps the proxyl labels will prove to be more suitable because of their greater stability in synthetic manipulations (Keana et al., 1976; Keana, 1979).

Table A.2 lists the various conditions required for making the different phospholipid labels discussed in this chapter; details of a typical synthesis are given later.

The cardiolipin spin label synthesis is given in detail by Cable et al., (1978) and is shown in Scheme E.

Although the yield of monolysocardiolipin was not as low in our hands as reported by Cable et al. (1978), reacylation produces at least four species as seen by TLC, the major one corresponding to the cardiolipin spin label. Reference to note 4d in Table A.2 should be made for chromatography on silicic acid columns.

Experimental

Phosphatidylcholine Spin Label (PCSL) Synthesis (Scheme C)

Completely dry fatty acid spin label, $I(m,n)$ (380 mg; 1 mmol) is dissolved in sodium-dried benzene (4 ml) together with N,N-carbonylimidazole (162 mg; 1 mmol) (Merck) and stirred for 30 min at room temperature or until no more CO_2 effervescence is seen. Lysophospha-tidylcholine [either from egg yolk lecithin (Lipid Products, Nutfield Ridge, Surrey, England)

Scheme C. Reaction sequence for synthesizing phosphatidylcholine spin labels (PCSL from fatty acid spin labels $I(m,n)$ and lysophosphatidylcholine.

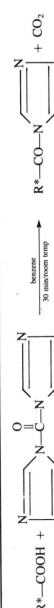

R*—COOH +

N,N-Carbonylimidazole

$I(m,n)$

benzene
30 min/room temp

R*—CO—N + CO₂

Imidazole fatty acid

R*—CO—N

+

HO—CH
H₂C—OOC—R′

H₂COPO(CH₂)₂N(CH₃)₃
O⁻

Lysolecithin

2–3 days
70°C

R*—O—CH
H₂C—OOC—R′

H₂COPO(CH₂)₂N(CH₃)₃
O⁻

PCSL

R′ = (usually) CH₃(CH₂)₁₄— , palmitic acid

R* = CH₃(CH₂)ₘ—C—(CH₂)ₙ— , $I(m,n)$, stearic acid spin label

115

Scheme D. Transphosphatidylation reaction of phosphatidylcholine spin labels mediated by phospholipase D.

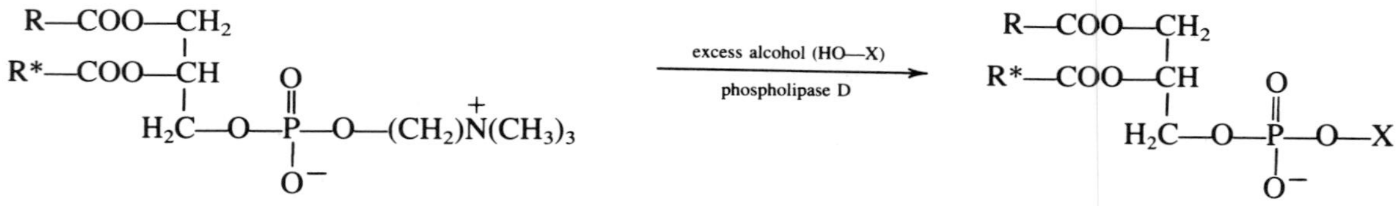

R—COO—CH$_2$
R*—COO—CH
H$_2$C—O—P—O—(CH$_2$)N(CH$_3$)$_3$
O=P
O$^-$

$\xrightarrow[\text{phospholipase D}]{\text{excess alcohol (HO—X)}}$

R—COO—CH$_2$
R*—COO—CH
H$_2$C—O—P—O—X
O=P
O$^-$

R   = acyl chain

R* = spin-labeled acyl chain, $I(m,n)$

X   = H → PASL

     = (CH$_2$)$_2$NH$_2$ → PESL

     = CH$_2$CHOHCH$_2$OH → PGSL

     = CH$_2$CH(NH$_2$)COOH → PSSL

     = CH$_3$ → Me-PASL

     = C$_2$H$_5$ → Et-PASL

Scheme 1. Reaction sequence for synthesis of the cardiolipin spin label from natural cardiolipin and fatty acid spin label, $I(m,n)$. From Cable et al. (1978). Crude venom is from *Trimersurus flavoviridis (habu)* and contains a phospholipase A_2 (available from Sigma Chemical Co.).

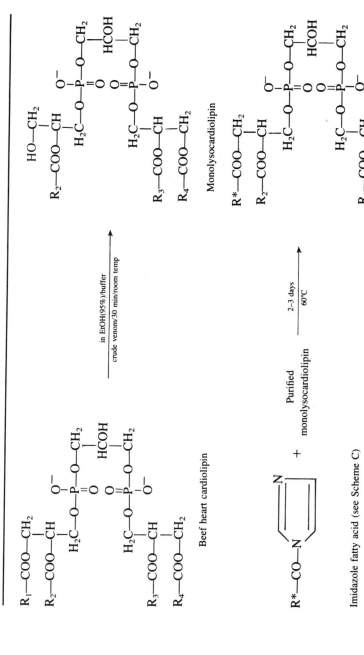

Beef heart cardiolipin

Monolysocardiolipin

Cardiolipin spin label

in EtOH(95%)/buffer
crude venom/30 min/room temp

Purified monolysocardiolipin

+

Imidazole fatty acid (see Scheme C)

2–3 days
60°C

R_1, R_2, R_3, R_4 = acyl chains; R^* = fatty acid spin label, $I(m,n)$; see Scheme C or palmitoyl-lyso-α-phosphatidylcholine (Fluka, or Sigma Chemical Co.)] (400 mg; 0.8 mmol), which has been dried under vacuum over P_2O_5, is then added to the imidazole fatty acid and stirred for 3 h at room temperature. After removing the solvent under reduced pressure, the yellowish mass is stirred at 60–70°C for 2–3 days under vacuum. Checking the reaction by TLC determines the best reaction time for different spin label isomers.

The reaction is quenched by adding water (4 ml). Carbon tetrachloride (20 ml), is added and then removed under reduced pressure; repetition of this process about five times leaves a yellow oil, which is then dissolved in $CHCl_3$-CH_3OH- 33% NH_4OH (200 : 15 : 1) (1–2 ml) and applied to a silicic acid column for purification.

The column (30 cm × 3 cm) is eluted with successively increasing polarity solvents made from CH_3Cl-CH_3OH- 33% NH_4OH. An initial yellow band eluted with (200 : 15 : 1) is discarded and the excess pure fatty acid eluted with (100 : 15 : 1). Pure PCSL is then eluted with (65 : 30 : 3) and rotary evaporated to dryness and stored at − 20°C. The labeled phosphatidylcholines tend to run a little faster on TLC than their unlabeled counterparts, and are detected by molybdate spray for phosphate, Dragendorff's reagent for choline, and on TLC plates containing fluorescent indicator for the spin label.

Phospholipase D-Mediated Exchange of PCSL to Give PASL, PESL, PGSL, PSSL, Me-PASL, and Et-PASL (see Scheme D)

Phosphatidylcholine spin label (see Scheme C) (100 mg, 1.1 mmol) is dissolved in alcohol-free diethyl ether (25 ml), added to buffer (25 ml; 0.1 M acetate/0.1 M $CaCl_2$, pH 5.6) and stirred at 35°C together with phospholipase D (EC 3.1.1.4; from white cabbage; Boehringer-Mannheim; Sigma Chemical Co.). Initially about 1 IU/ml of enzyme is added to the reaction mixture, followed by further similar additions as required over the total incubation period.

TABLE A.2 Summarized Details for Phospholipase D-Mediated Head Group Exchange of Phosphatidylcholine Spin Labels to Give Phospholipid Spin Labels with Different Head Groups

Label[a]	Alcohol[b]	Amount of Alcohol[c]	Remarks
PASL	HOH	100%	Normally a by-product of all other reactions; therefore can be separated from other reactions rather than made specifically.
			See note d below.
			Yields usually good.
PESL	HO-$(CH_2)_2NH_2$ Ethanolamine	34% wt/vol	Adjust pH of aqueous phase after adding ethanolamine, but before adding enzyme. Bring back to pH 5.6 with acetic acid.
			Use ninhydrin spray for TLC identification.
			Yields not very good; 5–15% depending on isomer.

TABLE A.2 (Continued)

Label[a]	Alcohol[b]	Amount of Alcohol[c]	Remarks
PGSL	HO-CH$_2$CHOHCH$_2$OH Glycerol	60% vol/vol	See note *d* below.
			Yields 15–35% depending on isomer.
			Racemic mixture of head group formed.
PSSL	HO-CH$_2$CH(NH$_2$)COOH L-Serine	46% wt/vol	Reaction temperature 45°C.
			PSSL and any remaining PCSL are not easily separated on silicic acid columns. Whatman Cellulose CM 52 can be used, but PSSL eluted with 3–5% methanol.
			See note *d,* below.
			Use ninhydrin spray for TLC identification.
			Yields 7–10%.
			L-Isomer in head group.
			Insoluble in EtOH.
Me-PASL	HO-CH$_3$ Methanol	12% vol/vol	Often found as by-product in reactions with other alcohols when methanol is impurity.
Et-PASL	HO-C$_2$H$_5$ Ethanol	12% vol/vol	

Adapted from Comfurius & Zwaal (1977).

[a]Abbreviations: PASL, phosphatidic acid; PESL, phosphatidylethanolamine; PGSL, phosphatidylglycerol; PSSL, phosphatidylserine; Me-PASL, methyl-phosphatidic acid; Et-PASL, ethyl-phosphatidic acid spin labels. See Fig. 15 for structures.

[b]Alcohols, as well as original phosphatidylcholine spin label and ether, should be free of methanol and ethanol, since Me-PASL and Et-PASL are formed more readily than other substituted head group labels.

[c]Amount of alcohol required refers to the aqueous phase, not total reaction volume.

[d]Charged lipids should be chromatographed on basic preequilibrated silicic acid columns (or preparative TLC plates), since they may elute as two species, charged and uncharged lipid.

The reaction is monitored by TLC against starting PCSL and unlabeled final product. Reaction time is usually found to be between 1 and 3 days, much longer than with unlabeled lipids.

The reaction is stopped when either no PCSL is left or no more desired lipid is formed at the expense of PASL, by adding 0.1 N HCl to give a pH of 2–3. After removing the ether phase with N$_2$, the lipid labels are extracted with CHCl$_3$ and applied to a silicic acid column (15 cm × 1 cm) for purification. Charged lipids are shaken against EDTA solution (equal volume, 0.1 M EDTA/0.1 M sodium acetate, pH 7–9) before reducing to a small volume to apply to the column.

Further details for each lipid type are given in Table A.2.

III. OTHER SPIN LABELS

Steroid Derivatives

The cholestane (Fig. 5) and androstane spin labels are synthesized essentially as shown in Scheme B from the commercially available keto products, cholestane-3-one and 4-androstane-17-hydroxy-3-one (Aldrich Co.), respectively. Both labels are available from Syva, Palo Alto, California.

Proxyl Labels

The proxyl long chain labels discussed in Sections IV.C and IV.E were first synthesized by Keana et al. (1976). Concise experimental details of a number of different derivatives are also given in a recent review by Keana (1979). Essentially the reaction sequence involves a Grignard reaction of a 1-bromoalkane with the commercially available nitrone (Aldrich Co.), followed by a second Grignard to give the long chain proxyl label, with protected functional group as shown in Scheme F. Removal of the protecting group realizes a label that can be further chemically modified, without loss of spin label intensity because of the greater stability of the proxyl ring.

TEMPO Spin Label

Exact details of the preparation of the 2,2,6,6-tetramethylpiperidine-N-oxyl (TEMPO) spin label (see Fig. 12) are given by Rozantsev (1970). This label is available from Syva, but is also readily produced by oxidation of 2,2,6,6,-tetramethyl-4-piperidine (Aldrich Co.; Eastman-Kodak). The oxidation can be made by dropping 2 g of the piperidine into 30 ml H_2O_2 (30%) containing phosphotungstic acid (60 mg) at 0°C with stirring. After 2–3 days at room temperature, the TEMPO is extracted with ether, the solvent dried and rotary evaporated at 0°C. The TEMPO is collected by sublimation as orange crystals (1.8 g).

Scheme F.   General synthetic route for proxyl-(5,5-dim ● yl-Δ^1-pyrroline N-oxide) fatty acid derivatives. Adapted from Keana (1979). For proxyl long chain alcohol, R′ is tetrahydropyranyl ether.

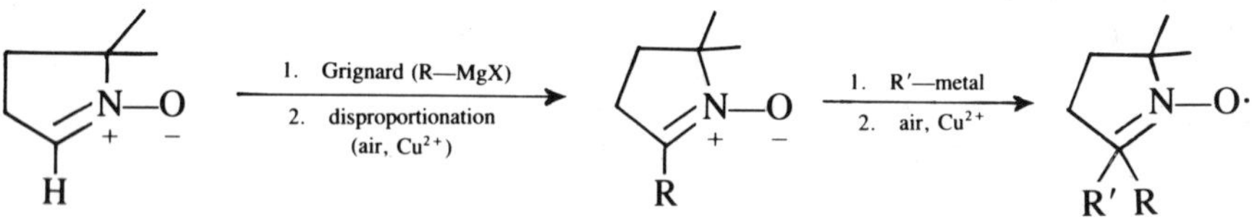

Nitrone

R = $(CH_2)_n CH_3$

R′ = $(CH_2)_m CO$(protecting group)

REFERENCES

Albert, A. D., & Litman, B. J. (1978). Independent Structural Domains in the Membrane Protein, Rhodopsin. *Biochemistry* **17**, 3893–3900.

Bakerman, S., & Wasemiller, G. (1967). Studies on Structural Units of Erythrocyte Membrane. I. Separation, Isolation and Partial Characterization. *Biochemistry* **6**, 1100–1113.

Baroin, A., Thomas, D. D., Osborne, B., & Devaux, P. F. (1977). Saturation Transfer Electron Paramagnetic Resonance on Membrane Bound Proteins. I. Rotational Diffusion of Rhodopsin in the Visual Receptor Membrane. *Biochem. Biophys. Res. Commun.* **78**, 442–447.

Barret, D. G., Sharom, F. J., Thede, A. E., & Grant, C. W. M. (1977). Isolation and Incorporation into Lipid Vesicles of a Concanavalin Receptor from Human Erythrocytes. *Biochim. Biophys. Acta* **465**, 191–197.

Berliner, L. J. (Ed.) (1976). *Spin-Labeling. Theory and Applications.* Vol I, Academic Press, New York-San Francisco-London.

Bienvenüe, A., Rousselet, A., Kato, G., & Devaux, P. F. (1977). Fluidity of the Lipids Next to the Acetylcholine Receptor Protein of *Torpedo* Membrane Fragments. Use of Amphiphilic Reversible Spin-Labels. *Biochemistry* **16**, 841–848.

Billington, T., & Nayudu, P. R. (1978). Studies on the Brush Border Membrane of Mouse Duodenum: Lipids. *Aust. J. Exp. Biol. Med. Sci* **1**, 25–29.

Birrell, G. B., Sistrom, W. R., & Griffith, O. H. (1978). Lipid-protein Associations in Chromatophores from the Photosynthetic Bacterium, *Rhodopseudomonas sphaeroides. Biochemistry* **17**, 3768–3773.

Bisson, R., Montecucco, C., Gutweniger, H., & Azzi, A. (1979). Cytochrome *c* Oxidase Subunits in Contact with Phospholipids. Hydrophobic Photolabelling with Azidophospholipids. *J. Biol. Chem.* **254**, 9962–9965.

Boss, W. F., Kelley, C. J., & Landsberger, F.R. (1975). A Novel Synthesis of Spin-Label Derivatives of Phosphatidylcholine. *Anal. Biochem.* **64**, 289–292.

Brotherus, J. R., Jost, P. C., Griffith, O. H., Keana, J. F. W., & Hokin, L. E. (1980). Charge Selectivity at the Lipid-Protein Interface of Membraneous Na, K-ATPase. *Proc. Natl. Acad. Sci. U.S.A.* **77**, 272–276.

Brotherus, J. R., Griffith, O. H., Brotherus, M. O., Jost, P. C., Silvius, J. R., & Hokin, L. E. (1981). Lipid-Protein Multiple Binding Equilibria in Membranes. *Biochemistry* **20**, 5261–5267.

Cable, M. B., Jacobus, J., & Powell, G. L. (1978). Cardiolipin: A Stereospecifically Spin-Labeled Analogue and Its Specific Enzymic Hydrolysis. *Proc. Natl. Acad. Sci. U.S.A.* **75**, 1227–1231.

Cable, M. B., & Powell, G. L. (1980). Spin-Labelled Cardiolipin: Preferential Segregation in the Boundary Layer of Cytochrome *c* Oxidase. *Biochemistry* **19**, 5679–5686.

Capaldi, R. A., & Briggs, M. (1976). The Structure of Cytochrome Oxidase. In *The Enzymes of Biological Membranes* (A. Martonosi, Ed.), Vol. 4, pp. 87–102, Wiley, London-New York-Sydney-Toronto.

Caspar, D. L. D., Goodenough, D. A., Makowsi, L., & Phillips, W. C. (1977). Gap-Junction Structures. I. Correlated Electron Microscopy and X-Ray Diffraction. *J. Cell Biol.* **29**, 605–628.

Cerletti, N., & Schatz, G. (1979). Cytochrome *c* Oxidase from Baker's Yeast. Photolabelling of Subunits Exposed to the Lipid Bilayer. *J. Biol. Chem.* **254**, 7746–7751.

Chignell, C. F., & Chignell, D. A. (1975). A Spin-Label Study of Purple Membranes from *Halobacterium halobium. Biochem. Biophys. Res. Commun.* **62**, 136–143.

Comfurius, P., & Zwaal, R. F. A. (1977). The Enzymatic Synthesis of Phosphatidylserine and Purification by CM-Cellulose Chromatography. *Biochim. Biophys. Acta* **488**, 36–42.

Daemen, F. J. M. (1973). Vertebrate Rod Outer Segment Membranes. *Biochim. Biophys. Acta* **300**, 255–288.

Davoust, J., Schoot, B. M., & Devaux, P. F. (1979). Physical Modifications of Rhodopsin Boundary Lipids in Lecithin-Rhodopsin Complexes: A Spin-Label Study. *Proc. Natl. Acad. Sci. U.S.A.* **76**, 2755–2759.

Davoust, J., Bienvenüe, A., Fellman, P., & Devaux, P. F. (1980). Boundary Lipids and Protein Mobility in Rhodopsin-Phosphatidylcholine Vesicles. Effects of Lipid Phase Transitions. *Biochim. Biophys. Acta* **596**, 28–42.

Davoust, J., & Devaux, P. F. (1980). Interaction Between ^{15}N-^{14}N Nitroxides. A Method to Determine the Specificity and Rates of Collisions Between Phospholipids and Proteins in Membranes; The Example of Rhodopsin-Lipid Recombinants. 9th International Conference on Magnetic Resonance in Biological Systems. Abstr. No. 133.

Deguchi, N., Jørgensen, P. L., & Maunsbach, A. B. (1977). Ultrastructure of the Sodium Pump. Comparison of Thin Sectioning, Negative Staining and Freeze-Fracture of Purified, Membrane-Bound (Na$^+$,K$^+$)-ATPase. *J. Cell Biol.* **75**, 619–634.

De Pont, J. J. H. H. M., van Prooijen-van Eeden, A., & Bonting, S. L. (1978). Role of Negatively Charged Phospholipids in Highly Purified (Na$^+$-K$^+$)-ATPase from Rabbit Kidney Outer Medulla. *Biochim. Biophys. Acta* **508**, 464–477.

Devaux, P. F., Davoust, J.,& Rousselet, A. (1981). Electron Spin Resonance Studies of Lipid–Protein Interactions in Membranes. Biochem. Soc. Symp. **46**, 207–222.

Durham, L. J., McLeod, D. J., & Cason, J. (1963). Methyl Hydrogen Hendecanoate. *Org. Syn. Coll.* **4**, 635–638.

Engelmann, D. M., Henderson, R., McLachlan, A. D., & Wallace, B. A. (1980). Path of the Polypeptide in Bacteriorhodopsin. *Proc. Natl. Acad. Sci. U.S.A.* **77**, 2023–2027.

Favre, E., Baroin, A., Bienvenüe, A., & Devaux, P. F. (1979). Spin-Label Studies of Lipid-Protein Interactions in Retinal Rod Outer Segment Membranes. Fluidity of the Boundary Layer. *Biochemistry* **18**, 1156–1162.

Feihn, W., & Hasselbach, W. (1970). The Effects of Phospholipase A on Calcium Transport and the Role of Unsaturated Fatty Acids in the ATPase Activity of Sarcoplasmic Vesicles. *Eur. J. Biochem.* **13**, 510–518.

Fleming, P. J., Dailey, H. A., Corcoran, D., & Strittmatter, P. (1978). Primary Structure of the Nonpolar Segment of Bovine Cytochrome b_5. *J. Biol. Chem* **253**, 5369–5372.

Freed, J. H. (1976). Theory of Slow-Tumbling ESR Spectra for Nitroxides. In *Spin-Labelling. Theory and Applications* (L. J. Berliner, Ed.), Vol. 1, pp. 53–132, Academic Press, New York-San Francisco-London.

Fretten, P., Morris, S. J., Watts, A., & Marsh, D. (1980). Lipid-Lipid and Lipid-Protein Interactions in Chromaffin Granule Membranes. A Spin-Label Study. *Biochim. Biophys. Acta* **598**, 247–259.

Fuller, S. D., Capaldi, R. A., & Henderson, R. (1979). Structure of Cytochrome Oxidase in Deoxycholate-Derived Two-Dimensional Crystals. *J. Mol. Biol.* **134**, 305–327.

Gaffney, B. J. (1976). Practical Considerations for the Calculation of Order Parameters for Fatty Acid or Phospholipid Spin Labels in Membranes. In *Spin-Labelling. Theory and Applications* (L. J. Berliner, Ed.), Vol. 1, pp. 567–571, Academic Press, New York-San Francisco-London.

Griffith, O. H. & Jost, P. C. (1976). Lipid Spin Labels in Biological Membranes. In *Spin-Labelling. Theory and Applications* (L. J. Berliner, Ed.), Vol. 1, pp. 454–523, Academic Press, New York-San Francisco-London.

Griffith, O. H., & Jost, P. C. (1978). Lipid-Protein Associations. In *Molecular Specialization and Symmetry in Membrane Function* (A. K. Solomon & M. Karnovsky, Eds.), pp. 31–60, Harvard University Press, Cambridge, MA.

Griffith, O. H., & Jost, P. C. (1979). The Lipid-Protein Interface in Cytochrome Oxidase. In *Cytochrome Oxidase* (B. Chance, T. E. King, K. Okunuki, Y. Orii, Eds.); pp. 207–218, Elsevier/North-Holland Biomedical Press, Amsterdam.

Grisham, M., & Barnett, R. E. (1973). The Rôle of Lipid-Phase Transitions in the Regulation of the (Sodium-Potassium) Adenosine Triphosphatase. *Biochemistry* **12**, 2635–2637.

Hartzell, C. R., Beinert, H., van Gelder, B. F., & King, T. E. (1978). In *Biomembranes. Methods in Enzymology* (S. Fleischer & L. Packer, Eds.), Vol. 53, Part D, pp. 54–66, Academic Press, New York-San Francisco-London.

Henderson, R., & Unwin, P. N. T. (1975). Three-Dimensional Model of Purple Membrane Obtained by Electron Microscopy. *Nature (London)* **257**, 28–32.

Henderson, R., Capaldi, R. A., & Leigh, J. H. (1977). Arrangement of Cytochrome Oxidase Molecules in Two-Dimensional Vesicle Crystals. *J. Mol. Biol.* **112**, 631–648.

Hesketh, J. R., Smith, G. A., Houslay, M. D., McGill, K. A., Birdsall, N. J. M., Metcalfe, J. C., & Warren, G. B. (1976). Annular Lipids Determine the ATPase Activity of a Calcium Transport Protein Complexed with Dipalmitoyl Lecithin. *Biochemistry* **15**, 4145–4151.

Hong, K., & Hubbell, W. L. (1972). Preparation and Properties of Phospholipid Bilayers Containing Rhodopsin. *Proc. Natl. Acad. Sci. U.S.A.* **69**, 2617–2621.

Hubbell, W. L., & McConnell, H. M. (1971). Molecular Motion in Spin-Labeled Phospholipids and Membranes. *J. Am. Chem. Soc.* **93**, 314–326.

Hyde, J. S. (1978). Saturation-Transfer Spectroscopy. In *Methods in Enzymology*, Vol. 49 (C. H. W. Hirs & S. N. Timasheff, Eds.), Academic Press, New York-San Francisco-London.

Jørgensen, P. L. (1975). Isolation and Characterization of the Components of the Sodium Pump. *Q. Rev. Biophys.* **7**, 239–274.

Jost, P. C., & Griffith, O. H. (1978a). Lipid-Protein Interactions: Influence of Integral Membrane Proteins on Bilayer Lipids. In *Biomolecular Structure and Function* (P. F. Agris, Ed.), pp. 25–65, Academic Press, New York.

Jost, P. C., & Griffith, O. H. (1978b). The Spin-Labelling Technique. In *Enzyme Structure. Methods in Enzymology* (C. H. W. Hirs & S. N. Timasheff, Eds.), Vol. 49, pp. 369–418, Academic Press, New York-San Francisco-London.

Jost, P. C., & Griffith. O. H. (1980). The Lipid-Protein Interface in Biological Membranes. *Ann. N.Y. Acad. Sci.* **38**, 391–407.

Jost, P., Libertini, L. J., Hebert, V., & Griffith, O. H. (1971). Lipid Spin Labels in Lecithin Multilayers. A Study of Motion Along Fatty Acid Chains. *J. Mol. Biol.* **59**, 77–98.

Jost, P. C., Griffith, O. H., Capaldi, R. A., & Vanderkooi, G. (1973a). Evidence for Boundary Lipid in Membranes. *Proc. Natl. Acad. Sci. U.S.A.* **70**, 480–484.

Jost, P. C., Capaldi, R. A., Vanderkooi, G., & Griffith, O. H. (1973b). Lipid-Protein and Lipid-Lipid Interactions in Cytochrome Oxidase Model Membranes. *J. Supramol. Struct.* **1**, 269–280.

Jost, P. C., Griffith, O. H., Capaldi, R. A., & Vanderkooi, G. A. (1973c). Identification and Extent of Fluid Bilayer Regions in Membranous Cytochrome Oxidase. *Biochim. Biophys. Acta* **311**, 141–152.

Jost, P. C., Nadakavukaren, K. K., & Griffith, O. H. (1977). Phosphatidylcholine Exchange Between the Boundary Lipid and Bilayer Domains in Cytochrome Oxidase-Containing Membranes. *Biochemistry* **16**, 3110–3114.

Jost, P. C., McMillen, D. A., Morgan, W. D., & Stoeckenius, W. (1978). Lipid-Protein Interactions in the Purple Membrane. In *Light Transducing Membranes* (D. Deamer, Ed.), pp. 141–154, Academic Press, New York-San Francisco-London.

Kang, S. Y., Gutowsky, H. S., Hsung, J. C., Jacobs, R., King, T. E., Rice, D., & Oldfield, E. (1979). Nuclear Magnetic Resonance Investigation of Cytochrome Oxidase-Phospholipid Interaction. A New Model for Boundary Lipid. *Biochemistry* **18**, 3257–3267.

Keana, J. F. W. (1979). New Aspects of Nitroxide Chemistry. In *Spin-Labeling. Theory and Applications.* (L. J. Berliner, Ed.), Vol. II, pp. 115–172, Academic Press, New York-San Francisco-London.

Keana, J. F. W. (1978). Newer Aspects of the Synthesis and Chemistry of Nitroxide Spin-Labels. *Chem. Rev.* **78**, 37–64.

Keana, J. F. W., Keana, S. B., & Beetham, D. (1967). A New Versatile Ketone Spin-Label. *J. Am. Chem. Soc.* **89**, 3055–3056.

Knowles, P. F., Marsh, D., & Rattle, H. W. E. (1976). *Magnetic Resonance of Biomolecules,* Wiley, London.

Knowles, P. F., Watts, A., & Marsh, D. (1979). Lipid Immobilization in Dimyristoylphosphatidyl-choline-Substituted Cytochrome Oxidase. *Biochemistry* **18**, 4480–4487.

Knowles, P. F., Watts, A., & Marsh, D. (1981). Spin-Label Studies of Headgroup Specificity in the Interaction of Phospholipids With Yeast Cytochrome Oxidase. *Biochemistry* **20**, 5888–5894.

Kusumi, A., Ohnishi, S., Ito, T., & Yoshizawa, T. (1978). Rotational Motion of Rhodopsin in the Visual Receptor Membrane as Studied by Saturation Transfer Spectroscopy. *Biochim. Biophys. Acta* **507**, 539–543.

Kushwaha, S. C., Kates, M., & Martin, W. G. (1975). Characterization and Composition of the Purple and Red Membrane from *Halobacterium cutirubrum*. *Can. J. Biochem.* **53**, 284–292.

LeMaire, M., Møller, J. V., & Tanford, C. (1976a). Retention of Activity by a Detergent-Solubilized Sarcoplasmic Ca^{2+}-ATPase. *Biochemistry* **15**, 2336–2342.

Le Maire, M., Jørgensen, K. E., Røigaard-Petersen, H., & Møller, J. V. (1976b). Properties of Deoxycholate Solubilized Sarcoplasmic Reticulum Ca^{2+}-ATPase. *Biochemistry* **15**, 5805–5812.

LeMaire, M., Møller, J. V., & Tardieu, A. (1981). Shape and Thermodynamic Parameters of a Ca^{2+}-Dependent ATPase. A Solution X-Ray Scattering and Sedimentation Equilibrium Study. *J. Mol. Biol.* **150**, 273–296.

Marsh, D., (1981). Electron Spin-Resonance: Spin-Labels. In *Membrane Spectroscopy, Molecular Biology, Biochemistry and Biophysics* (E. Grell, Ed.), Vol. 31, pp. 51–142, Springer-Verlag, Berlin-Heidelberg-New York.

Marsh, D. (1980b). Molecular Motion in Phospholipid Bilayers in the Gel Phase: Long Axis Rotation. *Biochemistry* **19**, 1632–1637.

Marsh, D., & Barrantes, F. J. (1978). Immobilized Lipid in Acetylcholine Receptor-Rich Membranes from *Torpedo marmorata*. *Proc. Natl. Acad. Sci. U.S.A.* **75**, 4329–4333.

Marsh, D., & Watts, A. (1981). ESR Spin Label Studies. In *Liposomes: From Physical Structure to Therapeutic Applications* (C. G. Knight, Ed.), Ch. 6, Elsevier/North-Holland Biomedical Press B.V., Amsterdam.

Marsh, D., Watts, A., Maschke, W., & Knowles, P. F. (1978). Protein-Immobilized Lipid in Di-myristoylphosphatidylcholine-Substituted Cytochrome Oxidase: Evidence for Both Boundary and Trapped Bilayer Lipid. *Biochem. Biophys. Res. Commun.* **81**, 397–402.

Marsh, D., Watts, A., & Barrantes, F. J. (1981). Phospholipid Chain Immobilization and Steroid Rotational Immobilization in Acetylcholine Receptor-Rich Membranes from *Torpedo marmorata*. *Biochim. Biophys. Acta* **645**, 97–101.

McConnell, H. M., & McFarland, B. J. (1972). The Flexibility Gradient in Biological Membranes. *Ann. N.Y. Acad. Sci.* **195**, 207–217.

Melchior, D. L., & Steim, J. M. (1976). Thermotropic Transitions in Biomembranes. *Ann. Rev. Biophys. Bioeng.* (L. J. Mullins, Ed.), **5**, 205–238.

Murphy, A. J. (1976). Cross-Linking of the Sarcoplasmic Reticulum ATPase Protein. *Biochem. Biophys. Res. Commun.* **70**, 160–166.

Nakamura, M., & Ohnishi, S.-I. (1972). Spin-Labelled Yeast Cells. *Biochem. Biophys. Res. Commun.* **46**, 926–932.

Nakamura, M., & Ohnishi, S.-I. (1975). Organization of Lipids in Sarcoplasmic Reticulum Membrane and Ca^{2+}-Dependent ATPase Activity. *J. Biochem.* **78**, 1039–1045.

O'Brien, J. S., & Sampson, E. L. (1965). Lipid Composition of Normal Human Brain—Gray Matter, White Matter and Myelin. *J. Lipid Res.* **6**, 537–544.

Osborne, H. B., Sardet, C., Michel-Villaz, M., & Carbre, M. (1978). Structural Study of Rhodopsin in Detergent Micelles by Small Angle Neutron Scattering. *J. Mol. Biol.* **123**, 177–206.

Ovchinnikov, Y. A., Abdulaev, N. G., Feigina, M. Yu., Kiselev, A. V., & Lobanov, N. A. (1979). The Structural Basis of the Functioning of Bacteriorhodopsin: An Overview. *FEBS Lett.* **100**, 219–224.

Ozols, J., & Gerard, C. (1977). Primary Structure of the Membranous Segment of Cytochrome b_5. *Proc. Natl. Acad. Sci. U.S.A.* **74**, 3725–3729.

Polnaszek, C. F., Marsh, D., & Smith, I. C. P. (1981). Simulation of the EPR Spectra of the Cholestane Spin Probe under Conditions of Slow Axial Rotation. Application to Gel Phase Phosphatidylcholine. *J. Magn. Reson.* **43**, 54–64.

Pontus, M., & Delmelle, M. (1975). Fluid Lipid Fraction in Rod Outer Segment Membrane. *Biochim. Biophys. Acta* **401**, 221–230.

Reynolds, J. A., & Karlin, A. (1978). Molecular Weight in Detergent Solution of Acetylcholine Receptor from *Torpedo californica*. *Biochemistry* **17**, 2035–2038.

Rice, D. M., Hsung, J. C., King, T. E., & Oldfield, E. (1979). Protein-Lipid Interactions. High Field Deuterium and Phosphorus Nuclear Magnetic Resonance Spectroscopic Investigation of the Cytochrome Oxidase-Phospholipid Interaction and the Effects of Cholate. *Biochemistry* **18**, 5885–5892.

Rouser, G., Nelson, G. J., Fleischer, S., & Simon, G. (1968). Lipid Composition of Animal Cell Membranes, Organelles and Organs. In *Biological Membranes* (D. Chapman, Ed.), pp. 20–69, Academic Press, London-New York.

Rousselet, A., Cartaud, J., & Devaux, P. F. (1979a). Importance des Interactions Protéine-Protéine dans la Maintien de la Structure des Fragments Excitables de l'Organe Électrique de *Torpedo marmorata*. *Compt. Rend.* **289**, 461–463.

Rousselet, A., Devaux, P. F., & Wirtz, K. W. (1979b). Free Fatty Acids and Esters can be Immobilized by the Receptor-Rich Membranes from *Torpedo marmorata*, but not Phospholipid Acyl Chains. *Biochem. Biophys. Res. Commun.* **90**, 871–877.

Rozantzev, E. G. (1970). *Free Nitroxide Radicals* (Transl. by B. J. Hazzard), Plenum, New York.

Sackmann, E., Träuble, H., Galla, H.-J., & Overath, P. (1973). Lateral diffusion, protein mobility, and phase transitions in *Escherichia coli* membranes. A spin-label study. *Biochemistry* **12**, 5360–5369.

Sardet, C., Tardieu, A., & Luzzati, V. (1976). Shape and Size of Bovine Rhodopsin: A Small-Angle X-Ray Scattering Study of a Rhodopsin-Detergent Complex. *J. Mol. Biol.* **105**, 383–407.

Scandella, C. J., Devaux, P. F., & McConnell, H. M. (1972). Rapid Lateral Diffusion of Phospholipids in Rabbit Sarcoplasmic Reticulum. *Proc. Natl. Acad. Sci. U.S.A.* **69**, 2056–2060.

Schindler, H.-J., & Seelig, J. (1973). ESR Spectra of Spin Labels in Lipid Bilayers. *J. Chem. Phys.* **59**, 1841–1850.

Schreier, S., Polnaszek, C. F., & Smith, I. C. P. (1978). Spin Labels in Membranes: Problems in Practice. *Biochim. Biophys. Acta* **515**, 375–436.

Schulte, T. H., & Marchesi, V. T. (1979). Conformation of Human Erythrocyte Glycophorin A and Its Constituent Peptides. *Biochemistry* **18**, 275–280.

Seelig, A., & Seelig, J. (1974). The Dynamic Structure of Fatty Acyl Chains in a Phospholipid Bilayer Measured by Deuterium Magnetic Resonance. *Biochemistry* **13**, 4839–4845.

Seelig, A., & Seelig, J. (1978). Lipid-Protein Interaction in Reconstituted Cytochrome *c* Oxidase–Phospholipid Membranes. *Hoppe-Seyler's Z. Physiol. Chem.* **359**, 1747–1756.

Segrest, J. P. (1977). The Erythrocyte: Topomolecular Anatomy of Mn-Glycoprotein. In *Mammalian Cell Membranes* (G. A. Jamieson & D. M. Robinson, Eds.), Vol. 3, pp. 1–26, Butterworths, London.

Shiga, T., Suda, T., & Maeda, N. (1977). Spin-Label Studies on the Human Erythrocyte Membrane. Two Sites and Two Phases for Fatty Acid Spin-Labels. *Biochim. Biophys. Acta* **466**, 231–244.

Simpkins, H., & Hokin, L. E. (1973). Studies on the Characterization of the Sodium-Potassium Adenosinetriphosphatase. XIII. On the Organization and Role of Phospholipids in the Purified Enzyme. *Arch. Biochem. Biophys.* **159**, 897–902.

Skou, J. C., & Nørby, J. G. (1979). *Na⁺-K⁺-ATPase. Structure and Kinetics,* Academic Press, New York-San Francisco-London.

Smith, I. C. P., & Butler, K. W. (1976). Oriented Lipid Systems as Model Membranes. In *Spin-labelling. Theory and Applications* (L. J. Berliner, Ed.), Vol. 1, pp. 411–448, Academic Press, New York-San Francisco-London.

Takemoto, J., & Huang Kao, M. Y. C. (1977). Effects of Incident Light Levels on Photosynthetic Membrane Polypeptide Composition and Assembly in *Rhodopseudomonas sphaeroides. J. Bacteriol.* **129**, 1102–1109.

Takeuchi, Y., Ohnishi, S.-I., Ishinga, M., & Kito, M. (1978). Spin-Labelling of *Escherichia coli* Membrane by Enzymatic Synthesis of Phosphatidylglycerol and Divalent Cation-Induced Interaction of Phosphatidylglycerol with Membrane Proteins. *Biochim. Biophys. Acta* **506**, 54–63.

Thomas, D. D., Dalton, L. R., & Hyde, J. S. (1976). Rotational Diffusion Studied by Passage Saturation Transfer Electron Paramagnetic Resonance. *J. Chem. Phys.* **65**, 3006–3024.

Tomita, M., Furthmayr, H., & Marchesi, V. T. (1978). Primary Structure of Human Erythrocyte Glycophorin A. Isolation and Characterization of Peptides and Complete Amino Acid Sequence. *Biochemistry* **17**, 4756–4770.

Träuble, H., & Sackmann, E. (1972). Studies of the Crystalline-Liquid-Crystalline Phase Transition of Lipid Model Membranes. III. Structure of a Steroid-Lecithin System Below and Above the Lipid-Phase Transition. *J. Am. Chem. Soc.* **94**, 4499–4510.

Unwin, P. N. T., & Zampighi, G. (1980). Structure of the Junction Between Communicating Cells. *Nature (London)* **283**, 545–549.

Waggoner, A. S., Kingzett, T. J., Rottschaeffer, S., Griffith, O. H., & Keith, A. D. (1979). A Spin-Labeled Lipid for Probing Biological Membranes. *Chem. Phys. Lipids* **3**, 245–253.

Warren, G. B., Toon, P. A., Birdsall, N. J. M., Lee, A. G., & Metcalfe, J. C. (1974). Reconstitution of a Calcium Pump Using Defined Membrane Components. *Proc. Natl. Acad. Sci. U.S.A.* **71**, 622–626.

Watts, A., & Marsh, D. (1981). Saturation Transfer ESR Studies of Molecular Motion in Phosphatidylglycerol Bilayers in the Gel Phase. *Biochim. Biophys. Acta* **642**, 231–241.

Watts, A., Marsh, D., & Knowles, P. F. (1978). Lipid-Substituted Cytochrome Oxidase: No Absolute Requirement of Cardiolipin for Activity. *Biochem. Biophys. Res. Commun.* **81**, 403–409.

Watts, A., Volotovski, I. D., & Marsh, D. (1979). Rhodopsin-Lipid Associations in Bovine Rod Outer Segment Membranes. Identification of Immobilized Lipid by Spin-Labels. *Biochemistry* **18**, 5006–5013.

Watts, A., Davoust, J., Marsh, D., & Devaux, P. F. (1981). Distinct States of Lipid Mobility in Bovine Rod Outer Segment Membranes. Resolution of Spin Label Results. *Biochim. Biophys. Acta* **643**, 673–676.

Watts, A., Volotovski, I. D., Pates, R., & Marsh, D. (1982). Spin Label Studies of Rhodopsin–Lipid Interactions. *Biophys. J.* (in press).

Winkler, H. (1976). The Composition of Adrenal Chromaffin Granules: An Assessment of Controversial Results. *Neuroscience* **1**, 65–80.

Yasunobu, K. T., Tanaka, M., Haniu, M., Sameshima, M., Reimer, N., Eto, T., King, T. E., Yu, C.-a., Yu, L., & Wei, Y.-h. (1979). Sequence Studies of Bovine Heart Cytochrome Oxidase Subunits. *Dev. Biochem.* **5**, 91–101.

Zingsheim, H. P., Neugebauer, D.-C., Barrantes, F. J., & Frank, J. (1980). Structural Details of Membrane-Bound Acetylcholine Receptor from *Torpedo marmorata. Proc. Natl. Acad. Sci. U.S.A.* **77**, 952–956.

Three

Nuclear Magnetic Resonance and Lipid-Protein Interactions

JOACHIM SEELIG, ANNA SEELIG, AND LUKAS TAMM

Biocenter of the University of Basel, Basel, Switzerland

ABBREVIATIONS

CSA, chemical shielding anisotropy
DEPC, dielaidoylphosphatidylcholine
DOPC, dioleoylphosphatidylcholine
ESR, electron spin resonance
NMR, nuclear magnetic resonance
S_{CD}, order parameter
T_1, spin-lattice relaxation time
τ_c, rotational correlation time
$\Delta\nu_Q$, quadrupole splitting

I. INTRODUCTION

The study of membranes by means of nuclear magnetic resonance (NMR) has proliferated during recent years, with a wealth of data from ^{1}H, ^{2}H, ^{13}C, ^{19}F, ^{14}N, and ^{31}P nuclei being reported. Almost all these techniques have also been employed to investigate the question of how lipids interact with proteins. This interaction can be highly specific, similar to an enzyme-substrate interaction, or it can result from a fairly nonspecific balance of hydrophobic and electrostatic forces involving large contact regions between lipids and proteins. Examples for both types of interaction can be found in the literature. Unfortunately, at the present stage of our knowledge, we have just scratched the surface of a multifaceted problem, and the idea of a critical overview appears to be rather preposterous at present. Furthermore, each nucleus has its own characteristic NMR properties, and a detailed discussion of the various NMR techniques is certainly beyond the scope of this chapter. Instead, we will limit ourselves to two recent NMR techniques that have been exploited with some success in the study of lipid-protein interaction, namely, deuterium (^{2}H) and phosphorus (^{31}P) NMR. Both methods provide insight into the spatial organization and the dynamic properties of the lipids and are very sensitive to changes in these parameters that are induced by lipid-protein association. A brief description of the two NMR methods is given first, followed by a discussion of results obtained with various reconstituted membrane systems. Since most of the examples are taken from our own work and since one or the other aspect has already been dealt with in a different context, a certain redundance and duplication of earlier literature is unavoidable.

II. DEUTERIUM MAGNETIC RESONANCE

Biological membranes are highly organized systems that are nevertheless fluid enough to allow considerable translational, rotational, and flexing movements of the constituent phospholipid and protein molecules. The fluid (liquid crystalline) membranes behave like optically uniaxial crystals with the optical axis perpendicular to the surface of the membrane. This means that the rotational and flexing movements are characterized, on the average, by cylindrical symmetry, with the bilayer normal as the axis of motional averaging. Compared to an isotropic solution, the movements of phospholipid molecules in membranes are thus anisotropic. Furthermore, the motional restrictions vary with the position in the membrane; for example, they differ in the hydrocarbon region from those in the polar region.

Deuterium magnetic resonance (^{2}H-NMR) is ideally suited to measure quantitatively the motional restrictions and to provide a detailed picture of the phospholipid conformations in biomembranes.

The principles of the method can be described as follows. By means of appropriate chemical synthesis some protons in a lipid molecule are replaced by deuterons. Since the van der Waals radii of the two isotopes are identical, this substitution leaves the membrane virtually unchanged. Other methods, such as spin label electron spin resonance (ESR) or fluorescence spectroscopy require the incorporation of bulky reporter groups into the membrane, which may lead to a distortion of the membrane. ^{2}H-NMR of selectively deuterated lipids thus offers the advantage of providing a nonperturbing probe technique. A second convenient feature of ^{2}H-NMR is the straightforward assignment of the signals. The natural abundance of deuterium in biological materials is negligible (0.016%); thus the ^{2}H-NMR signal of the selectively deuterated segment is the only resonance signal in the spectrum. This may be compared with proton and ^{13}C-NMR of biomembranes and derived liposomes, where a manifold of partially overlapping resonances is encountered. However, the most important advantage in the use of ^{2}H-NMR in the elucidation of membrane structure is the measurement of the segmental anisotropies, which can be derived from the so-called deuterium quadrupole splittings (for reviews see Seelig, 1977; Mantsch et al., 1977).

   In brief, the deuteron has a nuclear spin $I = 1$, which in an external magnetic field has three allowed orientations $m = +1, 0, -1$. The magnetic energy levels are equally spaced and, in the absence of quadrupole effects, the two allowed transitions $m = -1 \rightarrow m = 0$ and $m = 0 \rightarrow m = +1$ give rise to the same NMR signal. However, in addition to the magnetic moment, the deuteron nucleus also possesses an electric quadrupole moment. The interaction of the electric quadrupole with the electric field gradient of the surrounding bonding electrons modifies the magnetic energy levels and removes the degeneracy of the two allowed quantum transitions. Consequently, the deuterium signal splits into two resonances, since the two allowed transitions are now characterized by different energies. The frequency separation between the two signals is the residual deuterium quadrupole splitting, $\Delta\nu_Q$. For a C—D bond in a perfectly ordered environment (as, e.g., in a single crystal), the quadrupole splitting may reach a limiting value of 170 kHz (Burnett & Muller, 1971), whereas in fluid membranes $|\Delta\nu_Q|$ is usually found to be in the range of 1 kHz $< |\Delta\nu_Q| <$ 60 kHz.

   It should be noted that in isotropic solution the quadrupolar effects are averaged out due to the rapid tumbling of the molecules through all angles of space so that the ^{2}H-NMR spectrum then consists of a single line. On the other hand, the more restricted the angular excursions of the labeled segment, the larger the quadrupole separation becomes. Typical ^{2}H-NMR spectra of selectively deuterated lipid bilayers homogeneously oriented between planar glass plates are shown in the lower part of Fig. 1. The bilayer investigated in Fig. 1 is a lyotropic liquid crystal composed of a fatty acid, a deuterated long chain alcohol, and water (for details see Seelig & Niederberger, 1974). X-Ray diffraction shows that in this ternary mixture the lipid molecules are arranged in bilayers that are separated from each other by layers of water. Both amphiphilic components, the fatty acid as well as the alcohol, are integral constituents of the lipid region. A macroscopic ordering of this phase can be induced by forcing it between two closely spaced, parallel glass surfaces. Shearing forces and the contact with the glass surfaces arrange the microdomains in parallel layers on the supporting plates. The axis of motional averaging (director axis z') is always perpendicular to the bilayer surface. For the macroscopically oriented sample the director axis is identical with the normal to the glass plates.

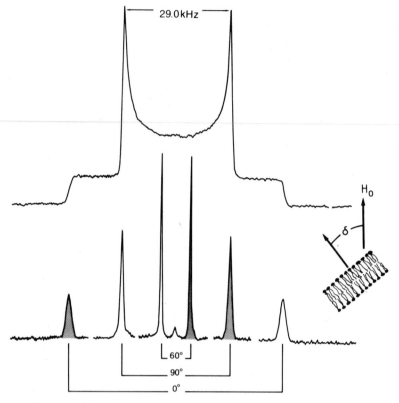

Figure 1. Deuterium NMR spectra (at 46 MHz) of a selectively deuterated soap-like bilayer. Sample composition: sodium octanoate, 34 wt %; [1,1−$^2H_2$]octanol, 30 wt %; water, 36 wt %. The lower part of the figure shows the ^{2}H-NMR spectra of planar oriented bilayers at various angles δ with respect to the magnetic field. The different shadings denote the two different transitions −1 ↔ 0 and 0 ↔ +1. A crossover of the two resonances occurs at the "magic angle" with δ = 54.73°. The quadrupole splitting $\Delta\nu_Q$ (δ) is the frequency separation of the two transitions. The upper spectrum is the ^{2}H-NMR spectrum of a randomly dispersed sample of the same bilayer as was used for the orientation study. The two most intense peaks in the "powder-type" pattern correspond to the δ = 90° orientation, while the weak outer wings reflect the δ = 0° orientation. Figure courtesy of Dr. J. L. Browning.

The deuterium quadrupole splitting $\Delta\nu_Q$ of a homogeneously oriented lipid bilayer obeys the relation (see Seelig, 1977)

$$\Delta\nu_Q\ (\delta) = {}^3\!/_2\ \frac{e^2qQ}{h}\ S_{CD}\ \frac{3\ \cos^2\!\delta\ -\ 1}{2} \tag{1}$$

where e^2qQ/h is the static deuterium quadrupole coupling constant ($\sim$ 170 kHz for paraffinic C—D bonds), S_{CD} is the order parameter of the C—D bond vector, and δ is the angle between the magnetic field and the bilayer normal.

First, it may be seen that the residual deuterium quadrupole splitting of a lipid bilayer depends strongly on the orientation of the planar strata with respect to the external magnetic field. The quadrupole splitting will collapse if the membranes are oriented at the so-called "magic angle" with $\delta = 54.73°$, since $(3 \cos^2 54.73 - 1) = 0$ for this orientation. On the other hand, the quadrupole splitting is largest if the bilayer normal is parallel to the magnetic field ($\delta = 0°$). Second, the quadrupole splitting depends on the molecular properties of the membrane as described by the order parameter S_{CD}, which is defined according to:

$$S_{CD} = \frac{1}{2} \overline{(3 \cos^2\theta - 1)} \qquad (2)$$

where θ is the instantaneous angle between the C—D bond vector and the bilayer normal, while the bar denotes a time-average over all motions that are fast compared to the static quadrupole splitting constant. The molecular interpretation of S_{CD} can be complicated at times, since it is actually a tensorial average that depends both on the geometry of the deuterated molecule and on the statistical fluctuations of this molecule within the bilayer. The term "order parameter" is therefore misleading to a certain extent.

The preparation of homogeneously oriented membranes is not easily achieved and remains the exception rather than the rule. Fortunately, ^{2}H-NMR is not limited to oriented samples but can be performed with coarse liposomes as well. The spectra of such nonoriented samples are often called "powder-type" spectra in the NMR literature. The shape of the powder-type spectrum is then the average over the resonance of all possible bilayer orientations weighted by their corresponding geometric probability (cf. Seelig, 1977). The upper part of Fig. 1 shows the "powder-type" spectrum of the lyotropic phase discussed above, while Fig. 2 represents the same type of spectrum for a true phospholipid bilayer, that is, randomly dispersed multilayers of 1,2-di[9,10-^{2}H$_2$] elaidoyl-sn-glycero-3-phosphocholine (DEPC). For practical purposes the most important feature of the "powder-type" spectra is the frequency spacing of the two most intense peaks. These peaks arise from bilayer microdomains that are oriented perpendicular to the magnetic field ($\delta = 90°$) and the spacing of the two peaks therefore is

$$\Delta\nu_{powder} = \frac{3}{4} \left(\frac{e^2qQ}{h}\right) S_{CD} \qquad (3)$$

The edges of the spectrum correspond to orientations of the bilayer normal parallel to the magnetic field ($\delta = 0$), and the separation of the edges is therefore exactly twice as large as that of the inner peaks. Therefore the deuterium order parameter S_{CD} can be determined easily from powder-type samples such as coarse dispersions of phospholipids in water.

Figure 2 further illustrates that the spectral intensity is much lower in the outer wings than at the $\delta = 90°$ orientations. Only recently has it been possible to actually measure the theoretically predicted line shape as shown in Fig. 2 (see Davis et al.,

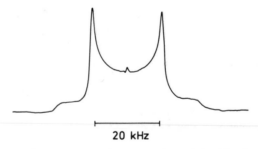

20 kHz

Figure 2. Deuterium NMR spectrum (at 41.3 MHz) of a multilamellar lipid dispersion of 1,2-di[9,10-^{2}H$_2$]elaidoyl-sn-glycero-3-phosphocholine (DEPC) at 24°C. Since the bilayer microdomains are distributed over all angles of space, this spectrum is also called a "powder"-type spectrum. Phase composition: DEPC, 50 wt %; H$_2$O, 50 wt % (cf. Seelig & Waespe-Šarčeviè, 1978).

1976), whereas in most earlier publications only the two most intense peaks could be detected. Obviously, this deficiency of the earlier spectra does not affect the evaluation of S_{CD}. On the other hand, it should be noted that S_{CD} cannot be evaluated from single-walled vesicles of small radius. For these systems the vesicle tumbling rate is fast compared to the residual quadrupole separation and the doublet will collapse into the singlet (cf. Stockton et al., 1976). Furthermore quadrupole effects are unique to nuclear spins with $I \geq 1$ and are not observed in more commonly employed NMR techniques such as proton, carbon-13 or phosphorus-31 NMR, since for these nuclei $I = \frac{1}{2}$.

III. PHOSPHORUS-31 NUCLEAR MAGNETIC RESONANCE

The lipids in membranes are predominantly phospholipids and ^{31}P-NMR is thus an attractive tool to study the motion of the phosphate segment. The anisotropy of the phosphate movement is reflected in the chemical shielding anisotropy (CSA) $\Delta\sigma$, the physical basis of which may be explained as follows. The magnetic field experienced by the phosphorus nucleus is reduced by the bonding electrons (chemical shielding). Since the electron density is not isotropic but depends on the bonding pattern, the chemical shielding will be smallest (largest) along the molecular axis with the lowest (highest) electron density, and the ^{31}P-NMR frequency will vary with the orientation of the phosphate segment with respect to the magnetic field. The chemical shielding anisotropy is most pronounced in the crystalline state where the resonance frequency for phospholipids may shift between -100 and $+100$ ppm (Kohler & Klein, 1976, 1977; Griffin, 1976; Herzfeld et al., 1978) while this effect is completely averaged out in isotropic solutions, where the molecules tumble rapidly through all angles of space. Biomembranes represent an intermediate situation. The movements of the phosphate segment are neither completely free nor perfectly anisotropic, with the result that the static chemical shielding anisotropy is only partially averaged out. This is demonstrated in Fig. 3, which contains the rotation pattern for a planar oriented phospholipid bilayer (Seelig & Gally, 1976).

Depending on the orientation of the bilayer with respect to the magnetic field, indicated by the rotation angle δ, the resonance frequency varies between -23 ppm for $\delta = 0°$ (magnetic field parallel to the bilayer normal) and $+12$ ppm for $\delta = 90°$ (magnetic field perpendicular to the bilayer normal). The chemical shielding anisotropy $\Delta\sigma$ is defined as the difference between these two extreme values and amounts to -35 ppm. The chemical shielding anisotropy in ^{31}P-NMR is comparable to the deuterium quadrupole splitting in ^{2}H-NMR. However, the quantitative interpretation of $\Delta\sigma$ is more complicated than the corresponding analysis of the deuterium quadrupole splitting $\Delta\nu_Q$, since $\Delta\sigma$ is determined by two independent order parameters compared to only one in the case of $\Delta\nu_Q$ (Niederberger & Seelig, 1976; Seelig, 1978):

$$\Delta\sigma = S_{11}(\sigma_{11} - \sigma_{22}) + S_{33}(\sigma_{33} - \sigma_{22}) \tag{4}$$

Here the σ_{ii} are the principal elements of the static chemical shielding tensor, whereas the S_{ii} are the order parameters of the corresponding principal axes. The principal elements can be determined easily from measurements of crystalline samples (Kohler & Klein, 1976, 1977; Griffin, 1976; Herzfeld et al., 1978). However, a quantitative analysis of $\Delta\sigma$ is still not possible without additional assumptions, since two independent order parameters should be evaluated from just one experimental result ($\Delta\sigma$). In spite of this theoretical difficulty, the measurement of the chemical shielding anisotropy has been of central importance in the understanding of the head group structure of phospholipids in membranes (see Seelig, 1978; Cullis & de Kruijff, 1979).

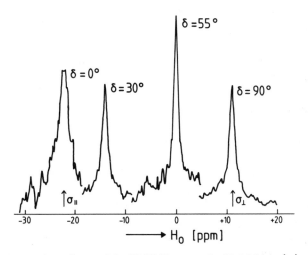

Figure 3. Orientation dependence of the ^{31}P-NMR spectra (at 36.4 MHz) of planar multilayers of 1,2-dipalmitoyl-sn-glycero-3-phosphoethanolamine: δ is the angle between the magnetic field and the normal to the plane of the membranes; temperature, 77°C; pH 5.5. Reproduced with permission from Seelig & Gally (1976).

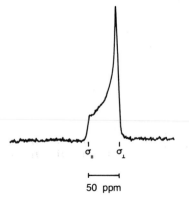

Figure 4. Proton-decoupled ^{31}P-NMR spectrum (at 121.4 MHz) of multilamellar vesicles of 1,2-dioleoyl-*sn*-glycero-3-phosphocholine measured at 14°C. The chemical shielding anisotropy is given by the separation of the edges of the spectrum. Reproduced with permission from Seelig et al. (1981).

The measurement of $\Delta\sigma$ is not limited to planar oriented samples, but $\Delta\sigma$ can also be obtained readily from random dispersions of phospholipid bilayers (Gally et al., 1975). A typical powder-type pattern of coarse liposomes is reproduced in Fig. 4. The important feature to note is that the edges of the powder-type spectrum correspond to $\sigma_\perp$ (high intensity shoulder, $\delta = 90°$) and $\sigma_\parallel$ (low intensity shoulder, $\delta = 0°$). Spectrum 4 was recorded under proton decoupling conditions. Without proton decoupling the lineshape of the powder pattern is seriously distorted, and a reliable evaluation of $\Delta\sigma$ is not possible under these circumstances.

IV. SPIN-LATTICE (T_1) RELAXATION TIME MEASUREMENTS

The information obtainable from ^{2}H- and ^{31}P-NMR is of two kinds. Measurement of the quadrupole splitting, $\Delta\nu_Q$, and the chemical shielding anisotropy, $\Delta\sigma$, provides information about the time-averaged *orientation* of the segments involved and about the amplitudes of their *angular fluctuations*. In contrast, measurement of deuterium and phosphorus T_1 relaxation times gives information on the *rate* of segmental motion. Both parameters have been used to characterize model membranes and biological membranes in terms of a somewhat loosely defined "membrane fluidity". The confusion arises because the term "fluid" sometimes refers to the disordered nature of the hydrocarbon chains (structural property), and in other cases to the rate of molecular motions (dynamic property). Since, however, it is not obvious that a disordered bilayer interior must be highly fluid under all circumstances, it is important to keep the two concepts apart.

Let us briefly discuss in a more quantitative way how the spin lattice relaxation time T_1 is related to the molecular motions. T_1 calculations can become very complicated for anisotropic rotations, and we therefore limit the discussion to the relatively simple case of an isotropic fluid, which in many cases is also a good approximation for an anisotropic system. It can then be shown that the T_1 relaxation time depends on a molecular correlation time τ_c according to

$$\frac{1}{T_1} \propto \frac{\tau_c}{(1 + \omega_o^2 \tau_c^2)} \tag{5}$$

where ω_o is the NMR frequency (in rad/s). It is obvious from inspection of Eq. (5) that the T_1 relaxation is most efficient (equivalent to a minimum in T_1) if the value of τ_c is such that $\omega_o\tau_c = 1$. If the molecular motions are slower or faster than ω_o, the T_1 times must become longer. Since the rate of the molecular fluctuations depends on the viscosity η and the temperature T (in °K) according to

$$\tau_c = \frac{4\pi r^3}{3} \frac{\eta}{kT} = \frac{\eta V}{kT} \tag{6}$$

where r and V are radius and volume of the segment, respectively, and kT is the thermal energy, a plot of the spin-lattice relaxation time T_1 versus the reciprocal temperature should exhibit a minimum if at a certain temperature the condition $\omega_o\tau_c \simeq 1$ is satisfied.

^{2}H-NMR spin lattice relaxation time studies on phospholipid bilayers and re-constituted membranes in the fluid phase consistently showed an increase of T_1 with increasing temperature, indicating that for specific experimental conditions employed the molecular motions were much faster than the resonance frequency ($\omega_o\tau_c \ll 1$, fast correlation time regime). Under these conditions the following T_1 formula has been derived for the reorientation rate of a C—D bond vector characterized by the order parameter S_{CD} (Brown et al., 1979):

$$\frac{1}{T_1} = \frac{3\pi^2}{2} \left(\frac{e^2qQ}{h}\right)^2 (1 + \frac{1}{2} S_{CD} - \frac{3}{2}S_{CD}^2)\, \tau_c \tag{7}$$

The dominant relaxation pathway in ^{2}H-NMR is quadrupole relaxation. This simplifies the interpretation of the T_1 relaxation times and has the further advantage of a relatively fast T_1 relaxation experiment, since the T_1 times of lipid bilayers are usually in the range of 10–100 ms.

The analysis of phosphorus T_1 relaxation times is generally more complicated. The most important mechanisms contributing to ^{31}P spin lattice relaxation are dipole-dipole relaxation (T_{1Dip}) and relaxation via chemical shielding anisotropy (T_{1CSA}). The two relaxation mechanisms can be differentiated by their dependence on the magnetic field strength H_0. Dipole-dipole relaxation is independent of H_0, whereas relaxation via chemical shielding anisotropy increases according to $1/T_{1CSA} \propto H_0^2$. It has been demonstrated that up to a field strength of 2.25 tesla the spin lattice relaxation time is dominated by dipolar interactions and that contributions from other mechanisms are negligible (Yeagle et al., 1975, 1977). Quantitatively, the dipolar contribution (T_{1Dip}) to the total relaxation time may be estimated according to (Doddrell et al., 1972):

$$\frac{1}{T_{1\ Dip}} = \frac{1}{10}\ |F_o|^2\ [J(\omega_H - \omega_P) + 3J(\omega_P) + 6J(\omega_H + \omega_P)] \tag{8}$$

with

$$J(\omega_i) = \frac{\tau_c}{1 + \omega_i^2\ \tau_c^2} \tag{9}$$

where $\omega_H \triangleq$ proton resonance frequency and $\omega_P \triangleq$ phosphorus resonance frequency. Equation 8 is strictly valid only for an isotropic motion characterized by a single correlation time τ_c. If a single proton at distance r from the phosphorus atom is involved in the relaxation process, F_o can be calculated according to

$$F_o = \frac{\hbar\ \gamma_H\ \gamma_P}{r^3} \tag{10}$$

where γ is the gyromagnetic ratio of the corresponding nucleus. The situation is more complex, however, for the phosphocholine head group where intra- as well as intermolecular dipole-dipole interactions are involved and furthermore the head group is flexible. It is then almost impossible to provide any realistic quantitative estimates of the amplitude factor F_o. The T_1 minimum of Eq. 8 does not occur exactly at $\tau_c = 1/\omega_o$ but at $\tau_c \simeq 0.7/\omega_o$.

A second relaxation mechanism is provided by relaxation via chemical shielding anisotropy (CSA). For isotropic reorientation the following expression is obtained for T_{1CSA} (Abragam, 1961):

$$\frac{1}{T_{1CSA}} = \frac{3}{10}\ \omega_P^2\ \delta\sigma^2\ \frac{\tau_c}{1 + \omega_P^2\ \tau_c^2} \tag{11}$$

The amplitude factor, $\delta\sigma$, is the chemical shielding anisotropy that is averaged by the molecular tumbling. In an anisotropic medium such as the phospholipid bilayer, only part of the total chemical shielding anisotropy is averaged, and it is difficult to predict $\delta\sigma$ a priori. In this respect the problem is similar to dipolar relaxation where F_o is unknown.

The two relaxation mechanisms are additive according to

$$\frac{1}{T_1} = \frac{1}{T_{1Dip}} + \frac{1}{T_{1CSA}} \tag{12}$$

At a field strength of 7 tesla ($\triangleq$ 300 MHz proton resonance frequency) both mechanisms contribute about equally to the relaxation rate of the phosphate segments of lipid bilayers (Seelig et al., 1981). The quantitative evaluation of phosphorus T_1 relaxation times in terms of a correlation time τ_c is therefore less straightforward than that of deuterium T_1 relaxation time, since the amplitude factors of the dipolar relaxation (F_o) and relaxation via chemical shielding anisotropy $(\delta\sigma)$ are generally unknown. However, it should be appreciated that knowledge of the amplitude factors is not essential if a *minimum* in the T_1 relaxation can be observed, since at the T_1 minimum the correlation time τ_c must be equal to the reciprocal resonance frequency of the phosphorus nucleus (i.e., $\tau_c = 1/\omega_P$).

V. THE PROBLEM OF TIME SCALES IN ESR AND NMR EXPERIMENTS

The principal method to study lipid-protein interactions has been spin label electron spin resonance (ESR) (see Jost et al., 1973; Jost & Griffith, 1980). The ESR spectra of spin-labeled membranes consistently showed the presence of two components. One spectral component was almost identical to the spectrum of a pure lipid bilayer, while the second spectral component was much broader and characteristic of a more restricted motion of the spin label. On the time-scale of ESR this second spectral component can be interpreted as an almost completely "immobilized" spin label. The ESR studies thus revealed the existence of two classes of spin-labeled phospholipids, that is, labels in immediate contact with the hydrophobic protein surface (the so-called boundary lipids) and those farther away from it. Since two spectral components were clearly resolved, the exchange rate of a spin label between the two environments had to be slow on the ESR time scale. This problem is equivalent to the classical two-site exchange in magnetic resonance (see Carrington & McLachlan, 1967). If a nuclear spin or an electron spin is jumping between two sites of different resonance frequencies ω_A and ω_B, two separate lines will be observed as long as the exchange rate is small. The lines will merge if the residence time τ of the spin at each site becomes shorter than

$$\tau \leqslant \frac{2\sqrt{2}}{\Delta\omega} = \frac{\sqrt{2}}{\pi\Delta\nu} \tag{13}$$

In the above-mentioned spin label studies the spectral differences between the pure lipid bilayer and the more immobilized boundary lipid were of the order of 2–10 gauss, corresponding to a frequency separation of 5.7–28 MHz. Thus it could be concluded that the exchange frequency was certainly less than 10^7–10^8 Hz. NMR methods operate at a much lower frequency due to the smaller nuclear magnetic moment. As an extreme and purely hypothetical case we envisage the possibility of a deuteron jumping between an isotropic fluid with a quadrupole splitting $\Delta\nu_Q = 0$ and a solid with $\Delta\nu_Q \simeq 100$ kHz. In order to observe an exchange-narrowed line, the exchange rate must be of the order of 10^5 Hz; that is, it can be two to three orders of magnitude slower than in the case of ESR.

In one of the first membrane reconstitution studies using cytochrome c oxidase and deuterated lipids, two different deuterium quadrupole couplings were indeed observed and were assigned to boundary lipid ($\Delta\nu_Q = 4.6$ kHz) and bulk lipid ($\Delta\nu_Q = 2.6$ kHz), respectively. From the spectral difference of about 2 kHz it was concluded that the rate of exchange of lipid between the free and restricted environments was slower than 10^4 Hz (Dahlquist et al., 1977). However, this experiment turned out not to be reproducible. In all further ^{2}H- and ^{31}P-NMR experiments with reconstituted membrane systems only a single spectral component was observed, placing the exchange rate between boundary lipid and bulk lipid above 10^4–10^5 Hz.

Thus the simplest explanation for the difference between ESR and ^{2}H-NMR experiments could be an exchange rate of the order of 10^5–10^7 Hz. This exchange rate would be fast enough to produce an exchange-narrowed ^{2}H-NMR spectrum, but would be slow compared to the ESR time scale, in order to account for different ESR spectra of boundary and bulk lipid. However, some alternative points of view have been proposed (Davoust et al., 1980; Chapman et al., 1979). For further discussion see Chapter 2, Section IV.A.

A different frequency range is addressed by the T_1 relaxation time measurements. As outlined above, T_1 relaxation is most efficient if the rate of the molecular fluctuations is close to the NMR resonance frequency of the nucleus under study (to be more precise, $\tau_c \simeq 1/\omega_i$ for minimum T_1). At a magnetic field strength of 7 tesla, the deuterium and phosphorus resonance frequencies are 46 and 121 MHz, respectively, and measurement of the T_1 relaxation times thus sheds light on molecular motions in the range of $> 10^9$ Hz even though the T_1 times themselves are of the order of 10 ms (^{2}H) to a few seconds (^{31}P). The frequency range accessible by T_1 measurements corresponds essentially to that of the intramolecular segmental reorientations of the phospholipid molecules.

Let us next assume that the phospholipids of a membrane are distributed between two different environments characterized by substantially different T_1 relaxation times but otherwise equal NMR parameters (e.g., same quadrupole splitting $\Delta\nu_Q$). If there is no exchange between the two populations, a T_1 relaxation time experiment will reveal a biphasic recovery of the magnetization after a 180° pulse and it will be possible to extract the two individual T_1's. However, if the residence time of the phospholipid molecules at the two sites is shorter than individual T_1's, only an averaged T_1 value can be observed. Thus exchange rates of 1–1000 Hz are already sufficient to efficiently average the differences in ^{2}H and ^{31}P T_1 relaxation times encountered in a two-site exchange, even though the molecular processes affecting the relaxation are many orders of magnitude faster.

The molecular motions in a lipid bilayer encompass a wide range of frequencies, and measurement of T_1 relaxation times filters out a relatively narrow frequency interval of 10^9–10^{12} Hz. By contrast, much slower motions are already sufficient to average the static deuterium quadrupole splitting constant ($e^2qQ/h = 170$ kHz). The occurrence of an axially symmetric line shape for the deuterium NMR spectra can be explained by motions with frequencies of only 10^6–10^7 Hz (Meirovitch & Freed, 1979), that is, fast compared to the static quadrupole splittings. However, such slow motions should have a profound effect on the line shape and the spin-spin relaxation time T_2. Support for the existence of slower motions comes from the observation of relatively broad resonance lines, which increase by almost an order of magnitude from pure lipid vesicles to a reconstituted membrane system. This effect cannot be rationalized in terms of the small changes generally observed in the T_1 relaxation times but require the assumption of motions with distinctly slower frequencies. The molecular nature and frequency range of these motions remain unclear at present.

VI. STRUCTURAL ASPECTS OF LIPID-PROTEIN INTERACTION AS DETECTED BY ^{2}H- AND ^{31}P-NMR

Two different approaches can be employed to study the interaction of lipids with membrane proteins by means of ^{2}H- and ^{31}P-NMR. One possibility is the purification and delipidation of membrane-bound proteins followed by reconstitution with selectively deuterated lipids; the other possibility is the biosynthetic incorporation of deuterated fatty acids or other deuterated substrates into biological membranes. In the latter case the intact biological membrane is compared with aqueous bilayer dispersions formed from the extracted lipids. A variety of transmembrane proteins have now been reconstituted in functional form into a matrix of deuterated lipids, most notably cytochrome c oxidase (Oldfield et al., 1978; Seelig & Seelig, 1978; Kang et al., 1979b) and Ca^{2+}-dependent ATPase (Rice et al., 1979; Seelig et al., 1981).

Figure 5 shows ^{2}H-NMR spectra of 1,2-di[9,10-^{2}H$_2$]oleoyl-sn-glycero-3-phosphocholine ([9,10-^{2}H$_2$]DOPC) and sarcoplasmic reticulum membrane vesicles exchanged to greater than 90% with [9,10-^{2}H$_2$]-DOPC. DOPC bilayers deuterated at the cis double bond exhibit a characteristic ^{2}H-NMR spectrum consisting of three distinct quadrupole splittings. The smallest $\Delta\nu_Q$ comes from the C-10 deuteron in

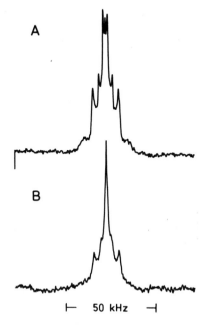

Figure 5. ^{2}H-NMR spectra at 46.1 MHz of functional reconstituted sarcoplasmic reticulum membrane vesicles (R-SR) exchanged with 1,2-di[9,10-^{2}H$_2$]oleoyl-sn-glycero-3-phosphocholine ([9,10-^{2}H$_2$]DOPC). Spectra were recorded with the quadrupole echo technique using quadrature phase detection: spectral width, 100 kHz; temperature, 4°C (liquid crystalline phase). (A) Pure [9,10-^{2}H$_2$]DOPC dispersed in water. (B) R-SR in buffer. Lipid-protein ratio L/P = 0.43 (wt/wt) corresponding to about 64 mol of phospholipid per mol of Ca^{2+} pump protein. The lipid was 99% DOPC, of which at least 90% was [9,10-^{2}H$_2$]DOPC. Reproduced with permission from Seelig et al. (1981).

the sn-2-oleic acyl chain, the intermediate from the C-10 deuteron in the sn-1 chain, while the largest is a superposition of the two remaining C-9 deuterons (Seelig & Waespe-Šarčevič, 1978; Seelig et al., 1981). The molecular origin of this spectrum is (1) the conformational inequivalence of the two fatty acyl chains and (2) a small tilting of the cis double bond with respect to the bilayer normal. The combination of these two effects leads to the motional nonequivalence of the four deuterons, which is reflected in the three different quadrupole splittings.

For sarcoplasmic reticulum membrane reconstituted with [9,10-^{2}H$_2$]DOPC we observe qualitatively the same spectrum as for pure DOPC bilayers, although the ratio of lipid to protein is 0.43 (wt/wt), corresponding to about 60 mol of phospholipid per mol of Ca^{2+} pump protein. Since it has been suggested that at least 30 lipids are needed to form a discrete and coherent lipid "annulus" around the protein (Metcalfe & Warren, 1977), the rather high protein concentration in the reconstituted sarcoplasmic reticulum (R-SR) membrane must allow at least every second lipid molecule to have direct contact with the protein surface. The remarkable result of the reconstitution study (Fig. 5B) thus is the relatively small change observed in the NMR parameters in spite of the high protein concentration. Compared to the spectrum from pure DOPC liposomes, the resonance lines in the spectrum from reconstituted sarcoplasmic reticulum membrane are broader and the quadrupole splittings are reduced. These effects are most conspicuous for the innermost splitting (C-10 deuteron of sn-2 chain) which is well resolved in pure DOPC but collapsed to a single broad line in R-SR.

The variation of the quadrupole splitting $\Delta\nu_Q$ and the chemical shielding anisotropy $\Delta\sigma$ of pure DOPC liposomes and R-SR membranes with temperature are summarized in Fig. 6. The $\Delta\nu_Q$ and $\Delta\sigma$ parameters are found to decrease monotonically with increasing temperature. Except for the quantitative differences, the deuterium and phosphorus spectra of pure DOPC and R-SR are quite similar. It is important to note that the ^{2}H-NMR spectrum of reconstituted sarcoplasmic reticulum membranes is characteristic of a single homogeneous lipid phase and does not provide evidence for two discrete and long-lived lipid environments around the Ca^{2+} pump protein (Seelig et al., 1981). The same result has been obtained for a variety of reconstituted cytochrome c oxidase systems employing saturated and unsaturated synthetic lecithins (Oldfield et al., 1978; Kang et al., 1979b; Seelig & Seelig, 1978; L. Tamm & J. Seelig, unpublished results).

The reduction in the deuterium quadrupole splitting has also been the common finding for the reconstituted membranes discussed above and for a variety of other reconstituted membrane systems (Rice et al., 1979). However, it could be argued that the results obtained are due in part to the biochemical manipulations involved in the purification, delipidation, and reconstitution of the membrane proteins. This problem is avoided by a more biological approach in which deuterated precursors such as deuterated fatty acids or deuterated glycerol are incorporated biosynthetically into a biological membrane. ^{2}H-NMR studies on intact biological membranes have now been reported for *Acholeplasma laidlawii* (Oldfield et al., 1972; Stockton et al., 1977; Smith et al., 1979; Rance et al., 1980) and for *Escherichia coli* (Davis et al., 1979; Gally et al., 1979, 1980, 1981; Kang et al., 1979a; Nichol et al., 1980).

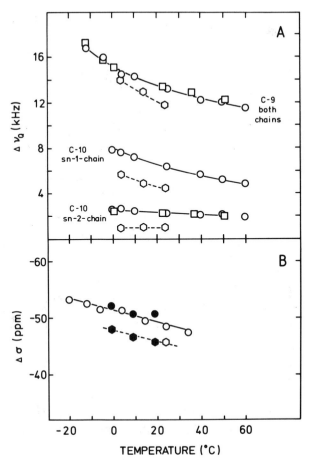

Figure 6. Reconstituted sarcoplasmic reticulum membranes exchanged with 1,2-di[9,10-²H₂]oleoyl-*sn*-glycero-3-phosphocholine: temperature dependence of ²H- and ³¹P-NMR parameters. (*A*) ²H-NMR. Temperature dependence of the residual quadrupole coupling constant, $\Delta\nu_Q$. O, 1,2-di[9,10-²H₂]oleoyl-*sn*-glycero-3-phosphocholine liposomes in excess water; □, 1-oleoyl-2-[9,10-²H₂]oleoyl-*sn*-glycero-3-phosphocholine; ⊙, R-SR membranes. The smallest splitting of R-SR is estimated and is certainly smaller than 1 kHz. (*B*)³¹P-NMR. Variation of the chemical shielding anisotropy $\Delta\sigma$ with temperature. Measurements were made at 121.4 MHz (open symbols) and at 36.4 MHz (solid symbols): O, ●, 1,2-dioleoyl-*sn*-glycero-3-phosphocholine in excess water (same sample as Fig. 5A); ⊙, ◉, R-SR membranes. The R-SR was the same sample as used in Fig. 5B. Reproduced with permission from Seelig et al. (1981).

Figure 7 shows the result of a ²H-NMR study of *E. coli* fatty acid auxotrophs grown on [9,10-²H₂]elaidic acid (Gally et al., 1980). The deuterium order parameters $|S_{CD}|$ of cells and inner and outer membranes are compared with those of the extracted lipids and are found to increase in the order of inner membrane < outer membrane < pure phospholipid bilayers. The differences between inner and outer membrane, are quite small and the order parameters of both membranes can be approximated by those of whole cells. The most interesting results of Fig. 7 are

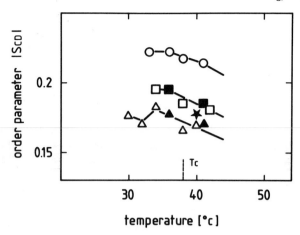

Figure 7. Temperature dependence of the deuterium order parameter $|S_{CD}|$ of elaidate-[9,10-^{2}H$_2$]-enriched *E. coli* membranes and derived liposomes: ○, liposomes in 0.1 M NaCl, 0.01 M NaPO$_4$, and 0.001 M EDTA, pH 7.0 (79 mol % phosphatidylethanolamine; 21 mol % phosphatidylglycerol). The gel-liquid crystal phase transition is centered around 38°C as revealed by differential thermal analysis. Outer membrane (□) and inner membrane (△) in Mg^{2+}- free buffer. Outer membrane (■) and inner membrane (▲) in the presence of 0.01 M MgCl$_2$ and no EDTA. Whole cells (★) in the presence of 0.01 M MgCl$_2$. Reproduced with permission from Gally et al. (1980).

the findings that the hydrocarbon chains are little affected by the presence of protein and that they are better ordered in a bilayer without protein than with protein. Similar results have also been obtained for the other biological membrane systems studied. As a general conclusion it follows from these studies that the *average lipid conformation is not much changed in the presence of membrane protein,* at least not at the level of the hydrocarbon chains and the glycerol backbone (for details of this lipid conformation see Seelig & Seelig, 1980, Gally et al., 1981). At the level of the polar head group the situation could be different, however. Only phosphorus NMR data have been reported at present, whereas ^{2}H-NMR studies with selectively deuterated phospholipid head groups, which would be more sensitive to specific interactions, are just in their beginning (L. Tamm & J. Seelig, unpublished work).

The small reductions in the deuterium quadrupole splittings, $\Delta\nu_Q$, are the recurrent result of almost all reconstitution studies and investigations of intact membranes. This reduction can be explained by assuming that the protein molecules have an irregular, rugged surface due to the protruding amino acid side groups. In the fluid membrane the flexible chains of the lipid molecules will fill the space between these side groups to create a densely packed lipid-protein bilayer. This has the consequence that the already distorted hydrocarbon chains become even more distorted in the presence of protein, the amplitudes of the angular excursions of the methylene segment increase and the deuterium order parameter decreases (Seelig & Seelig, 1978). As an alternative explanation it has been suggested that in common with soluble proteins, integral proteins in membranes have a "fluidlike" outer region, which provides an approximate fluid mechanical match with the liquid crystalline phospholipid membrane (Bloom, 1979).

VII. DYNAMIC ASPECTS OF LIPID-PROTEIN INTERACTION

This section is devoted to a short discussion of T_1 relaxation times in reconstituted membrane systems. Unfortunately, there is a dearth of experimental data and the only systematic studies available are those on reconstituted sarcoplasmic membranes (R-SR) (Seelig et al., 1981) and to a lesser extent, those on reconstituted cytochrome c oxidase membranes (L. Tamm & J. Seelig, unpublished work). The result of a phosphorus T_1 relaxation time experiment (at $\sim$ 121 MHz) on sarcoplasmic reticulum membranes reconstituted with DOPC is reproduced in Fig. 8. No anisotropy in the T_1 relaxation time was observed; that is, the T_1 relaxation time was the same for all parts of the "powder-type" spectrum, and the relaxation process could be described by a single exponential.

The temperature dependence of the phosphorus T_1 relaxation time (at $\sim$ 121 MHz) was characterized by a distinct minimum as is shown in Fig. 9. For pure DOPC this minimum occurred at 277°K with a relaxation time of $\sim$ 1.0 s, whereas for R-SR the minimum was shifted toward higher temperatures (287°K) and had a longer T_1 of 1.2 s. These phosphorus measurements are the first example of a T_1 minimum for a lipid bilayer. The observation of a T_1 minimum immediately allows

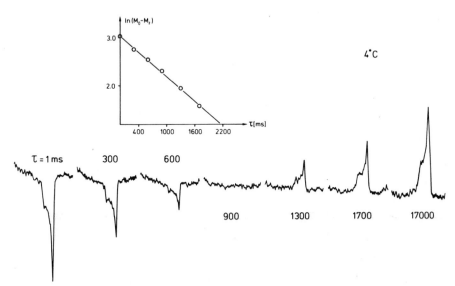

Figure 8. Measurement of the phosphorus T_1 relaxation time at 121.4 MHz using the inversion recovery technique: R-SR membrane vesicles exchanged with 1,2-di[9,10-²H₂]oleoyl-sn-glycero-3- phosphocholine (same sample as Fig. 5B), 31.25 kHz spectral width, proton-decoupled spectra using inverse gated-decoupling conditions, 300 scans per spectrum; T_1 = 1.30 s; temperature, 4°C. Reproduced with permission from Seelig et al. (1981).

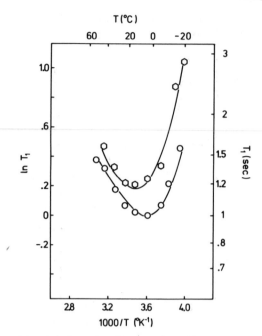

Figure 9. Variation of the phosphorus T_1 relaxation time with temperature: $\circ$, DOPC liposomes measured at 121.4 MHz resonance frequency; $\mathbb{O}$, sarcoplasmic reticulum membrane exchanged with DOPC measured at 121.4 MHz (same sample as Fig. 5B). Reproduced with permission from Seelig et al. (1981).

a quantitative evaluation of the correlation time, since $\tau_c \simeq 1/\omega_o$ at the minimum. Hence we conclude from Fig. 9 that the correlation time in pure DOPC bilayers at 277°K is about 1 ns. In reconstituted sarcoplasmic membrane vesicles the T_1 minimum is shifted by 10°C toward higher temperatures, which means that the high frequency motions dominating the T_1 process are slowed by 10–20% in the presence of Ca^{2+} pump protein.

Deuterium T_1 relaxation times for [9,10-^{2}H$_2$]DOPC and sarcoplasmic reticulum membranes reconstituted with [9,10-^{2}H$_2$]DOPC are shown in Fig. 10. The T_1 relaxation time increases with increasing temperature, from which it can be concluded that the rate of reorientation of the *cis* double bond falls into the short correlation time regime ($\omega_o\tau_c \ll 1$, $\omega_o = 2.9 \times 10^8$ rad/s). Figure 10 also shows that the deuterium T_1 relaxation times are shorter in R-SR than in pure DOPC vesicles at ambient temperature. It is the reciprocal of the T_1 relaxation time that is proportional to the motional parameter (i.e., the rotational correlation time τ_c). For DOPC bilayers at 24°C one measures $\Delta\nu_Q = 13.2$ kHz and $T_1 = 13.8$ ms, leading to $S_{CD} \sim 0.10$ and $\tau_c = 0.17$ ns (see Eq. 7). For R-SR at the same temperature, the T_1 relaxation time is 11.0 ms with $\Delta\nu_Q = 11.7$ kHz; hence $S_{CD} \sim 0.09$ and $\tau_c = 0.21$ ns. Thus the analysis of the deuterium T_1 relaxation time yields the same conclusion as reached above for the phosphate group, namely, the rate of segmental reorientation is slowed by about 20% due to the presence of protein.

Taken together, the ^{2}H- and ^{31}P-NMR relaxation time studies demonstrate unambiguously that motions with frequencies in the relatively narrow interval of 10^9–10^{10} Hz have a high probability of occurrence in the bilayer head group region as well as in the hydrocarbon region. These motions must be identified as intramolecular segmental reorientations. In addition, the data show that the high frequency motions are only slightly slowed (by 10–20% at most) in the presence of protein, even at very high protein contents. The deuterium T_1 relaxation time studies on reconstituted cytochrome c oxidase membrane have led to similar results as shown in Fig. 10 and generally support the above conclusions.

Evaluation of the NMR correlation times in terms of an apparent "microviscosity" using Eq. 6 yields values of $\eta \sim 0.1$–0.5 poise. In contrast, estimates of membrane "microviscosities" derived from translational diffusion constants of lipids and rotational or translational diffusion constants of membrane proteins yield values of the order of 1–10 poise (Edidin, 1974; Cherry, 1979). This suggests that the fast internal modes of motion of a phospholipid molecule are much less influenced by

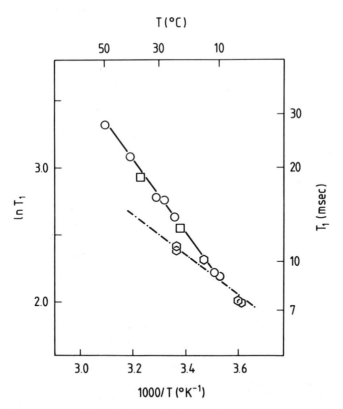

Figure 10. Arrhenius plots of the deuterium T_1 relaxation times (at 46.1 MHz): ○, 1,2-di[9,10-^{2}H$_2$]oleoyl-sn-glycero-3-phosphocholine dispersed in excess water; □, 1-oleoyl-2-[9,10-^{2}H$_2$]oleoyl-sn-glycero-3-phosphocholine in excess water; ○, sarcoplasmic reticulum membrane vesicles reconstituted with 1,2-di[9,10-^{2}H$_2$]oleoyl-sn-glycero-3-phosphocholine dispersed in buffer (same sample as in Fig. 5B). Reproduced with permission from Seelig et al. (1981).

membrane proteins than are the translocation and also, perhaps, the reorientation of the molecule as a whole. The differences of the various methods also illustrate the inherent deficiencies of the concept of "microviscosity" as applied to ordered membrane systems.

REFERENCES

Abragam, A. (1961). *The Principles of Nuclear Magnetism,* Oxford University Press, London.

Bloom, M. (1979). Squishy Proteins in Fluid Membranes. *Can. J. Phys.* **57,** 2227–2230.

Brown, M. F., Seelig, J., & Haeberlen, U. (1979). Structural Dynamics in Phospholipid Bilayers from Deuterium Spin Lattice Relaxation Time Measurements. *J. Chem. Phys.* **70,** 5045–5053.

Burnett, L. J., & Muller, B. H. (1971). Deuteron Quadrupole Coupling Constants in Three Deuterated Paraffin Hydrocarbons: C_2D_6, C_4D_{10}, C_6D_{14}. *J. Chem. Phys.* **55,** 5829–5831.

Carrington, A., & McLachlan, A. D. (1967). *Introduction to Magnetic Resonance,* Harper & Row, New York.

Chapman, D., Gomez-Fernandez, J. C., & Goni, F. M. (1979). Intrinsic Protein-Lipid Interactions. *FEBS Lett.* **98,** 211–223.

Cherry, R. J. (1979). Rotational and Lateral Diffusion of Membrane Proteins. *Biochim. Biophys. Acta* **559,** 289–327.

Cullis, P. R., & de Kruijff, B. (1979). Lipid Polymorphism and the Functional Roles of Lipids in Biological Membranes. *Biochim. Biophys. Acta* **559,** 399–420.

Dahlquist, F. W., Muchmore, D. C., Davis, J. H., & Bloom, M. (1977). Deuterium Magnetic Resonance Studies of the Interaction of Lipids with Membrane Proteins. *Proc. Natl. Acad. Sci. U.S.A.* **74,** 5435–5439.

Davis, J. H., Jeffrey, K. R., Bloom, M., Valic, M. I., & Higgs, T. P. (1976). Quadrupolar Echo Deuteron Magnetic Resonance Spectroscopy in Ordered Hydrocarbon Chains. *Chem. Phys. Lett.* **42,** 390–394.

Davis, J. H., Nichol, C. P., Weeks, G., & Bloom, M. (1979). Study of the Cytoplasmic and Outer Membranes of *E. coli* by Deuterium Magnetic Resonance. *Biochemistry* **18,** 2103–2112.

Davoust, J., Bienvenue, A., Fellmann, P., & Deveaux, P. F. (1980). Boundary Lipids and Protein Mobility in Rhodopsin-Phosphatidylcholine Vesicles. Effect of Lipid Phase Transition. *Biochim. Biophys. Acta* **596,** 28–42.

Doddrell, D., Glushko, V., & Allerhand, A. (1972). Theory of Nuclear Overhauser Enhancement and $^{13}C-^1H$ Dipolar Relaxation in Proton-Decoupled Carbon-13 NMR Spectra of Macromolecules. *J. Chem. Phys.* **56,** 3683–3689.

Edidin, M. (1974). Rotational and Translational Diffusion in Membranes. *Annu. Rev. Biophys. Bioeng.* **3,** 179–201.

Gally, H. U., Niederberger, W., & Seelig, J. (1975). Conformation and Motion of the Choline Head Group in Bilayers of Dipalmitoyl-3-*sn*-phosphatidylcholine. *Biochemistry* **14,** 3647–3652.

Gally, H. U., Pluschke, G., Overath, P., & Seelig, J. (1979). Structure of *Escherichia coli* Membranes. Phospholipid Conformation in Model Membranes and Cells as Studied by Deuterium Magnetic Resonance. *Biochemistry* **18,** 5605–5610.

Gally, H. U., Pluschke, G., Overath, P., & Seelig, J. (1980). Structure of *Escherichia coli* Membranes. Fatty Acyl Chain Order Parameters of Inner and Outer Membrane and Derived Liposomes. *Biochemistry* **19,** 1638–1643.

Gally, H. U., Pluschke, G., Overath, P., & Seelig, J. (1981). Structure of *Escherichia coli* Membranes. Glycerol Auxotrophs as a Tool for the Analysis of the Phospholipid Head Group Region by Deuterium Magnetic Resonance. *Biochemistry* **20,** 1826–1831.

Griffin, R. G. (1976). Observation of the Effect of Water on the ^{31}P-Nuclear Magnetic Resonance Spectra of Dipalmitoyllecithin. *J. Am. Chem. Soc.* **98**, 851–853.

Herzfeld, J., Griffin, R. G., & Haberkorn, R. A. (1978). Phosphorus-31 Chemical Shift Tensors in Barium Diethyl Phosphate and Urea–Phosphoric Acid: Model Compounds for Phospholipid Head Group Studies. *Biochemistry* **17**, 2711–2718.

Jost, P. C., Griffith, O. H., Capaldi, R. A., & Vanderkooi, G. (1973). Evidence for Boundary Lipids in Membranes. *Proc. Natl. Acad. Sci. U.S.A.* **70**, 480–484.

Jost, P. C., & Griffith, O. H. (1980). The Lipid-Protein Interface in Biological Membranes. *Ann. N.Y. Acad. Sci.* **348**, 391–407.

Kang, S. Y., Gutowsky, H. S., & Oldfield, E. (1979a). Spectroscopic Studies of Specifically Deuterium Labelled Membrane Systems. Nuclear Magnetic Resonance Investigation of Protein-Lipid Interactions in *Escherichia coli* Membranes. *Biochemistry* **18**, 3268–3272.

Kang, S. Y., Gutowsky, H. S., Hsung, H. C., Jacobs, R., King, T. E., Rice, D., & Oldfield, E. (1979b). Nuclear Magnetic Resonance Investigation of the Cytochrome Oxidase–Phospholipid Interaction. A New Model for Boundary Lipid. *Biochemistry* **18**, 3257–3267.

Kohler, S. J., & Klein, M. P. (1976). ^{31}P Chemical Shielding Tensors of Phosphorylethanolamine, Lecithin and Related Compounds: Application to Head Group Motion in Membranes. *Biochemistry* **15**, 967–973.

Meirovitch, E., & Freed, J. H. (1979). Slow Motional Lineshapes for Very Anisotropic Diffusion: $I = 1$ Nuclei. *Chem. Phys. Lett.* **64**, 311–316.

Metcalfe, J. C., & Warren, G. B. (1977). Lipid-Protein Interaction in a Reconstituted Calcium Pump. In *International Cell Biology* (B. R. Brinkley, & K. R. Porter, Eds.), pp. 15–23, Rockefeller University Press, New York.

Nichol, C. P., Davis, J. H., Weeks, G., & Bloom, M. (1980). Quantitative Study of the Fluidity of *Escherichia coli* Membranes Using Deuterium Magnetic Resonance. *Biochemistry* **19**, 451–457.

Niederberger, W., & Seelig, J. (1976). Phosphorus-31 Chemical Shift Anisotropy in Unsonicated Phospholipid Bilayers. *J. Am. Chem. Soc.* **98**, 3704–3706.

Oldfield, E., Chapman, D., & Derbyshire, W. (1972). Lipid Mobility in *Acholeplasma* Membranes Using Deuteron Magnetic Resonance. *Chem. Phys. Lipids* **9**, 69–81.

Oldfield, E., Gilmore, R., Glaser, M., Gutowsky, H. S., Hsung, J. C., Kang, S. Y., King, T. E., Meadows, W., & Rice, D. (1978). Deuterium Nuclear Magnetic Resonance Investigation of the Effects of Proteins and Polypeptides on Hydrocarbon Chain Order in Model Membrane Systems. *Proc. Natl. Acad. Sci. U.S.A.* **75**, 4657–4660.

Rance, M., Jeffrey, K. R., Tulloch, A. P., Butler, K. W., & Smith, I. C. P. (1980). Orientational Order of Unsaturated Phospholipids in the Membranes of *Acholeplasma laidlawii* as Observed by Deuterium NMR. *Biochim. Biophys. Acta* **600**, 245–262.

Rice, D. M., Meadows, M. D., Scheinman, A. O., Goni, F. M., Gomez-Fernandez, J. C., Moscarello, M. A., Chapman, D., & Oldfield, E. (1979). Protein-Lipid Interactions. A Nuclear Magnetic Resonance Study of Sarcoplasmic Reticulum Ca^{2+}, Mg^{2+}-ATPase, Lipophilin, and Proteolipid Apoprotein-Lecithin Systems and a Comparison with the Effects of Cholesterol. *Biochemistry* **18**, 5893–5903.

Seelig, J. (1977). Deuterium Magnetic Resonance: Theory and Application to Lipid Membranes. *Q. Rev. Biophys.* **10**, 353–418.

Seelig, J. (1978). Phosphorus-31 Nuclear Magnetic Resonance and the Head Group Structure of Phospholipids in Membranes. *Biochim. Biophys. Acta* **505**, 105–141.

Seelig, J., & Niederberger, W. (1974). Deuterium Labeled Lipids as Structural Probes in Liquid Crystalline Bilayers. *J. Am. Chem. Soc.* **96**, 2069–2072.

Seelig, J. & Gally, H. U. (1976). Investigation of Phosphatidylethanolamine Bilayers by Deuterium and Phosphorus-31 Nuclear Magnetic Resonance. *Biochemistry* **15**, 5199–5204.

Seelig, A., & Seelig, J. (1978). Lipid-Protein Interaction in Reconstituted Cytochrome *c* Oxidase Phospholipid Membranes. *Hoppe Seyler's Z. Physiol. Chem.* **359,** 1747–1756.

Seelig, J., & Waespe-Šarčevič, N. (1978). Molcular Order in cis- and trans-Unsaturated Phospholipid Bilayers. *Biochemistry* **17,** 3310–3315.

Seelig, J., & Seelig, A. (1980). Lipid Conformation in Model Membranes and Biological Membranes. *Q. Rev. Biophys.* **13,** 9–61.

Seelig, J., Tamm, L., Hymel, L., & Fleischer, S. (1981). Deuterium and Phosphorus NMR and Fluorescence Depolarization Studies of Functional Reconstituted Sarcoplasmic Reticulum Membrane Vesicles. *Biochemistry* **20,** 3922–3932.

Smith, I. C. P., Butler, K. W., Tulloch, A. P., Davis, J. H., & Bloom, M. (1979). The Properties of Gel State Lipid Membranes of *Acholeplasma laidlawii* as Observed by Deuterium Nuclear Magnetic Resonance. *FEBS Lett.* **100,** 57–61.

Stockton, G. W., Johnson, K. G., Butler, K., Tulloch, A. P., Boulanger, Y., Smith, I. C. P., Davis, J. H., & Bloom, M. (1977). Deuterium NMR Study of Lipid Organisation in *Acholeplasma laidlawii* Membranes. *Nature (London)* **269,** 267–268.

Stockton, G. W., Polnaszek, C. F., Tulloch, A. P., Hasan, F., & Smith, I. C. P. (1976). Molecular Motion and Order in Single-Bilayer Vesicles and Multilamellar Dispersions of Egg Lecithin and Lecithin-Cholesterol Mixtures. A Deuterium Nuclear Magnetic Resonance Study of Specifically Labeled Lipids. *Biochemistry* **15,** 954–966.

Yeagle, P. L., Hutton, W. C., Huang, C., & Martin, R. B. (1975). Head Group Conformation and Lipid Cholesterol Association in Phosphatidylcholine Vesicles. A ^{31}P {^{1}H} Nuclear Overhauser Effect Study. *Proc. Natl. Acad. Sci. U.S.A.* **72,** 3477–3481.

Yeagle, P. L., Hutton, W. C., Huang, C., & Martin, R. B. (1977). Phospholipid Head Group Conformations. Intermolecular Interactions and Cholesterol Effects. *Biochemistry* **16,** 4344–4349.

Four

Photochemical Cross-Linking in Studies of Lipid-Protein Interactions

ROBERT J. ROBSON, RAMACHANDRAN RADHAKRISHNAN, ALONZO H. ROSS, YOHTAROH TAKAGAKI, AND H. GOBIND KHORANA

Departments of Biology and Chemistry, Massachusetts Institute of Technology, Cambridge, Massachusetts U.S.A.

ABBREVIATIONS

CIDNP, chemically induced dynamic nuclear polarization
DMAP, N,N-dimethyl-4-aminopyridine
INA, 5-iodonapthyl-1-azide
PC-PLEP, phosphatidylcholine-phospholipid exchange protein
TPD, 3-trifluoromethyl-3-phenyl diazirine

I. INTRODUCTION

Studies of protein-lipid interactions are important to the understanding of structural and functional properties of membranes. Integral membrane proteins may be expected to contain two types of domains. One type of protein domain may interact with the hydrophobic fatty acyl chains of phospholipids by virtue of being embedded in the bilayer (e.g., the bilayer anchoring portions of cytochrome b_5, glycophorin A, and influenza hemagglutinin). The second type of protein domain may interact to some extent with the polar head groups of phospholipids and in most cases protrude out of the bilayer and into the aqueous medium. Extrinsic membrane proteins, such as spectrin, actin, and certain apolipoproteins, may interact predominantly with the phospholipid polar head groups.

For the peptide chains embedded in the phospholipid bilayer, it has generally been assumed that their ability to exist in a lipid bilayer depends on the presence of hydrophobic amino acid side chains juxtaposed to the lipid fatty acyl chains. However, there is very little specific information on the topography or structures of such chains within the hydrophobic milieu.

Progress is being made in the determination of the amino acid sequences of membrane proteins (Table 1). Although the primary structure determination is a necessary first step, additional approaches will be required for the determination of secondary and tertiary structures, including detailed knowledge of the phospholipid-protein interactions. Recently, a variety of physicochemical approaches have been developed for the study of phospholipid-protein interactions in the hydrophobic regions of membrane proteins. Some of these approaches are discussed in other chapters of this book.

The goal of most organochemical approaches is the formation of covalent linkages between certain added reagents and specific sites on the exposed or embedded regions of proteins or phospholipids. Broadly, two concepts have been used. In the first, reagents capable of specific chemical reactions with functional groups (amino, carboxyl, sulfhydryl, hydroxyl) in the polypeptide chains are used. Within this group, reagents may be classified as membrane permeant (lipophilic) or non-permeant (hydrophilic). One permeant reagent is phenylisothiocyanate (Sigrist & Zahler, 1978). Non-permeant reagents include the *bis*-alkyl imidate (Hartman & Wold, 1966), the tartyl diazides (Lutter et al., 1974), and various sulfhydryl reagents, including *bis*-maleimides (see, e.g., Wold, 1972). Cross-linking by these reagents occurs if functional groups are suitably placed within the molecule.

In the second general organochemical approach, introduced by Westheimer and colleagues (Singh et al., 1962), chemical cross-linking does not depend on reactive functional groups on the proteins but merely on the proximity of the reagent to the polypeptide chains. The aim of this approach as applied to membrane studies is the determination of contacts or interactions between polypeptides and lipid fatty acyl chains by the formation of cross-links between points of nearest contact (Chakrabarti & Khorana, 1975). As such, reagents are needed that can cross-link even to such hydrocarbons as the amino acid side chains of valine, leucine, or isoleucine, as well as the alkyl chains of phospholipids. All approaches so far to develop such highly reactive and non-selective reagents have made use of light-sensitive (photolabile) reagents that form highly reactive intermediates upon irradiation at the proper wavelengths. This photolabeling approach, the subject of this chapter, utilizes reagents that are chemically stable in the absence of light and have the capacity to undergo photodecomposition to a short-lived, highly reactive intermediate that can cross-link to form stable, covalent bonds. In general, the reactive intermediates are carbenes, nitrenes, and free radicals. Free radicals, however, can undergo chain reactions and are therefore unsuitable.

This chapter reviews the requirements, characteristics, and physical properties of photolabile cross-linking reagents in general, cross-linking studies with membrane-permeant lipophilic nitrene and carbene precursors, and cross-linking studies using phospholipids carrying photolabels at the fatty acyl termini and in the polar head group. Recent reviews of photolabeling of biological systems include those of Chowdhry and Westheimer (1979) and Bayley and Knowles (1977), and a series of papers from a recent symposium of the New York Academy of Sciences (Tometsko & Richards, 1980).

II. PHOTOSENSITIVE REAGENTS AND PHOTOGENERATED SPECIES

The characteristics necessary for an effective photolabeling reagent are as follows: (1) reasonable thermal and chemical stability in the usual laboratory light and conditions, (2) photolysis upon irradiation at wavelengths that cause negligible photodamage to the proteins and lipids in the system, (3) capability of cross-linking based solely or mostly on proximity, (4) absence of undesirable intramolecular reactions or rearrangements, and (5) ready availability of the reagent, preferably

TABLE 1 Viral, Prokaryote, and Eukaryote Membrane Proteins of Known Sequence

Protein (Origin)	Size[a]	Method[b]	References
Virus			
Bacteriophage ZJ-2 B-protein	50 AA	M. Ed.	Snell & Offord (1972)
Bacteriophage fd coat protein (gene VIII)	50 AA	M. & L. Ed.	Nakashima & Konigsberg (1974)
Bacteriophage f1 coat protein	50 AA	L. Ed.	Bailey et al. (1977)
Bacteriophage M13 coat protein (gene VIII)	50 AA	DNA	Van Wezenbeck et al. (1980)
Vesicular Stomatitis Virus glycoprotein G	M 62,500[c]	DNA	Rose et al. (1980)
Semliki Forest Virus glycoprotein E1	438 AA	DNA	Garoff et al. (1980)
Semliki Forest Virus glycoprotein E2	422 AA	DNA	Garoff et al. (1980)
Fowl Plague Virus hemagglutinin	563 AA	DNA	Porter et al. (1979)
Human influenza A hemagglutinin HA-2	221 AA	DNA	Porter et al. (1979)
Avian Sarcoma Virus oncogene (src) pp60[src]	530 AA	DNA	Czernilofsky et al. (1980)
Avian Sarcoma Virus glycoprotein gp37	M 20,500[c]	DNA	Czernilofsky et al. (1980)
Prokaryote			
Bacteriorhodopsin (*H. halobium*)	248 AA	L. & S. Ed.	Khorana et al. (1979)
D-Alanine carboxypeptidase (*B. stearothermophilus*)	M 46,500[c]	L. Ed.	Waxman & Strominger (1981)
D-Alanine carboxypeptidase (*B. subtilis*)	M 50,000[c]	L. Ed.	Waxman & Strominger (1981)
DCCD-binding protein of ATPase complex (*E. coli*)	79 AA	S. Ed.	Wachter et al. (1980)
DCCD-binding protein of ATPase complex (*T. thermophilum* PS-3)	72 AA	S. Ed.	Hoppe & Sebald (1980)
Lactose permease (M-protein, lacY) (*E. coli*)	417 AA	DNA	Buchel et al. (1980)
Outer membrane protein I (porin) (*E. coli* B/r)	340 AA	M. Ed.	Chen et al. (1979)
Outer membrane protein II* (ompA) (*E. coli* K-12)	325 AA	M. Ed.	Chen et al. (1980)

Eukaryote

	Size	Method	Reference
Cytochrome b_5 (horse, porcine, bovine, rabbit liver)	133 AA	L. Ed.	See text
Cytochrome *c* oxidase subunit II (beef heart mitochondria)	227 AA	L. Ed.	Steffens & Buse (1979)
Cytochrome *c* oxidase subunit VIIIa (beef heart mitochondria)	47 AA	L. Ed.	Buse & Steffens (1978)
DCCD-binding protein of F_o-ATPase (*S. cerevisiae* mitochondria)	76 AA	S. Ed.	Sebald et al. (1979)
DCCD-binding protein of F_o-ATPase (*N. crassa* mitochondria)	81 AA	S. Ed.	Sebald et al. (1980)
DCCD-binding protein of F_o-ATPase (spinach chloroplast)	81 AA	S. Ed.	Sebald & Wachter (1980)
Glycophorin A (human erythrocyte)	131 AA	M. Ed.	See text
Immunoglobulin M (IgM) membrane bound μ-chain (mouse myeloma cell)	—	DNA	Rogers et al. (1980)
Isomaltase of sucrase-isomaltase complex (rabbit intestine)	M 150,000[c]	L. Ed.	Frank et al. (1980)

[a]The size of the protein is expressed in number of amino acids (AA) or molecular weight (M).

[b]Methods for the determination of primary sequence abbreviated as follows: DNA, deduced from DNA sequencing; L. Ed., automated liquid phase Edman degradation; S. Ed., automated solid phase Edman degradation; M. Ed., manual Edman degradation.

[c]The sequence of the presumed membranous portion is known, but not that of the entire protein.

in radiolabeled form. In order to know which amino acid side chains or phospholipids are adjacent to the photolabile probe, it is imperative that the reactive species react extremely rapidly. Lifetimes on the order of a few microseconds or less are desirable. The reagents investigated most often have been precursors of carbenes and nitrenes.

A. Carbenes

The common precursors for carbenes are diazo compounds and the three- membered cyclic isomers called diazirines. The general structures of these compounds are shown in Table 2. Carbenes are neutral bivalent carbon intermediates: the carbon atom is covalently linked to two other groups and possesses two non-bonding orbitals with two electrons whose spins may be antiparallel (singlet state) or parallel (triplet state). Carbenes react very rapidly with a variety of chemical functions. Physical techniques, such as low temperature absorption spectroscopy, electron spin resonance (ESR) spectroscopy, and chemically induced dynamic nuclear polarization (CIDNP) have been used to monitor the formation and detection of carbenes. Pulsed-laser spectrophotometric techniques have been used to measure the rate of intersystem crossing for the singlet to triplet carbene, and to evaluate the rate constants for reaction of both carbene species with olefins, hydrocarbons, and alcohols. Laser flash photolysis of diazofluorene produced a singlet and the long-lived triplet (Zupancic & Schuster, 1980). The lifetime of the triplet in the absence of a trapping agent was determined to be circa 30 μs. The short-lived transient was shown to be the singlet by its rapid conversion by methanol to an ether. The half-life for intersystem crossing of the singlet carbene was 17 ±2 ns. For further detailed analysis of this system, the reader is referred to Zupancic and Schuster (1980), and references cited therein.

Carbenes are potentially capable of insertion into C—H, O—H, and N—H bonds, addition to carbon-carbon double bonds and aromatic systems, coordination to nucleophilic centers to give carbanions, and hydrogen abstraction to give two free radicals that may subsequently couple. These possible fates for carbenes are shown schematically in Table 3.

TABLE 2   The Formation of Carbenes from Diazo Compounds and Diazirines

$$RR'C{=}\overset{+}{N}{=}\overset{-}{N} \longrightarrow RR'C\!: \ + \ N_2$$

(Diazo compound)   (Carbene)

$$RR'C\overset{\diagup N}{\underset{\diagdown N}{\|}} \longrightarrow RR'C\!: \ + \ N_2$$

(Diazirine)   (Carbene)

TABLE 3   Possible Fates for Carbenes

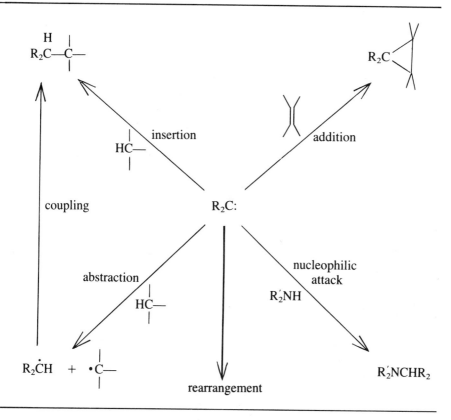

Specific reactions that a carbene may undergo depend on a number of factors. Among these are the R groups covalently bonded to the carbenic carbon, the spin state at the time of reaction, and temperature. For example, both singlet and triplet carbenes can undergo net C—H insertion reactions, but the mechanisms are qualitatively different. Singlet carbenes insert into C—H bonds in essentially a one-step process, whereas triplet carbenes first abstract a hydrogen atom to produce a triplet radical pair that results in an insertion product only after intersystem crossing. The differences in the reactivity of singlet and triplet carbenes can be found in tentative correlations made by Turro (1980).

To distinguish between singlet and triplet carbenes, one can analyze the products resulting from the addition to ethylenes that are nonsymmetrical along the C=C axis. This has become Skell's qualitative test for the spin states of the carbenes (Woodworth & Skell, 1959; Skell & Klebe, 1960). Singlet carbenes add stereospecifically to ethylenes, whereas triplet carbenes add non-stereospecifically (via a triplet diradical intermediate). This is shown in Fig. 1.

A number of covenient precursors for the photochemical generation of carbenes are shown in Table 4. The primary requirement for a carbene capable of inter-

Singlet carbenes:

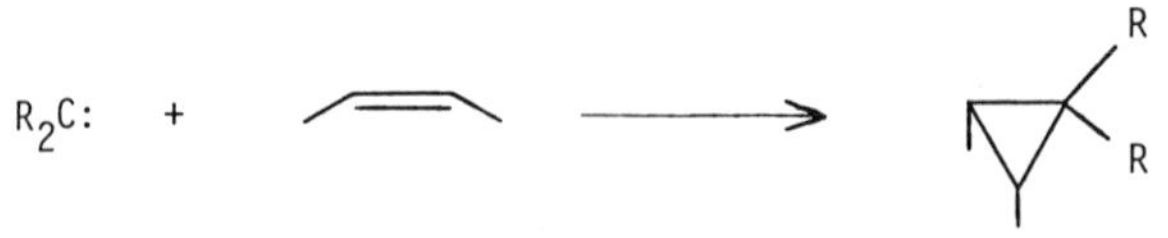

Triplet carbenes:

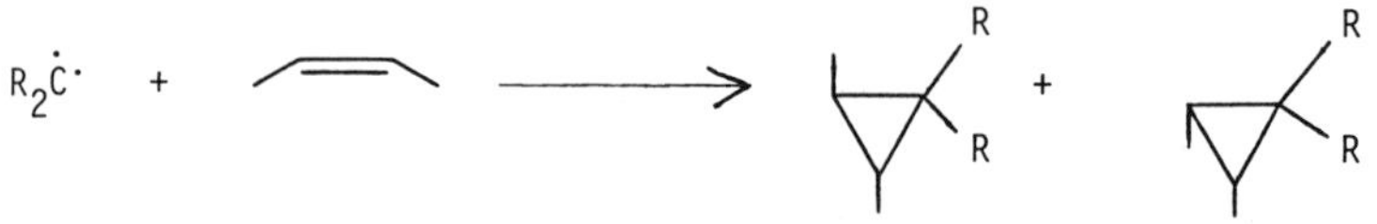

Figure 1. Reactions of singlet and triplet carbenes with ethylenes that are not symmetrical along the C=C axis. The products of singlet carbenes lead to the *syn* isomer exclusively, whereas triplet carbenes give mixtures of *syn* and *anti* isomers.

molecular insertion is that it must not have a hydrogen atom on the carbon adjacent to the carbenic center. Otherwise, hydrogen migration will occur, leading to an unreactive olefin. Diazoacetates and ketones (Table 4) meet this requirement. A major complicating side reaction, Wolff rearrangement, occurs on photolysis of diazoacyl compounds, and this is illustrated with the use of diazoacetate to map the active site of chymotrypsin, the first application of photolabeling (Fig. 2). Diazoacetates and ketones are not very stable, particularly at low pH. Diazomalonates (Table 4) (Vaughan & Westheimer, 1969) represented an improvement with regard to stability. Wolff rearrangement, however, was not reduced significantly.

Substitution of a group on the diazo-containing carbon with a trifluoromethyl moiety, as in ethyl trifluorodiazopropionate (Table 4), markedly stabilizes the diazo compound (stable for 24 hr in 1 M HCl, Chowdhry et al., 1976). On photolysis, less rearrangement occurred than with other known diazoacyl reagents. Photolysis in methanol gave insertion into the O—H bond of the solvent to the extent of about 85%, and there was only 15% reaction via the Wolff rearrangement (Chowdhry et al., 1976):

$$CF_3\overset{\overset{\displaystyle N_2}{\|}}{C}COOC_2H_5 \xrightarrow[\text{CH}_3\text{OH}]{254\text{ nm }h\nu} CF_3CHOCH_3COOC_2H_5 \quad (75\%)$$

$$+ \; CF_3CHOCH_3COOCH_3 \quad (\;8\%)$$

$$+ \; CF_3CHOC_2H_5COOCH_3 \quad (15\%)$$

TABLE 4 Useful Carbene Precursors

Type of Compound	Formula	Stability in the Dark at pH 7	Probability for Undesirable Internal Rearrangements
Diazoacetate	$-\text{OCCH}=\overset{+}{\text{N}}=\overset{-}{\text{N}}$ (with C=O)	fair	high
Diazoketones	$-\text{CCH}=\overset{+}{\text{N}}=\overset{-}{\text{N}}$ (with C=O)	fair	high
Diazomalonates	$-\text{C}-\text{C}=\overset{+}{\text{N}}=\overset{-}{\text{N}}$, COOR (with C=O)	fair	moderate
Trifluorodiazopropionate	$-\text{OCC}=\overset{+}{\text{N}}=\overset{-}{\text{N}}$, CF$_3$ (with C=O)	good	low
Aryl diazomethane	$\text{ArCH}=\overset{+}{\text{N}}=\overset{-}{\text{N}}$	poor	low
Aryl diazirine	$\underset{\text{ArC}}{\overset{\text{H}}{\diagdown}}\!\!\begin{array}{c}\text{N}\\ \| \\ \text{N}\end{array}$	good	low
Aryl trifluoroalkyl diazirine	$\underset{\text{Ar}-\text{C}}{\overset{\text{CF}_3}{\diagdown}}\!\!\begin{array}{c}\text{N}\\ \| \\ \text{N}\end{array}$	good	low

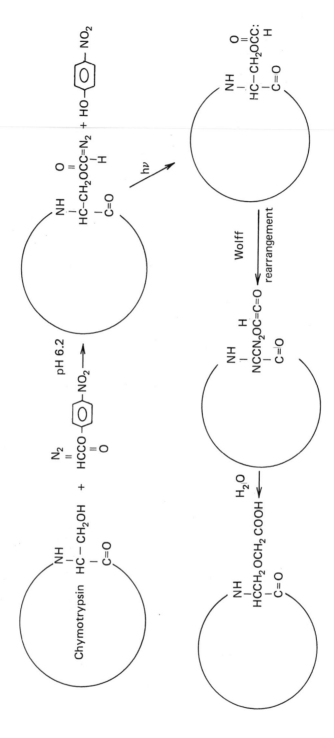

Figure 2. Photoaffinity labeling of chymotrypsin showing formation of the acyl enzyme intermediate, photolysis, Wolff rearrangement of the resulting carbene, and hydrolysis of the ketene. The small amount of cross-linking of chymotrypsin by the carbene is not shown.

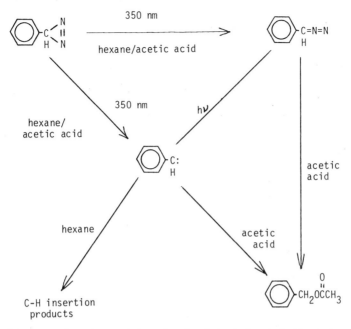

Figure 3. Primary products that result from the photolysis of phenyl diazirine in hexane-acetic acid solutions (Smith & Knowles, 1975). Estimates for percent products are shown for benzylacetate and C—H insertion products.

Aryl diazomethanes are chemically too reactive to be of much use for photoaffinity studies (Bayley & Knowles, 1977). However, their cyclic isomers, aryl diazirines, are stable for hours in 1 M acetic acid and are stable for months at −20°C as dilute solutions in hydrocarbon solvents. Photolyis proceeds readily near 350 nm, with some isomerization to the corresponding acid-labile diazo compound. Photolysis of aryl diazirines in hexane–acetic acid led to a mixture of benzyl acetate and products resulting from insertion of the phenyl carbene into the C—H bonds of hexane (Smith & Knowles, 1975) (Fig. 3).

3-Trifluoromethyl-3-phenyl diazirine (TPD) has recently been synthesized (Brunner et al., 1980) and has potentially desirable characteristics. The diazirine was rapidly photolyzed at wavelengths above 300 nm, leading to 65% carbene and 35% diazo isomer. The carbene quantitatively inserted into the O—H bond of methanol and reacted to the extent of at least 50% with cyclohexane (Fig. 4). The diazoisomer was very stable in cyclohexane or dioxane containing 1 M acetic acid and, in fact, was completely photolyzed to the carbene before more than 2% of the diazo compound had chemically reacted.

B. Nitrenes

The precursors for nitrenes are usually azides (see Table 5). Nitrenes are nitrogen analogues of carbenes that are electrically neutral, univalent nitrogen intermediates, the nitrogen atom being covalently linked to one other group. Nitrenes contain

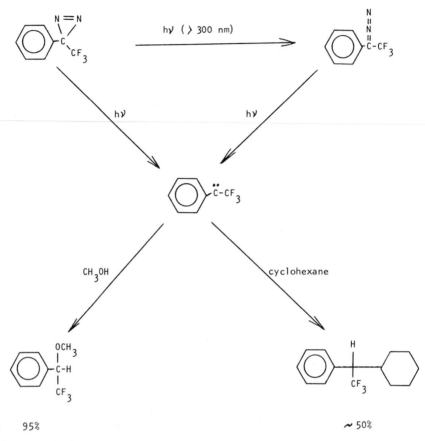

Figure 4. The photolysis of trifluoromethylphenyl diazirine leads to the carbene directly, or more slowly through photolysis of the diazo compound formed by the light-catalyzed isomerization of the diazirine to the diazo compound. The carbene has been shown to insert quantitatively into the O—H bond of methanol and to the extent of 50% into cyclohexane.

electrons in nonbonding orbitals whose spins may be antiparallel or parallel, as in carbenes. In general, nitrenes are not nearly as reactive as carbenes and are, therefore, much more reluctant to undergo insertion reactions.

Flash photolysis in solution and subsequent trapping experiments provide an idea of lifetimes. The flash photolysis of ethyl azidoformate appears to give the nitrene. From trapping experiments using cyclohexene, a crude range of $3 \times 10^{-7} < \tau < 10^{-5}$ s was determined for the lifetime τ of the nitrene (Berry et al., 1963). Several aryl azides have been photolyzed in organic glasses at 77°K in an attempt to trap the nitrenes. This intermediate was stable at liquid nitrogen temperature but disappeared on warming (Reiser & Frazer, 1965). Only 1-azidoanthracene was studied by flash photolysis; the experiments were carried out at room temperature in ethanol (5×10^{-4} M). The lifetime of the absorption at 342 nm indicated that the half-life of the nitrene was between 3 and 10 s (Reiser et al., 1966).

TABLE 5 The Formation of Nitrenes from Azides

$$R\overset{+}{\text{---}N}\overset{-}{=}N\overset{}{=}N \quad \xrightarrow{\ h\nu\ } \quad R\text{---}\ddot{N}: \ + \ N_2$$

(Azide)   (Nitrene)

Some convenient precursors for the photochemical generation of nitrenes are shown in Table 6. Possible fates of nitrenes are shown in Table 7. The most frequently used nitrene source is an aryl azide. As with carbenes, if the carbon adjacent to the nitrene is bonded to a hydrogen atom, then a molecular rearrangement becomes dominant. One problem common to aryl nitrenes is a ring expansion following reaction of a nucleophile on an azirine intermediate:

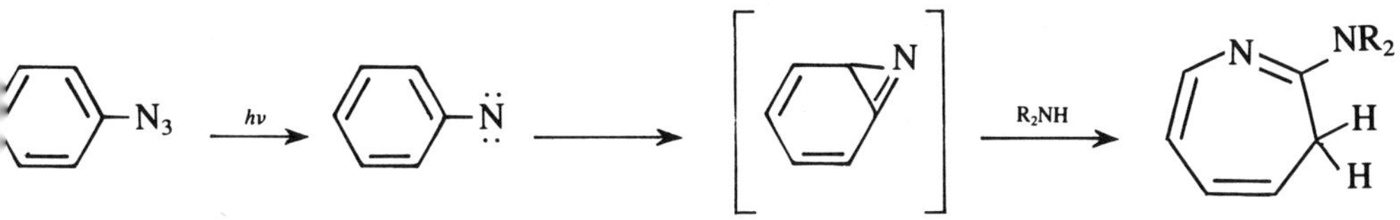

McRobbie et al. (1976) demonstrated that even when alternative pathways exist, singlet nitrenes react preferentially by insertion into O—H or N—H bonds. Nitrophenyl nitrenes have substantially shorter lifetimes and more reactivity than phenyl nitrenes in polystyrene matrices (Reiser et al., 1968). The use of nitrophenyl nitrenes is also advantageous because long wavelength light can be used for irradiation, thus reducing the risk of damage to the biological systems under study.

A recent comprehensive study of the photolysis of nine different *sec-* and *tert-*alkyl azides showed no evidence of typical nitrene processes such as aromatic

TABLE 6 Useful Nitrene Precursors

Type of Compound	Formula	Stability in the Dark at pH 7	Probability for Undesirable Internal Rearrangements
Alkyl azides	$R_3C\overset{+}{\text{---}N}\overset{-}{=}N\overset{}{=}N$	good	high
Aryl azides	$Ar\overset{+}{\text{---}N}\overset{-}{=}N\overset{}{=}N$	good	moderate
Nitro aryl azides	*m-* or *p-*$NO_2Ar\overset{+}{\text{---}N}\overset{-}{=}N\overset{}{=}N$	good	moderate

From Bayley and Knowles (1977).

TABLE 7   Possible Fates for Nitrenes

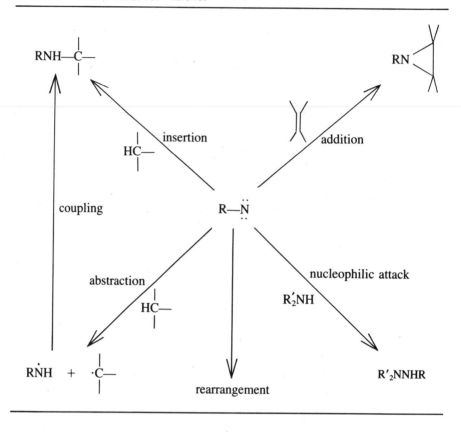

substitution or aliphatic C—H insertion (Kyba & Abramovitch, 1980). Instead, the products consisted of imines derived from 1,2-shifts of groups on the alkyl carbon to the azido nitrogen atom.

III.   LIPOPHILIC MEMBRANE-PERMEANT REAGENTS IN CROSS-LINKING STUDIES

Small, lipid-soluble photoreactive reagents can be used to label the hydrophobic sections of membrane lipids or proteins. As stated previously, the photogenerated intermediates from these reagents should insert into C—H bonds, even in the presence of other groups. Phospholipid vesicles can serve as useful models in the study of photochemical cross-linking to lipids because C—H insertion into acyl chains can be easily measured. Cross-linking to lipids can similarly be used as models for cross-linking into the hydrophobic regions of membrane proteins.

A. Cross-Linking to Lipids

The first reported experiments of photolabeling in the hydrophobic core were those of Klip and Gitler (1974) using radioactive 1-azidonaphthalene and 1-azido-4-iodobenzene (see Table 8). Extensive levels of attachment to the lipids resulted, and chemical degradation showed cross-linking to be predominantly to the fatty acyl chains. However, products were not characterized and may significantly or solely involve additions to the double bonds of the fatty acyl chains. Infavorable absorption maxima and problems with artifacts with these two compounds led to the development of 5-iodonaphthyl-1-azide (INA) (Bercovici & Gitler, 1978). INA was synthesized with high specific activity, had favorable absorption maxima, and partitioned greater than 99.9% into membranes (Gitler & Bercovici, 1980). Photolyses in liposomes composed of egg phosphatidylcholine, egg phosphatidylethanolamine, bovine phosphatidylserine, or phosphatidylinositol resulted in significant covalent attachment to the phospholipids (Gitler & Bercovici, 1980). However, the water-soluble nitrene scavenger glutathione (Bayley & Knowles, 1978a) decreased the INA bound to the lipids by 30%, suggesting frequent excursions to the bilayer surface. The products were again not characterized chemically and may contain substantial amounts of adducts to the double bonds of the fatty acyl chains. This label has also been used in probing the hydrophobic segments of integral membrane proteins (see below).

Bayley and Knowles (1978a, 1978b) carried out a detailed comparison of the nitrene and carbene probes by using phenyl azide and phenyl diazirine in well-defined model membrane systems constructed from phosphatidylcholines containing saturated or unsaturated fatty acyl chains. Photolysis of phenyl diazirine resulted in cross-linking to the extent of 5% (percentage of reagent attached to the phospholipid after photolysis) to dimyristoylphosphatidylcholine, and approximately 10% to dioleoylphosphatidylcholine in two seperate experiments, neither of which was diminished by the addition of glutathione. Products were characterized by gas chromatography–mass spectrometry and shown to result from C—H insertions into both the saturated and unsaturated fatty acyl chains and from addition products of the carbene to the double bonds of unsaturated fatty acyl chains. Photolysis of phenyl azide gave only 0.25% reaction with dimyristoylphosphatidylcholine and 0.75% with dioleoylphosphatidylcholine. The differences in efficiencies between phenyl carbene and phenyl nitrene can be partly explained by the longer lifetime (lower reactivity) of the nitrene, as well as greater partitioning into the membrane of the carbene precursor. A nitrophenyl azide may have been a better choice for a more reactive nitrene, but the precursor may not partition as favorably in the membrane.

Another hydrophobic membrane-soluble carbene precursor is spiro[adamantane-4,4′-diazirine] (adamantane diazirine) (Bayley & Knowles, 1978b, 1980). It partitions into membranes better than phenylazide or phenyldiazirine, but not as well as INA. Photolysis at 350 nm at temperatures above the lipid phase transitions in dimyristoyl- or dioleoylphosphatidylcholine sonicated vesicles, followed by labeling analysis, showed approximately 3 and 5% labeling, respectively.

TABLE 8   Lipophilic Membrane-Permeant Reagents

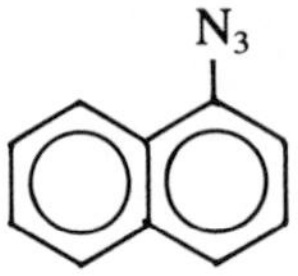

1-azidonaphthylene

1-azido-4-iodobenzene

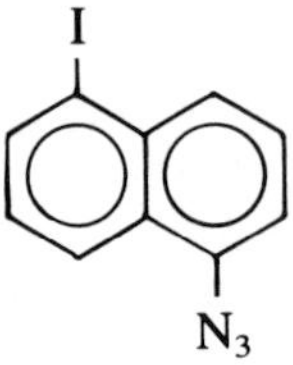

5-iodonaphthyl-1-azide

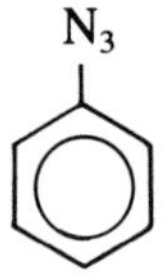

phenyl azide

phenyl diazirine

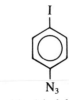

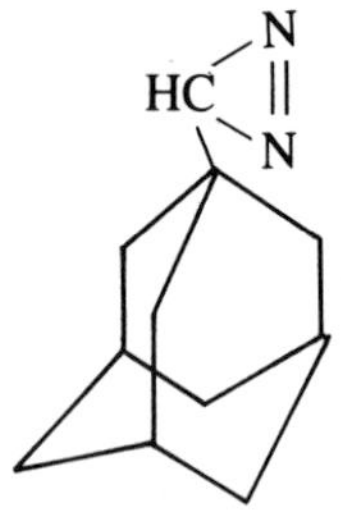

adamantane diazirine

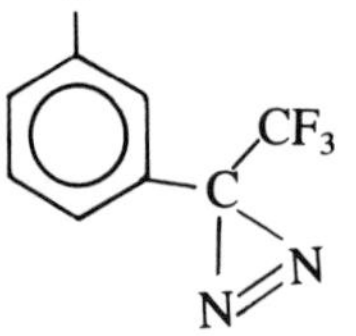

3-trifluoromethyl-3-(*m*-iodophenyl) diazirine

B. Cross-Linking to Proteins

The goal of using membrane-soluble photoactivatable probes in the study of membrane proteins is to label only those domains that are in contact with the hydrocarbon chains of the bilayer.

The first hydrophobic photolabeling of membrane proteins was reported by Klip and Gitler (1974), who used the above-described aromatic azides in the radiolabeled form as the nitrene precursors. Following photolysis, analysis showed that cross-linking occurred mainly to the protein domains adjacent to the lipids (Klip & Gitler, 1974). Later, INA was developed as the reagent (Bercovici & Gitler, 1978). Using several different proteins and isolated membranes, covalent incorporation of the radioactively labeled reagent into proteins was demonstrated in human erythrocytes, erythrocyte ghosts, sarcoplasmic reticulum (Bercovici & Gitler, 1978), purified (Na^+,K^+)-ATPase (Karlish et al., 1977), and brush border membranes (Sigrist-Nelson et al., 1977). The addition of nitrene scavengers such as p-aminobenzoic acid or glutathione did not significantly decrease the extent of cross-linking to the proteins. Cross-linked glycophorin was isolated from human erythrocyte membranes on photolysis in the presence of INA. Labeling analysis indicated that, as expected, the bulk (80–90%) of the radioactivity was in the trypsin-insoluble hydrophobic segment (Kahane & Gitler, 1978) that had been concluded by Furthmayr (1977) to be embedded in the bilayer. INA cross-linking to (Na^+,K^+)-ATPase followed by trypsinization resulted in a 12,300 dalton polypeptide containing all the radioactivity (Karlish et al., 1977), indicating that the membrane-embedded portion is within this fragment. Cross-linking to (Ca^{2+},Mg^{2+})-ATPase followed by trypsinization yielded two labeled peptides (Bercovici & Gitler, 1978; Gitler & Bercovici, 1980).

The careful work of Bayley and Knowles (1978a) showed that in the presence of phospholipid vesicles phenyl nitrene is scavenged in the aqueous solution and that C—H insertions do not occur in saturated phosphatidylcholine. Membrane labeling experiments with azidobenzene are thus questionable.

Bayley and Knowles (1978b) have used the above-described hydrophobic carbene precursor adamantane diazirine for the labeling of hydrophobic regions in membranes (Goldman et al., 1979; Farley et al., 1980). Three well-characterized membrane proteins were labeled in the native membrane, including glycophorin A, the heavy chains of the human histocompatibility antigens HLA-A2 and -B7, and the influenza virus hemagglutinin HA_2. All the radioactivity was found attached to polypeptide fragments believed from the results of other techniques to span the lipid bilayer. For example, photolysis of the histocompatibility antigens HLA-A2 and −B7 followed by purification and papain cleavage showed that most of the radioactivity lies in or close to the transmembrane sequence near the C-terminus of these antigens (Goldman et al., 1979). Analogous results were obtained with glycophorin A and the hemagglutinin HA_2.

3-Trifluoromethyl-3-phenyl diazirine may offer an advantage in membrane protein labeling experiments, since the diazo isomerization product is stable under experimental conditions (Brunner et al., 1980). The radioiodinated analogue has apparently been used in membrane labeling experiments with sucrose-isomaltase and human erythrocyte ghosts (Brunner, 1981).

IV. PHOSPHOLIPIDS CONTAINING PHOTOLABELS IN THE FATTY ACYL CHAINS: SYNTHESIS

An alternative approach to the exogenous addition of lipid-soluble photolabile reagents utilizes prior covalent attachment of photoreactive groups to fatty acyl chains at a defined position (Chakrabarti & Khorana, 1975; Gupta et al., 1977) and the subsequent synthesis of mixed 1,2- diacylphosphatidylcholines using these fatty acyl groups. The structures of the phospholipids thus synthesized are shown in Fig. 5.

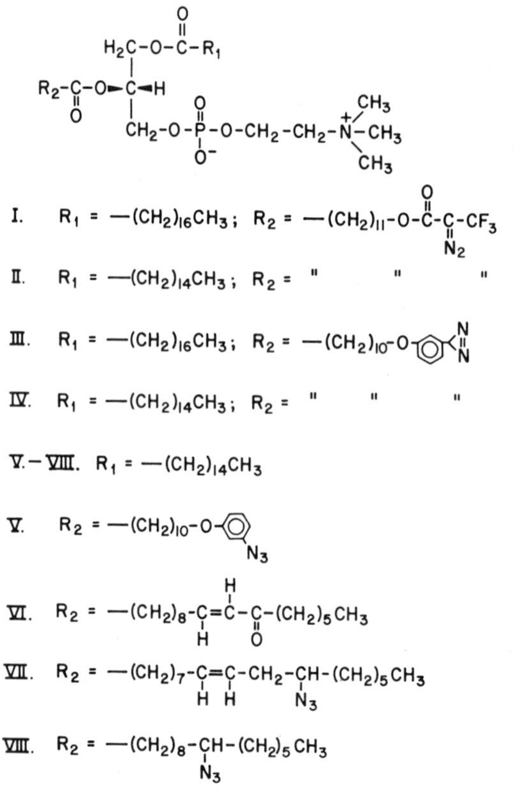

Figure 5. Synthetic mixed 1,2-diacylphosphatidylcholines carrying different photoactivatable groups in the fatty acyl chains on the *sn*-2 position of the glycerol backbone. The phospholipids were prepared from lysolecithins by acylation with fatty acyl anhydrides containing photoactivatable groups in the hydrocarbon chains. Taken with permission of Academic Press from Radhakrishnan et al. (1980).

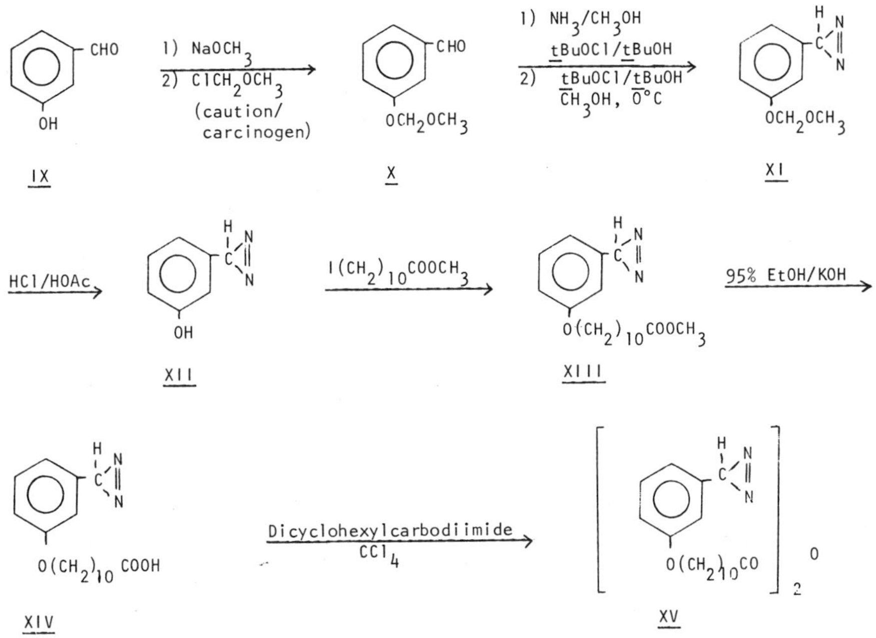

Figure 6. The synthesis of a fatty acid and its anhydride carrying a 3-diazirinophenoxy group in the terminus of the fatty acyl chain.

A. Fatty Acids

A photoreactive azido group can easily be introduced at any desired position along a hydrocarbon chain, and in initial work, a variety of fatty acids of this type were synthesized (Chakrabarti & Khorana, 1975) (e.g., **VII** and **VIII** in Fig. 5). Alternatively, photosensitive carbene precursors such as phenoxydiazirine or trifluorodiazopropionate were prepared and condensed, respectively, with the readily available ω-halo or ω-hydroxy fatty esters to give the desired fatty acid esters (e.g., **I** and **III** in Fig. 5).

The procedure used for the preparation of a fatty acid containing an aromatic diazirine as the photoreactive moiety is illustrated in Fig. 6 (Radhakrishnan et al., 1981). The phenolic hydroxyl of *m*-hydroxybenzaldehyde was protected with chlorodimethyl ether, and the aldehyde function was converted to a 3H-diazirine (Smith & Knowles, 1975). This was used to prepare a phenyl diazirine that was subsequently hydrolyzed to the free acid. The anhydride was prepared by the method of Selinger and Lapidot (1966). Details of this synthetic procedure are given in Section I of the Appendix at the end of this Chapter.

Another photolabel located at the terminus of the fatty acyl chains incorporated β-trifluoro-α-diazopropionate as the carbene precursor (see Fig. 7). The synthetic details will not be described, but in general the procedure involved the preparation of trifluorodiazoethane from trifluoroethylamine, followed by reaction with phos-

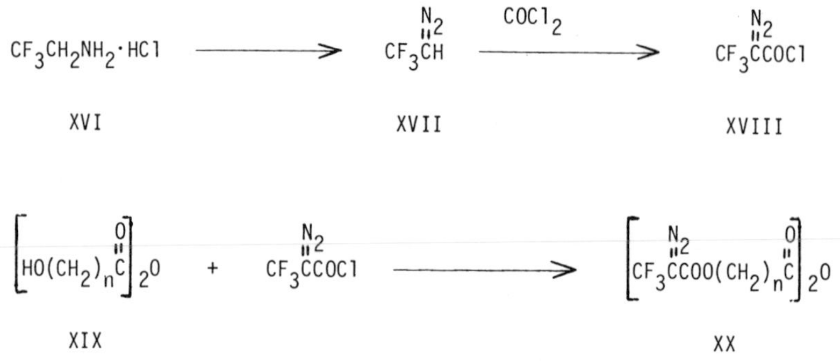

$$CF_3CH_2NH_2 \cdot HCl \longrightarrow CF_3\overset{N_2}{CH} \overset{COCl_2}{\longrightarrow} CF_3\overset{N_2}{\underset{O}{C}}COCl$$

XVI   XVII   XVIII

$$\left[HO(CH_2)_n\overset{O}{\underset{\parallel}{C}} \right]_2 O \quad + \quad CF_3\overset{N_2}{\underset{\parallel}{C}}COCl \longrightarrow \left[CF_3\overset{N_2}{\underset{\parallel}{C}}COO(CH_2)_n\overset{O}{\underset{\parallel}{C}} \right]_2 O$$

XIX   XX

Figure 7. The synthesis of the fatty acid anhydride containing a trifluorodiazopropionate group at the termini of the fatty acyl chain. The synthesis of trifluorodiazopropionyl chloride is that of Chowdhry et al. (1976), and the synthesis of the anhydride and phospholipid is that of Gupta et al. (1977).

gene to give **XVIII** (Chowdhry et al., 1976). This was then coupled to ω-hydroxy fatty acid anhydrides.

B. Phospholipids

The anhydrides of the fatty acids carrying the photoreactive group were used to acylate lysophosphatidylcholines using the powerful catalyst N,N-dimethyl-4-aminopyridine (DMAP) (Steglich & Höfle, 1969). This acylation procedure developed for phospholipids (Gupta et al., 1977) gives 40–80% yields, even on small scale preparations, and has been used to prepare these compounds in radiolabeled form as well. The procedure used is illustrated in Fig. 8 for the preparation of the mixed acyl lecithin carrying an aromatic diazirine as the photolabile moiety (Radhakrishnan et al., 1981).

V.   LIPID-LIPID CROSS-LINKING USING PHOSPHOLIPIDS CARRYING PHOTOREACTIVE LABELS IN THE ACYL CHAINS

Determination of the molecular details of protein folding within the bilayer is a reasonable goal using phospholipids containing photoreactive groups in the acyl chains if the carbenes formed insert into saturated phospholipids and if a relationship between the depth of the probe into the bilayer and the region of cross-linking can be demonstrated.

A.   Cross-Linking to Phospholipids

The first requirement of the synthetic phospholipid analogues is the capability of forming sealed vesicles. The formation of sealed sonicated vesicles using these synthetic phospholipids was shown by the trapping of [^{14}C]glucose when the latter was included in the sonication buffer. For photolysis, either sonicated vesicles or multilamellar liposomes composed of the phospholipids were photolyzed at 25°C using 366 nm light. A monochrometer was usually used to limit the range to ±20 nm, and a filter was used to remove all light less than 300 nm. For cross-linking analysis, the products of photolysis were separated by gel exclusion chromatography using Sephadex LH-20 with $CHCl_3$/MeOH as the eluent.

Extensive photolytic studies with vesicles prepared from azido phospholipids showed that the nitrene intermediates formed few, if any, intermolecular insertion products into C—H bonds (Gupta et al., 1979a). This was determined by the presence of phospholipid dimers on Sephadex LH-20 gel filtration chromatography. For instance, when mixed vesicles composed of 1,2- [^{14}C]palmitoyl-sn-glycero-3-phosphorylcholine and 1-palmitoyl-2-[10-azidopalmitoyl]-sn-glycero-3-phosphorylcholine were photolyzed at 300 nm, only 0.6% of radioactive phospholipid was shown to form intermolecularly linked dimers. The amount of cross-linking was too low to be useful in studies with membrane proteins.

Vesicles were formed from photoreactive phospholipids with a trifluorodiazopropionyloxy or diazirinophenoxy group at the fatty acyl termini. Photolysis yielded products formed by intermolecular cross-linking reactions, and these were analyzed after separation on gel permeation columns. Two representative separations of reaction products from the two types of phospholipid are shown in Figs. 9 and 10. The earlier elution of the first main peaks shows that these peaks consist of materials with molecular weights higher than those of the starting materials. The average yield of cross-linked products is 40 ±5% for the diazo phospholipid and 50–60%

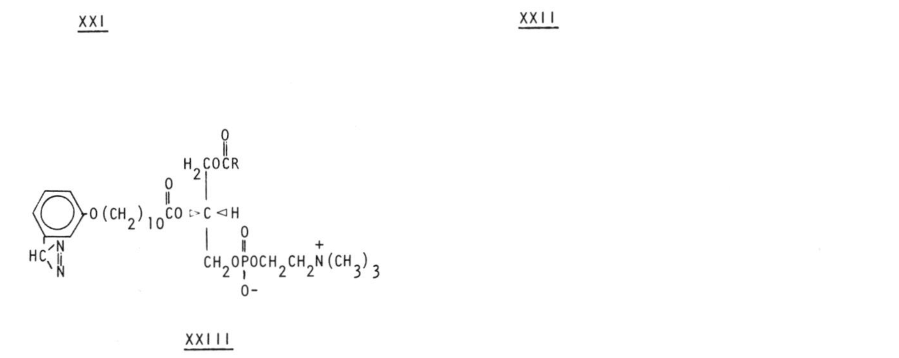

Figure 8. The synthesis of a mixed acyl phosphatidylcholine carrying a diazirinophenoxy group in the fatty acyl chain of the sn-2 position of the glycerol backbone.

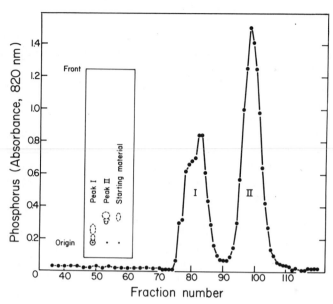

Figure 9. The elution pattern from a Sephadex LH-20 column (2.5 × 100 cm) of the products obtained from the photolysis of 1-palmitoyl-2-ω-(trifluorodiazopropionyl)lauryl-*sn*-glycero-3-phosphorylcholines. Peak I, the intermolecularly cross-linked products, accounted for 38% of the total phosphate-containing material. Peak II results from intramolecular insertion products and side reactions of the carbene intermediate. Reproduced with permission of Academic Press from Radhakrishnan et al. (1980).

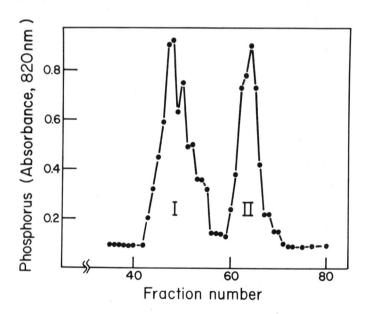

Figure 10. A Sephadex LH-20 column elution profile (2.5 × 100 cm) of the products obtained from the photolysis of 1-palmitoyl-2-ω-(diazirinophenoxy)undecanoyl phosphatidylcholine. Peak I accounted for 55% of the total phosphate-containing material and peak II the remainder. Reproduced with permission of Academic Press from Radhakrishnan et al. (1980).

for the diazirine phospholipid. Several lines of evidence establish that these products arise by a covalent attachment of the carbene to the saturated fatty acyl chain of a second phospholipid molecule.

When vesicles formed from a 1 : 1 mixture of the diazo phospholipid and radioactive acceptor dipalmitoylphosphatidylcholine were photolyzed, cross-linked material amounted to 10–15% of the acceptor. The cross-linked phospholipid dimer was further characterized by three degradative methods: phospholipase A_2 and C treatments, and transesterification by sodium methoxide. Four products were obtained by phospholipase A_2 treatment, and the structures for these compounds are shown in Fig. 11. All the cross-linked material was sensitive to cleavage by phospholipase A_2: two sets of lysolecithins, palmitic acid, and cross-linked palmitic acids resulted. Digestion by phospholipase C, an enzyme specific for the cleavage of head groups, resulted in the formation of radiolabeled diglycerides, but no radioactivity could be detected in the aqueous phase. The dimeric phospholipid adduct was also subjected to transesterification using sodium methoxide in methanol. Gel exclusion chromatography gave methyl palmitate and the cross-linked fatty acid diester, as expected (see Fig. 11). Finally, cross-linked fatty esters derived from the trifluorodiazopropionyl and the diazirinophenoxy phospholipids were analyzed by mass spectrometry and shown to have the general structures in Fig. 12. This confirmed that carbenes generated at the ω-positions of the fatty acyl termini restrict themselves to cross-linking within the bilayer.

Having demonstrated the occurrence of photocatalyzed intermolecular insertion reaction into saturated fatty acyl chains, it was important to see if there was a relationship between the position(s) of cross-linking in the acceptor chain and the location of the carbene precursor in the other chain. The usefulness of this approach would be increased if a relationship could be found between the location of the carbene precursor along the chain and the sites of cross-linking. This has been

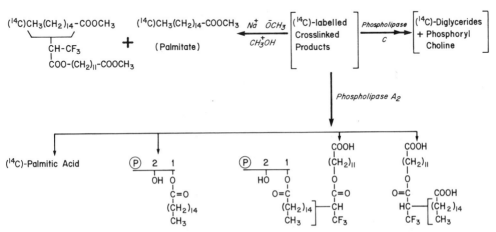

Figure 11.   The methods of degradation used for the analysis of cross-linked phospholipids. Only the [C^{14}]palmitic acid and products containing this radioactive residue are shown; ⓟ stands for phosphorylcholine.

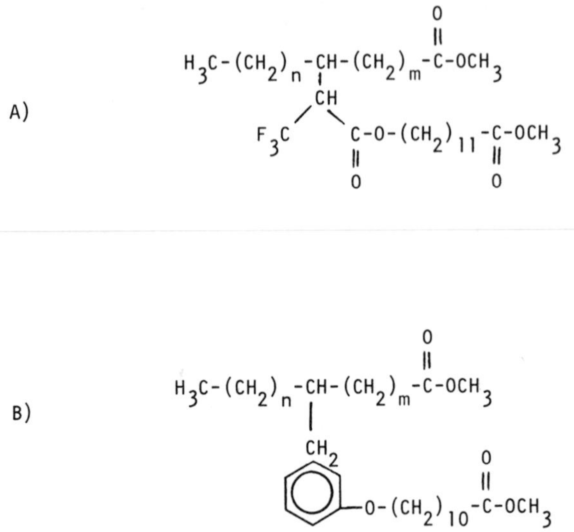

A)

$$H_3C-(CH_2)_n-\underset{\underset{CH}{|}}{CH}-(CH_2)_m-\overset{\overset{O}{\|}}{C}-OCH_3$$

$$F_3C \diagdown \quad \overset{\|}{\underset{O}{C}}-O-(CH_2)_{11}-\overset{\overset{O}{\|}}{C}-OCH_3$$

B)

$$H_3C-(CH_2)_n-\underset{|}{CH}-(CH_2)_m-\overset{\overset{O}{\|}}{C}-OCH_3$$

$$CH_2$$

$$-O-(CH_2)_{10}-\overset{\overset{O}{\|}}{C}-OCH_3$$

Figure 12. The general structures of the cross-linked fatty acid esters obtained using (A) the diazo and (B) the phenyldiazirine phospholipid systems.

demonstrated for the two series by analysis of the extent of cross-linking as measured by mass spectrometric fragmentation of cross-linked fatty esters.

The major mode of mass spectral fragmentation of the cross-linked diazirine-derived dimeric fatty esters was a cleavage at the benzylic site (Fig. 13). The benzylic ion appeared with m/e at 306 instead of 305 due to a γ-hydrogen abstraction. The mass spectrum also showed a peak at 307, a contribution from the natural abundance of the ^{13}C isotope. Utilizing the useful feature of γ-hydrogen abstraction, the acceptor chain was selectively di-deuterated along various points in the chain. The cross-linked product showed an enhanced intensity of the peak at m/e 307 when the cross-linking occurred at the carbon-bearing deuterium (Radhakrishnan et al.,

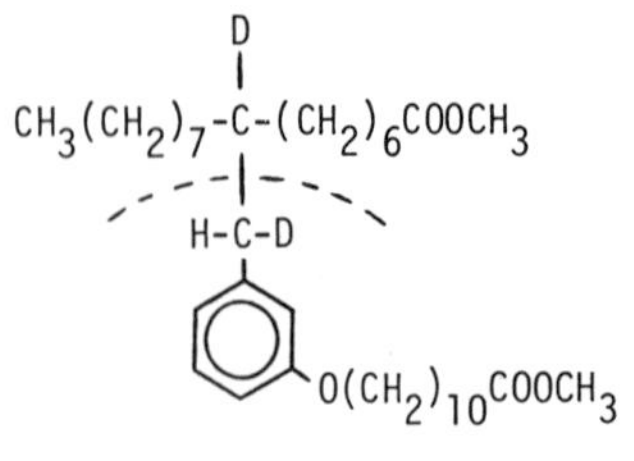

$$CH_3(CH_2)_7-\underset{\underset{H-C-D}{|}}{\overset{\overset{D}{|}}{C}}-(CH_2)_6COOCH_3$$

$$O(CH_2)_{10}COOCH_3$$

$C_{36}H_{60}D_2O_5$ M^+ 576.472

Figure 13. The major mode of mass spectral fragmentation observed in cross-linked diazirino fatty esters.

1980). Analysis of the abundance of the mass spectral ions showed a broad distribution of cross-linking along the hydrocarbon chain with an increase in the amount of cross-linking toward the methyl terminus (R. Radhakrishnan, C. E. Costello, & H. G. Khorana, unpublished data). A similar study with the phospholipid containing the trifluorodiazo-propionyl derivative at the fatty acyl termini using nondeuterated samples also showed a distribution in the cross-linking (Gupta et al., 1979b). These data confirmed that photolabel incorporated into the ω-positions of the fatty acid chains resides primarily in the middle of the bilayer.

Alternate phospholipid photolabels designed for a similar goal of utilizing phospholipid analogues containing carbene or nitrene precursors on the fatty acyl chains has been reported by Brunner and Richards (1980) (Fig. 14). Photosensitive aryl diazirine or aryl azide groups attached via a disulfide bridge to a fatty acyl chain were incorporated into phospholipids. Reduction of the disulfide bond after photolabeling generates free sulfhydryl groups on the attached label, which can be used to separate modified peptides or lipids to aid in subsequent analyses. These phospholipids were photolyzed in mixed vesicles containing dipalmitoylphosphatidylcholine or 1-palmitoyl-2-oleoylphosphatidylcholine. The carbene-generating probe labeled dipalmitoylphosphatidylcholine in yields of 15–20%, while labeling with the nitrene probe was about 10 times less efficient. By contrast, these probes were found to react with the unsaturated lipid with almost equal efficiency.

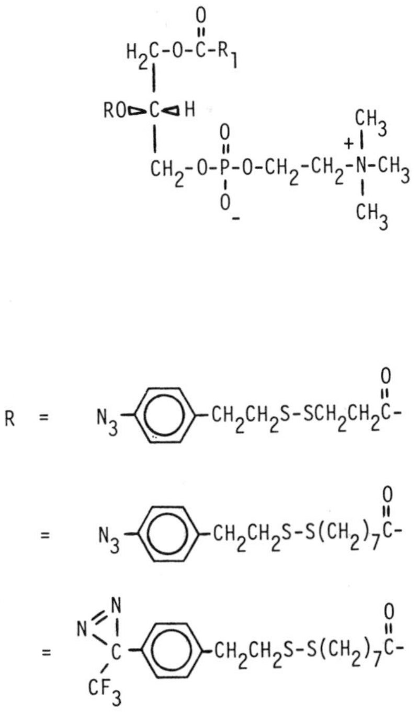

Figure 14. Phosphatidylcholine analogues containing the azidophenyl or 3-trifluoromethyl-3-phenyl diazirine photosensitive groups and an S—S bond (Brunner & Richards, 1980). These analogues were synthesized by acylation of lysophosphatylcholine with the anhydrides of the photosensitive fatty acids by the method of Gupta et al. (1977).

B. Cross-Linking to Cholesterol

Although the molecular mechanism of cholesterol function is poorly understood, extensive studies in model and biological membranes have indicated that cholesterol can affect the packing of the lipid hydrocarbon chains in the bilayer (Demel & de Kruijff, 1976). Photolabeling techniques were used to study cholesterol effects on lipid chain packing in liposomes. Sonicated vesicles composed of 33 mol % cholesterol in vesicles of the photosensitive phospholipids were photolyzed at 350 nm at 40°C. The same methodology outlined above for the separation of cross-linked phospholipid was used to monitor the intermolecular reactions. Photo-cross-linking resulted in approximately 12–15% of the high molecular weight product based on the cholesterol used in the experiment. The cross-linked products were characterized by their chromatographic behavior, cleavage by phospholipase A_2, base-catalyzed transesterification, and mass spectral measurements. The cross-linking was shown not to involve the 3-β-hydroxyl group of cholesterol. Presently, methods are being developed to determine the precise site(s) of cross-linking.

VI. LIPID-PROTEIN CROSS-LINKING USING PHOSPHOLIPIDS CARRYING PHOTOREACTIVE LABELS IN THE ACYL CHAINS

In studies of lipid-protein interactions, filters should be used to eliminate short wavelengths of ultraviolet light and long wavelengths of infrared radiation that may damage the protein system. Oxygen should also be excluded to eliminate light-catalyzed photooxidation. The chemical stability of the cross-links to a critical factor in the analysis of sites of cross-linking. For exact points of cross-linking, the bonds should be stable to conditions necessary to sequence the peptide fragment by Edman sequencing techniques.

A. Glycophorin A

Glycophorin A is a 31,000 dalton glycoprotein from the human erythrocyte (Furthmayr, 1977) and contains 60% carbohydrate by weight. Tomita et al. (1978) showed by protein sequencing that glycophorin A has a tripartite structure: two hydrophilic segments, the carbohydrate-containing N-terminus and the proline-rich C-terminus. The complete amino acid sequence is shown in Table 9. In the middle there exists a hydrophobic segment composed of about 23 uncharged amino acid residues. The carbohydrates of the N-terminus lie outside the erythrocyte (Steck & Dawson, 1974), and the C-terminus is exposed only on the inner surface of the erythrocyte membrane (Cotmore et al., 1977). Schulte and Marchesi (1979) concluded from circular dichroism measurements that the hydrophobic segment may be α-helical.

   Glycophorin A was reconstituted with radiolabeled photolabile diazirinophenoxy phospholipid (**XXIII**) and excess nonphotolabile lipid using the cholate dialysis

TABLE 9   Amino Acid Sequence of Glycophorin A[a]

C	C	C					C	C	C	C	C	C

(Leu)   (Glu)

(Ser)Ser-Thr-Thr(Gly)Val-Ala-Met-His-Thr-Thr-Thr-Ser-Ser-Ser-Val-Ser-Lys-Ser-Tyr-
  1   10   20

     C   C C   C

Ile-Ser-Ser-Gln-Thr-Asn-Asp-Thr-His-Lys-Arg-Asp-Thr-Tyr-Ala-Ala-Thr-Pro-Arg-Ala-
                      30   40

        C   C   C

His-Glu-Val-Ser-Glu-Ile-Ser-Val-Arg-Thr-Val-Tyr-Pro-Pro-Glu-Glu-Glu-Thr-Gly-Glu-
                      50   60

Arg-Val-Gln-Leu-Ala-His-His-Phe-Ser-Glu-Pro-Glu-Ile-Thr-Leu-Ile-Ile-Phe-Gly-Val-
                      70   80

Met-Ala-Gly-Val-Ile-Gly-Thr-Ile-Leu-Leu-Ile-Ser-Tyr-Gly-Ile-Arg-Arg-Leu-Ile-Lys-
                      90   100

Lys-Ser-Pro-Ser-Asp-Val-Lys-Pro-Leu-Pro-Ser-Pro-Asp-Thr-Asp-Val-Pro-Leu-Ser-Ser-
                      110   120

Val-Glu-Ile-Glu-Asn-Pro-Glu-Thr-Ser-Asp-Gln
                      130

From Tomita et al. (1978).

[a]"C" above an amino acid represents covalently bound carbohydrate.

procedure of Kagawa and Racker (1971). The vesicles were photolyzed, the products separated by gel exclusion chromatography, subjected to proteolytic digestion, and the cross-linked fragment analyzed by sequential Edman degradation. Glutamic acid-70 was the major site of cross-linking.

In additional experiments, intact erythrocyte membranes were photolyzed with externally added 1-[^{14}C]palmitoyl-2-diazirinohexanoyl phosphatidylcholine. Although the protein was not sequenced, the photoreactive lipid reacted almost exclusively with a segment corresponding to amino acids 62–81.

A schematic model of glycophorin A in the erythrocyte membrane is shown in Fig. 15. The model is based on the following: (1) cross-linking of **XXIII** to glutamic acid-70, (2) cross-linking by phospholipids carrying photolabels in the head groups occur in the fragments containing amino acids 62–81 and 82–96 (see Section VII), (3) residues 62–96 were interpreted to be α-helical from circular dichroism studies by Schulte and Marchesi (1979) and by nuclear magnetic resonance NMR studies (Cramer et al., 1980; Egmond et al., 1979), and (4) the thickness of the erythrocyte lipid bilayer is about 45 Å (Rand & Luzzati, 1968), which would imply that about 30 amino acid residues in an α-helix would be required to cross the membrane.

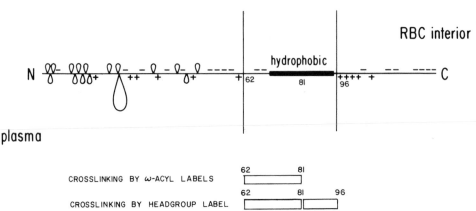

Figure 15. Schematic representation of glycophorin in the bilayer and sites of cross-linking by diazirine-containing phospholipids. The ω-acyl labels react only in the segment 62–81. The head group labels react in both segments 62–81 and 82–96. The oligosaccharide chains are represented by loops. The glutamic/aspartic acid and arginine/lysine are indicated by − and +, respectively.

B.   Cytochrome b_5

Cytochrome b_5 is an amphipathic membrane protein with a large hydrophilic N-terminus containing the catalytic site and a hydrophobic segment of 30–40 amino acids at the C-terminus that anchors the protein (Ozols, 1974). The hydrophilic and hydrophobic portions fold independently of each other (Visser et al., 1975).

The amino acid sequence of the hydrophilic portions of cytochrome b_5 from several species have been reported (Ozols et al., 1976), and more recently the primary structures of the membranous segments of four species have also been determined (Ozols & Gerard, 1977a, 1977b; Fleming et al., 1978; Takagaki et al., 1980). The amino acid sequence of the membranous segment of rabbit liver cytochrome b_5 is shown in Table 10. Cytochrome b_5 is known to bind to membranes in two different modes termed the "loosely bound" and "tightly bound" forms (Enoch et al., 1979). The former is generated by spontaneous, reversible incor-

TABLE 10   Partial Amino Acid Sequence[a] of Rabbit Liver Cytochrome b_5

Leu-Ser-Lys-Pro-Met-Glu-Thr-Leu-Ile-Thr-Thr-Val-Asp-<u>Ser-Asn-Ser-Ser-Trp-Trp-Thr-</u>
91   100   110

<u>Asn-Trp-Val-Ile-Pro-Ala-Ile-Ser-Ala-Leu-Ile-Val-Ala-Leu-Met-Tyr-Arg</u>-Leu-Tyr-Met-
                                          120   130

Ala-Asp-Asp
133

From Takagaki et al. (1980).
[a]The underlined region represents cross-linked amino acids in "loosely bound" form (see text).

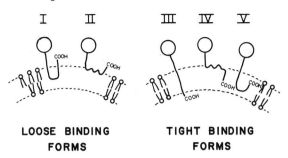

Figure 16. Models for the orientation of the membranous segment of cytochrome b_5 in the membrane. Originally proposed by Enoch et al. (1979).

poration of cytochrome b_5 to phospholipid vesicles. This form is susceptible to digestion by carboxypeptidase Y, indicating exposure of the C-terminal on the same side of the lipid bilayer as the hydrophilic segment. The "tightly bound" form in vesicles is formed when vesicles are made by detergent reconstitution methods, does not undergo the transfer between vesicles, and is not susceptible to digestion by carboxypeptidase Y. The C-terminal anionic residues are necessary in stabilizing the "tightly bound" form (Dailey & Strittmatter, 1981). Possible configurations of the membranous segment for these two forms are shown in Fig. 16 (Enoch et al., 1979). Photolabile phospholipid was used to distinguish among these possibilities (Takagaki et al., 1979, 1981).

Rabbit liver cytochrome b_5 was reconstituted in both the "loosely bound" and "tightly bound" forms into vesicles containing **XXIII** or **I**. After photolysis, uncross-linked phospholipid was separated from cytochrome b_5, and the association of phospholipid with cytochrome b_5 was established as covalent. Proteolytic and chemical fragmentation indicated that sites of cross-linking were confined within the hydrophobic C-terminal region from Glu- 96 to Asp- 133. Sequence analysis of the cross-linked cytochrome b_5 obtained from the "loosely bound" form showed a broad distribution of radioactivity starting from Ser- 104 to Tyr- 126 with a peak approximately at Pro- 115. This result favors a "looping up" model (Fig. 16, I) over a "membrane surface floating" model (Fig. 16, II). The sequence analysis of the cross-linked cytochrome b_5 obtained from the "tightly bound" form showed a similar distribution of the sites of cross-linking, favoring a "transmembrane model" (Fig. 16, III) over others.

C. Identification of the Lipid Binding Site of the Phosphatidylcholine Exchange Protein

Phosphatidylcholine-phospholipid exchange protein (PC-PLEP), also referred to as phosphatidylcholine transfer protein (see Volume 1, Chapter 4), acts as a specific carrier of phosphatidylcholine between membranes. At the lipid-water interface, PC-PLEP exposes a binding site that interacts specifically with the polar head group of phosphatidylcholine (Kamp et al., 1977). The primary structure of this protein has been determined very recently (Moonen et al., 1980; Akeroyd et al., 1981).

PC-PLEP forms a one-to-one molar complex with phosphatidylcholine. Electron spin resonance studies on this complex with 2-acyl spin-labeled phosphatidylcholine indicated a strongly immobilized probe unaffected by the presence of ascorbate, evidence that supports the buried nature of phosphatidylcholine in the exchange protein (Devaux et al., 1977; Machida & Ohnishi, 1978). To obtain further knowledge on the lipid binding site of this protein, the photolabeling approach was tried by Wirtz et al. (1980) with lipids carrying 7-(4-azido-2-nitrophenoxy) [^{14}C]heptanoic acid or 11-(m-diazirinophenoxy)[^{14}C]undecanoic acid in the sn-2 positions of phosphatidylcholines.

Incubation of phosphatidylcholine-PLEP with vesicles composed of the azido or diazirine lecithins formed the 1 : 1 complex, as with endogenous lipids (Demel et al., 1973; Kamp et al., 1975). Isolation of the complex, followed by photolysis, resulted in covalent association of the reactive species with the protein as determined by SDS-polyacrylamide slab gel electrophoresis. The yield of incorporation was approximately 30% for both compounds. The photolyzed complex was reduced with a thiol reagent, carboxymethylated, citraconylated, and then digested by *Staphylococcus aureus* protease. The radioactivity was primarily in a fragment that had lysine as the N-terminal (see Table 11). This fragment was subjected to digestion with clostripain, a proteolytic enzyme that specifically cleaves Arg—X bonds (Gilles et al., 1979). Products were separated, and the cross-linking region was narrowed to a sequence containing the residues 26–31 (residues 171–176, Table 11). The interesting feature is that this segment contains an extremely hydrophobic cluster of amino acids -Val-Phe-Met-Tyr-Tyr-Phe-. An analysis of the secondary structure of the binding site according to the method of Chou and Fasman predicts a β-sheet structure for this hexapeptide. According to this analysis, the tetrapeptide on either side of the hexapeptide (residues 21–24, 33–36) would form β-turns. A hexapeptide in β-sheet structure has an estimated length of 14 Å. This suggests that the binding site has the proper dimensions to accommodate the sn-2 fatty acyl chain of naturally occurring phosphatidylcholines (Wirtz et al., 1980).

TABLE 11   Partial Amino Acid Sequence[a] of Phospholipid Exchange Protein

Lys-Ser-Gly-Val-Ile-Arg-Val-Lys-His-Tyr-Lys-Gln-Arg-Leu-Ala-
  150   160

   (21)   (24)   (26)   (31)   (33)
Ile-Gln-Ser-Asp-Gly-Lys-Arg-Gly-Ser-Lys-Val-Phe-Met-Tyr-Tyr-Phe-Asp-Asn-Pro-Gly-
  170   180

(36)
Gly-Gln-Ile-Pro-Ser-Trp-Val-Ile-Asn-Trp-Ala-Ala
  190

From K. Wirtz, private communication (see also Akeroyd et al., 1981).
[a]The underlined hexapeptide contained the radioactivity from the photolabile phospholipid; the numbers in parentheses that appear above the sequence refer to positions in this fragment.

D. Gramicidin A

Multilamellar liposomes were prepared from 90 mol % of dipalmitoylphosphatidylcholine, 5 mol % of [^{14}C]gramicidin A and 5 mol % of phospholipid analogues containing carbene or nitrene precursors described above (Brunner & Richards, 1980). After photolysis at 45–50°C and separation on thin layer chromatography (TLC) two radioactive spots were detected. One spot corresponded to unreacted gramicidin A and the other to cross-linked gramicidin A. The nitrene-generating probe modified gramicidin two to three times more efficiently then did the carbene-generating probe (approximately 14 and 4%, respectively). The results of amino acid analysis and Edman degradation were interpreted as both probes predominantly reacting with tryptophan residues, although because of the manner in which analyses were performed, photodegradation of tryptophan has not been ruled out.

VII. PHOTOAFFINITY LABELS IN THE PHOSPHOLIPID HEAD GROUP

In recent years, a number of membrane-bound enzymes have been shown to exhibit varying degrees of specificities with respect to the polar head groups of phospholipids (Coleman, 1973). For example, the activity of the inner mitochondrial enzyme β-hydroxybutyrate dehydrogenase shows an almost absolute requirement for the choline headgroup (Grover et al., 1975; Gazzotti et al., 1975). One possible approach to probe this specificity and also to gain information on the amino acid residues at or near the lipid-water interface of membrane proteins is to use phospholipids carrying photolabels in the head group. Because of the hydration in the head group region, one might expect the reactive species generated to be extensively quenched by reaction with water. On the other hand, if there were to be an interaction between the polar head groups and the hydrophilic amino acid residues that lie at the membrane interface, there could be possibility of photo-cross-linking among these components. Therefore, a series of phospholipids with photoactivatable carbene or nitrene precursors located in the headgroup region was synthesized The structures of these derivatives are shown in Fig. 17 (Chakrabarti & Khorana, 1975; Radhakrishnan et al., 1981). The phosphatidylcholine analogue **XXVII** was synthesized by the scheme presented in Fig. 18. Pyridine-3-carboxaldehyde is converted to pyridyl-3-diazirine (Smith & Knowles, 1975), quaternized with ethylene oxide, and then attached to the head group via either a chemical or an enzymatic coupling method. The complete scheme is given in detail in Section II of the Appendix.

A. Cross-Linking to Phospholipids

When **XXVII** was mixed with dimyristoylphosphatidylcholine (mol ratio 1 : 9) and trace amounts of di[^{14}C]palmitoylphosphatidylcholine, sonicated, photolyzed, and analyzed by the usual procedure involving Sephadex LH-20 gel exclusion chromatography, approximately 2.5% of the radiolabeled lipid was found to be cross-linked. This cross-linking has been shown to be predominantly (at least 85%) in

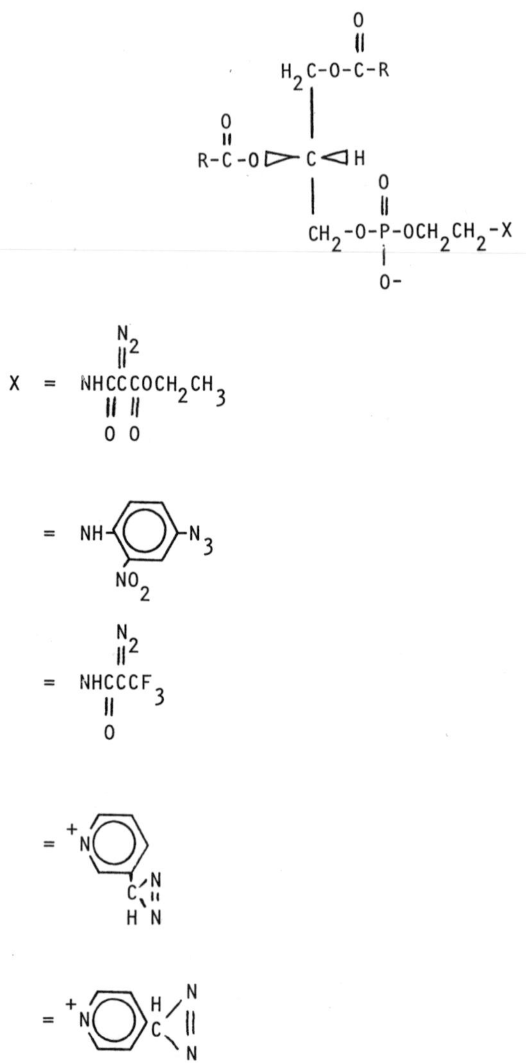

XXIV

XXV

XXVI

XXVII

XXVIII

Figure 17. Structures of phosphatidylethanolamine and phosphatidylcholine analogues carrying photoreactive polar head groups: The first **XXIV** and **XXV** were synthesized by Chakrabarti and Khorana (1975); **XXVI–XXVIII** by Robson and Khorana (Radhakrishnan et al., 1981).

the head group region, as expected (R. J. Robson & H. G. Khorana, unpublished results).

B. Cross-Linking to Proteins

The diazomalonyl phospholipid derivative (**XXIV**) has been used by Huang and Law (1981) to study the active site of phospholipase A_2 from snake venom. Tryptic degradation followed by cyanogen bromide cleavage narrowed the point(s) of cross-

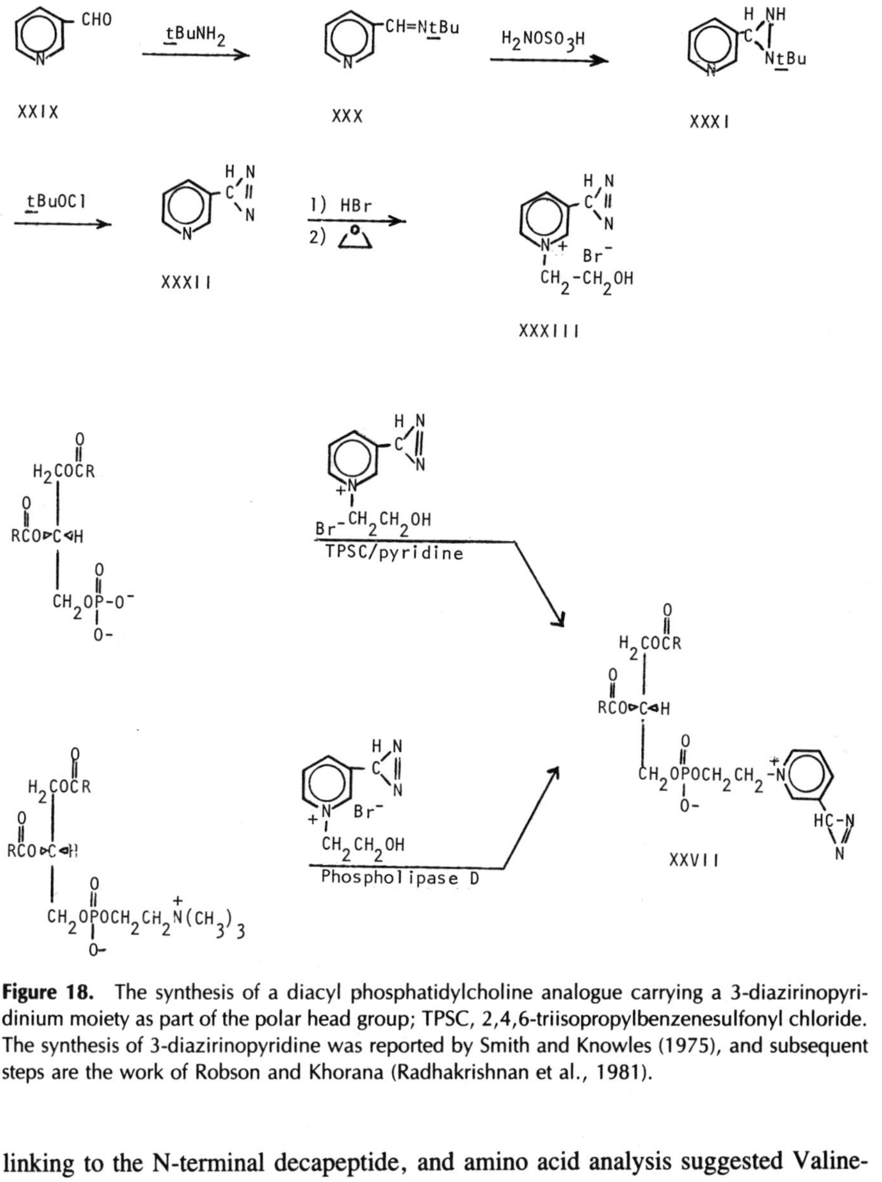

CHO
tBuNH₂ →
CH=NtBu
H₂NOSO₃H →
(XXXI structure)

XXIX   XXX   XXXI

tBuOCl →
(XXXII structure)
1) HBr
2) △O
→
(XXXIII structure)
Br⁻
CH₂-CH₂OH

XXXII   XXXIII

H₂COCR (O)
RCO►C◄H (O)
CH₂OP-O⁻ (O)
O-

(pyridinium diazirine structure)
Br⁻ CH₂CH₂OH
TPSC/pyridine →

H₂COCR (O)
RCO►C◄H (O)
CH₂OPOCH₂CH₂-N⁺
O-
HC-N
N

XXVII

H₂COCR (O)
RCO►C◄H (O)
CH₂OPOCH₂CH₂N⁺(CH₃)₃
O-

(pyridinium diazirine structure)
Br⁻
CH₂CH₂OH
Phospholipase D →

Figure 18. The synthesis of a diacyl phosphatidylcholine analogue carrying a 3-diazirinopyridinium moiety as part of the polar head group; TPSC, 2,4,6-triisopropylbenzenesulfonyl chloride. The synthesis of 3-diazirinopyridine was reported by Smith and Knowles (1975), and subsequent steps are the work of Robson and Khorana (Radhakrishnan et al., 1981).

linking to the N-terminal decapeptide, and amino acid analysis suggested Valine-3 as the cross-linked amino acid.

In studies with bacteriorhodopsin, the protein was reconstituted into vesicles containing the pyridinium diazirine phospholipid **XXVII** and dioleoylphosphatidylcholine by using the cholate dialysis procedure described by Huang et al. (1980). The vesicles were photolyzed at 30°C at 334 or 366 nm (using a ±20 nm monochrometer plus a low wavelength filter). The protein peak from an LH-60 gel permeation column contained approximately 1.5–2.0% of cross-linked radioactive phospholipid. A control experiment with bacteriorhodopsin and the diazirine phospholipid without photolysis showed very little association of phospholipid with the protein (0.07%). Further characterization of the site or sites of cross-linking should provide useful information on the peptide segments that are exposed into aqueous medium.

In studies with glycophorin A, **XXVII** cross-linked to polypeptide segments containing residues 62–81 and 82–96. In conjunction with the photolabels at the fatty acyl termini, these results are consistent with the membrane model for glycophorin A shown in Fig. 15.

VIII. SUMMARY

This chapter discusses the feasibility of studying lipid-protein interactions using photolabeling techniques. Considering the requirements of reactivity, reagents capable of generating nitrene and carbene intermediates are used as probes. Three types of reagents have been proposed and investigated in this laboratory and elsewhere: (1) hydrophobic, membrane-permeant reagents, (2) phospholipids containing carbene or nitrene precursors as integral parts of lipid molecules and located at the termini of the fatty acyl chains, and (3) phospholipids carrying these precursors in the polar head group region. All are prepared in radioactive form, which makes it possible to determine the sites of cross-linking. Studies with model systems using phospholipid vesicles showed that with membrane-soluble reagents it is necessary for the reagents to favorably partition into the bilayer interior and to be sufficiently reactive to insert into C—H bonds. In general, carbene precursors have been shown to be reagents of choice, whether as permeant reagents or as an integral part of phospholipid molecules.

Attachment of a carbene precursor to the termini of one of the fatty acyl chains of phospholipids offers particular advantages over strictly membrane permeant reagents. Localization of the probe within the bilayer is virtually guaranteed, and perturbation of the natural membrane environment is undoubtedly less than with bulky reagents added from outside (such as adamantane diazirine and iodonaphthyl azide). However, synthesis of these phospholipids may be more difficult and expensive. The utility of these modified phospholipids was shown by the photolysis of lipids containing diazirinophenoxy or trifluorodiazopropionyloxy groups and subsequent analysis of the cross-linked products. Degradative analysis showed that cross-linking occurred to a significant extent and was predominanty confined to the middle half of the bilayer. With phospholipids carrying the photolabel in the polar head group region, analysis of photolysis products in model lipid systems showed that cross-linking was restricted to the polar head group of acceptor phospholipid molecules.

The results with model lipid systems thus gave encouragement to an analysis of more complex membrane systems containing membrane proteins. Integral membrane proteins like glycophorin A and cytochrome b_5 were reconstituted with phospholipids containing carbene precursors at the fatty acyl termini. Cross-linking, followed by chemical and enzymatic analysis, showed essentially one site of cross-linking for glycophorin A (Glu-70) but a broad distribution of cross-linking in the case of cytochrome b_5 localized in the region containing the hydrophobic amino acids (residues 104–126). Binding of one mole of the photosensitive phospholipid with phospholipid exchange protein, followed by photolysis and analysis, showed that cross-linking was localized in a hydrophobic hexapeptide in the presumed fatty acyl binding region of this protein.

Phospholipids carrying the photolabels in the polar head group have been used to study the topography of bacteriorhodopsin, glycophorin A, and phospholipase A_2. Further experiments are necessary before the utility of these particular photo-labels is established conclusively.

ACKNOWLEDGMENTS

This work has been generously supported by grants NIH (AI-11479, GM-28289) and NSF (PCM 78-13713). Drs. A. Ross and R. Robson were supported by the Damon Runyon-Walter Winchell Cancer Fund (DRG-144-F and DRG-301-F, respectively).

APPENDIX: SYNTHESES

I. Procedure to Yield a Mixed Acyl Phosphatidylcholine Carrying an Aromatic diazirine in the Fatty Acyl Chain (See Figs. 6 and 8.)

m-(Methoxymethyleneoxy)benzaldehyde (X)

To a solution of *m*-hydroxy benzaldehyde (**IX**) (80 g, 0.66 mole) in anhydrous methanol (250 ml) is added sodium methoxide (40.5 g, 0.75 mole) in 250 ml of anhydrous methanol over a period of 1 h. The methanolic solution is evaporated after 30 min at a moderate vacuum (20 mm Hg). The residue is dried by repeated evaporation of anhydrous benzene. The yellow solid phenoxide thus obtained is dried *in vacuo* and then suspended in 250 ml of anhydrous dimethylformamide and cooled in an ice bath. To this suspension is added chlorodimethyl ether (*caution:* carcinogen) (64 g, 0.8 mol), slowly with stirring. The reaction is exothermic, and after the complete addition of the reagent, the mixture is stirred for another 4 h. TLC (petroleum ether : ether, 4 : 1, vol/vol) confirms the completion of the reaction. Dilution with 200 ml of ethyl ether precipitates sodium chloride, which is removed by filtration. Most of the organic solvents are removed by evaporation at a moderate vacuum. The residue is dissolved in ethyl ether (500 ml), and after washing with water (2 × 200 ml), the combined organic layer is dried and evaporated. The crude product is purified by silica gel chromatography. Elution with 5% ethyl ether in petroleum ether gives the title compound as a colorless oil. The yield is 80 g (72%). IR (neat): 1700 cm^{-1} (aldehyde).

m-(Methoxymethyleneoxy)phenyl-3H-diazirine (XI)

The procedure is adopted from the general methods used for aryl diazirines (Smith & Knowles, 1975; Schmitz, 1962). To 225 ml of anhydrous methanol at −40°C is added 75 ml of liquid ammonia to make an approximate 10 N solution. A solution of 30 ml of tert-butyl hypochlorite in 30 ml of tert-butanol is added dropwise to the above methanolic solution over a period of 1 h. The mixture is stirred magnetically during this period to effect the complete formation of chloramine. The protected benzaldehyde (41 g, 0.25 mole) is added and the mixture allowed to stand at −40°C for 1 h and then allowed to warm to room temperature gradually. Total reaction period varies from 24 to 36 h. Ammonia is evaporated, solvents are removed, and the crude product is dried in vacuo. The residue is suspended in methanol (200 ml) at 0°C and tert-butyl hypochlorite (12 ml) in tert-butanol (30 ml) is added dropwise over 20 min. After stirring at 0°C for 2 h, the reaction mixture is poured into a sodium metabisulfite solution (300 g in 1500 ml) and extracted into petroleum ether (4 × 150 ml). The organic solvent is removed under moderate vacuum at no greater than 5°C, and the crude diazirine thus obtained is purified by silicic acid column chromatography. The fastest running product is the desired diazirine and is obtained pure by elution with petroleum ether or 5% ethyl ether in petroleum ether. Preparative high performance liquid chromatography (Water Prep 500 system) using a silica column is employed for large scale purification using 1–2% ethyl ether in isooctane as the eluent. Using either method of purification, 4.1 g of pure diazirine is obtained as a pale yellow liquid. Stored in the freezer (−20°C), this compound is stable for months and is relatively stable to visible light, acids and bases. IR (Nujol): 1580 cm^{-1} (N$=$N); UV (cyclohexane): λ_{max} 362 nm (ε = 276); NMR (CDCl$_3$): δ 5.6–7.2 (m, 4H), δ 2.3 (s, 1H).

m-(Hydroxy)phenyldiazirine (XII)

A solution of the protected phenyldiazirine (534 mg, 3.0 millimoles) in 7 ml of glacial acetic acid is treated with 3 ml of 1 N hydrochloric acid at room temperature. The reaction is monitored by TLC (petroleum ether : ether, 4 : 1, vol/vol). After 3 h (or when the reaction is complete), the mixture is slowly added to a cooled solution of sodium bicarbonate (17 g in 500 ml of water). After the reaction subsides, the phenolic diazirine is extracted with ethyl ether (3 × 100 ml). The organic layer is dried under moderate vacuum and then purified by silica gel chromatography. Elution with 5% ethyl ether in petroleum ether removes any unreacted starting material. Elution with 10% ethyl ether in petroleum ether gives m-(hydroxyphenyl)diazirine as a colorless syrup (270 mg, 70% yield). The syrup may slowly turn brown due to oxidation but can be used in the alkylation reaction without any reduction in yield. IR (Nujol): 1580 cm^{-1} (N$=$N); UV (cyclohexane): λ_{max} 358 nm (ε = 265); NMR (CDCl$_3$): δ 5.6–7.2 (m, 5H; one of which disappears after D$_2$O treatment); δ 2.4 (s, 1H).

Methyl m-(3H-diazirino)phenoxy undecanoate (XIII)

The phenolic diazirine (400 mg, 3.0 millimoles) in 5 ml of anhydrous methanol is treated with a solution of sodium methoxide (80 mg, 3.3 millimoles) in methanol (2 ml). After standing for 20 min, methanol is evaporated under moderate vacuum. The gummy residue is dried with benzene and then finally in vacuo. The yellow solid phenoxide thus obtained is suspended in hexamethylphosphoramide (5 ml) and treated with methyl 11-iodo-undecanoate (980 mg, 3 millimoles) in hexamethylphosphoramide (5 ml). The reaction, stirred at room temperature, is complete in 6 h (TLC, using petroleum ether : ether, 4 : 1, vol/vol). The reaction mixture is partitioned between water and petroleum ether (100 ml each), and the aqueous phase is extracted with petroleum ether (2 × 50 ml). The combined organic layers are dried and concentrated. The residue is chromatographed on a silicic acid column. Elution with 5% ethyl ether in petroleum ether gives methyl m-(3H-diazirino)phenoxy undecanoate as a colorless syrup (700 mg, 70% yield). The product is pure as judged by high pressure liquid chromatography (monitored by UV absorption). A μPorasil column (0.4 × 30 cm) is used, eluting with petroleum ether : benzene (60 : 40, vol/vol) at a flow rate of 2 ml/min. The retention time is 9.0 min. IR (neat): 1750 cm^{-1} (C=O) and 1590 cm^{-1} (N=N); UV (cyclohexane): λ_{max} 358 nm (ε = 286); NMR (CDCl$_3$): δ 6.3–7.2 (m, 3H), δ 5.65 (d, 1H, J = 2 Hz), δ 3.5 (s, 3H), 3.8 (t, 2H, J = 7 Hz).

ω-(m-3H-Diazirino)phenoxy undecanoic acid (XIV)

A solution of methyl m-(3H-diazirino)phenoxy undecanoate (332 mg, 1.0 millimole) in 5% potassium hydroxide in 95% ethanol (3 ml) is stirred at room temperature, and the potassium salt of the fatty acid precipitates. After 5 h, the mixture is diluted with water, and the solution is neutralized with 1 N HCl until pH 5.0. The precipitated acid is extracted into ether (2 × 50 ml). The organic layer is washed once with water (1 × 20 ml). The etheral solution is dried over anhydrous sodium sulfate and evaporated under aspirator vacuum. The fatty acid is crystallized from petroleum ether containing minimal diethyl ether. Pale yellow crystals (200 mg, 63% yield) are obtained, mp 73-5°C. UV (cyclohexane): λ_{max} 360 nm (ε = 282).

1-Palmitoyl-2-(m-3H-diazirino)phenoxy- undecanoyl-sn-glycero-3-phosphorylcholine (XXIII)

Palmitoylphosphatidylcholine is prepared by published procedures (Gupta et al., 1977). To a solution of palmitoylphosphatidylcholine (348 mg. 0.5 millimole) in 80 ml of ethyl ether : methanol (99 : 1, vol/vol) is added 45 ml of an aqueous solution containing 40 mM calcium chloride, 20 mM Tris-HCl, pH 8.0, and 7 mg of *Crotalus adamanteus* rattlesnake venom. The reaction flask is sealed and stirred vigorously for approximately 5 h, or until the reaction is complete as determined by TLC (chloroform : methanol : water 65 : 25 : 4, vol/vol). The solvents are removed by a gentle stream of nitrogen, and 2-lysophosphatidyl-choline is extracted by the procedure of Bligh and Dyer (1959). Purification may be necessary on a Sephadex LH-20 column (chloroform : methanol, 1 : 1, vol/vol). The yield from di-palmitoylphosphatidylcholine is nearly quantitative (90–100%).

After removal of solvents by rotary evaporation (absolute ethanol may be added to prevent excessive foaming), the residue is rendered anhydrous by azeotroping with anhydrous benzene. To the dry lyso compound thus obtained is added the anhydride **XV** (0.63 millimole) and DMAP (61 mg, 0.5 millimole), followed by 10 ml of freshly distilled anhydrous chloroform. The flask is flushed with nitrogen, sealed, and stirred for 2 days. The solvent is removed under reduced pressure and the residue extracted by the method of Bligh and Dyer. The substitution of 1 N HCl for the water in the step effectively extracts all the DMAP. The phospholipid can be purified either on a silica gel or Sephadex LH-20 column (chloroform : methanol, 1 : 1, vol/vol). TLC (chloroform : methanol : water, 65 : 25 : 4, vol/vol) confirms the homogeneity of the product. The yield is approximately 80%.

II.   Procedure to Yield a Diacyl Phosphatidylcholine Analogue Carrying an Aromatic Diazirine in Place of the Trimethylammonium Group (See Fig. 18.)

1,2-Dimyristoyl-sn-glycero-3-phosphoric acid-(4-diazirinopyridinium)-ethyl ester (XXVII)

4-Pyridyl-3H-diazirine (XXXII)

The procedure for the synthesis of 3-(3-pyridyl)-3H-diazirine is that of Smith and Knowles (1975) and is used for both positional isomers, with modifications in methods of purification. Freshly distilled pyridine-4-carboxaldehyde (17.2 g, 0.16 mole) is dissolved in t-butylamine (30 ml, 0.29 mole) and allowed to stand for 3–4 h. Addition of anhydrous benzene (100 ml) followed by solvent removal at moderate vacuum gives a light yellow oil. This is dissolved in absolute ethanol (100 ml), triethylamine (40 ml), and water (40 ml). After cooling to −10° C, hydroxylamine-O-sulfonic acid (24 g, 0.21 millimole) is added over 30 min with vigorous stirring, and stirring is continued for an additional 2 h. t-Butyl hypochlorite (24 ml, 0.21 mole) is added over 30 min and the solution stirred at 4°C for 16 h. Sodium bicarbonate is added to bring the pH to 7. After the addition of water (100 ml), the product is extracted into methylene chloride (4 × 100 ml). The product is purified on silicic acid columns using gradients of acetone in petroleum ether, followed by a second column using ethyl ether in methylene chloride. IR (neat): 1595 cm^{-1} (N=N); UV (cyclohexane): λ_{max} = 344 nm (ε = 220); NMR (CCl$_4$): 2.07 (1H, s, diazirine), 6.75 (2H, d, 3-H), 8.43 (2H, d, 2-H). The yield is 15–20% after purification. Caution should be exercised as these compounds are rather volatile and can cause skin irritations.

N-β-Hydroxyethyl-4-diazirinopyridinium bromide XXXIII

For the quaternization of the 4-diazirinopyridine, equimolar amounts of 48% HBr and the diazirinopyridine (350 mg) are mixed at −20°C and the solvent removed at room temperature under high vacuum. Water (2 ml) is added, followed by ethylene oxide (4 mole excess, 0.65 ml). A dry ice–acetone cold finger is attached and the solution stirred at room temperature 4–6 h. After the solvent is removed, the product is crystallized from ethanol/ether. NMR (D$_2$O): δ 2.8 (1H, s, diazirine), δ 4.1 (2H, d), δ 4.7 (2H, d), δ 7.7 (2H, d), δ 8.8 (2H, d). UV (water): λ_{max} = 331 nm. Alternatively and with lower yields, alkylation can be performed with iodoethanol in acetonitrile or acetone.

1,2-Dimyristoyl-sn-glycero-3-phosphoric acid-(4-diazirinopyridinium)-ethyl ester XXVII

1,2-Dimyristoyl-sn-glycero-3-phosphoric acid is prepared by phospholipase D-catalyzed hydrolysis of 1,2-dimyristoyl-sn-glycero-3-phosphorylcholine. Dimyristoylphosphatidylcholine (500 mg) is dissolved in 25 ml ether. An aqueous solution of cabbage phospholipase D (50 mg) in 90 ml of 17 mM sodium acetate (pH 5.6) and 20 ml of 225 mM calcium chloride is added. The flask is sealed and stirred at 25–30°C for 1 day. The organic layer is removed, the aqueous layer extracted 3 times with ether, the organic layers are combined, and the solvent removed.

   1,2-Dimyristoyl-sn-glycero-3-phosphoric acid (0.2 millimole) and N-β-hydroxymethyl-4-diazirinopyridinium halide (0.2 millimole) are dried by the repeated addition and evaporation of dry pyridine. 2,4,6-Triisopropylbenzenesulfonyl chloride (0.4 millimole) is added, followed by dry pyridine (2 ml). After 12 h, the reaction is quenched by the addition of an equal volume of 50% pyridine in water. After solvent removal, ether (2 ml) is added, the solution filtered, and the filtrate applied to a silica gel column. Elution is effected with increasing amounts of methanol in methylene chloride, the yield being 70–90%.

Alternatively, N-β-hydroxyethyl-4-diazirinopyridinium bromide is exchanged for choline in a reaction catalyzed by phospholipase D. The reaction is two-phase, the upper phase being the phospholipid in ether, the lower phase containing the exchangeable alcohol, enzyme, and calcium in an aqueous buffer. The ether phase consists of dimyristoylphosphatidylcholine in ethyl ether (20 mg/ml). The aqueous phase consists of 100 mM Tris-HCl (pH 8.0), 100 mM calcium chloride, 10% N-β-hydroxyethyl-4-diazirinopyridinium halide, and 0.1–1.0 mg/ml phospholipase D. The ether and aqueous phases are rapidly stirred for several hours, monitoring the formation of the modified phospholipid by TLC (chloroform : methanol : water, 65 : 25 : 4, vol/vol) and stopping the reaction before significant phosphatidic acid is formed. The product is purified by silica gel chromatography with methanol and methylene chloride as eluents. The yield is 50–70%.

REFERENCES

Akeroyd, R., Moonen, P., Westerman, J., Puijk, W. C., & Wirtz, K. W. A. (1981). The Complete Primary Structure of the Phosphatidylcholine Transfer Protein from Bovine Liver. *Eur. J. Biochem.* **114,** 385–391.

Bailey, G. S., Gillet, D., Hill, F., & Petersen, G. B. (1977). Automated Sequencing of Insoluble Peptides Using Detergent: Bacteriophage f 1 Coat Protein. *J. Biol. Chem.* **252,** 2218–2225.

Bayley, H., & Knowles, J. R. (1977). Photoaffinity Labeling. *Methods Enzymol.* **46,** 69–114.

Bayley, H., & Knowles, J. R. (1978a). Photogenerated Reagents for Membrane Labeling. 1. Phenylnitrene Formed Within the Lipid Bilayer. *Biochemistry* **17,** 2414–2419.

Bayley, H., & Knowles, J. R. (1978b). Photogenerated Reagents for Membrane Labeling. 2. Phenylcarbene and Adamantylidene Formed Within the Lipid Bilayer. *Biochemistry* **17,** 2420–2423.

Bayley, H., & Knowles, J. R. (1980). Photogenerated Reagents for Membranes: Selective Labeling of Intrinsic Membrane Proteins in the Human Erythrocyte Membrane. *Biochemistry* **19,** 3883–3892.

Bercovici, T., & Gitler, C. (1978). 5-[^{125}I]Iodonaphthyl Azide, a Reagent to Determine the Penetration of Proteins into the Lipid Bilayer of Biological Membranes. *Biochemistry* **17,** 1484–1489.

Berry, R. S., Cornell, D., & Lwowski, W. (1963). Flash-Photolytic Decomposition of Gaseous Alkyl Azidoformates. *J. Am. Chem. Soc.* **85,** 1199.

Bligh, E. G., & Dyer, W. J. (1959). A Rapid Method of Total Lipid Extraction and Purification. *Can. J. Biochem. Physiol.* **37,** 911–917.

Brunner, J., & Richards, F. M. (1980). Analysis of Membranes Photolabeled with Lipid Analogues: Reaction of Phospholipids Containing a Disulfide Group and a Nitrene or Carbene Precursor with Lipids and with Gramicidin A. *J. Biol. Chem.* **255,** 3319–3329.

Brunner, J. (1981). Labeling the Hydrophobic Core of Membranes. *Trends Biochem. Sci.* **6,** 44–46.

Brunner, J., Senn, H., & Richards, F. M. (1980). 3-Trifluoromethyl-3-Phenyldiazirine. *J. Biol. Chem.* **255,** 3313–3318.

Büchel, D. E., Gronenborn, B., & Müller-Hill, B. (1980). Sequence of the Lactose Permease Gene. *Nature (London)* **283,** 541–545.

Buse, G., & Steffens, G. J. (1978). Studies on Cytochrome *c* Oxidase. II. The Chemical Constitution of a Short Polypeptide from the Beef Heart Enzyme. *Hoppe-Seyler's Z. Physiol. Chem.* **359,** 1005–1009.

Chakrabarti, P., & Khroana, H. G. (1975). A New Approach to the Study of Phospholipid-Protein Interactions in Biological Membranes. Synthesis of Fatty Acids and Phospholipids Containing Photosensitive Groups. *Biochemistry* **14**, 5021–5033.

Chen, R., Krämer, C., Schmidmayr, W., & Henning, V. (1979). Primary Structure of Major Outer Membrane Protein I of *Escherichia coli* B/r. *Proc. Natl. Acad. Sci. U.S.A.* **76**, 5014–5017.

Chen, R., Schmidmayr, W. Krämer, C., Chen-Schmeisser, V., Henning V. (1980). Primary Structure of Major Outer Membrane Protein II* (omp A protein) of *Escherichia coli* K-12. *Proc. Natl. Acad. Sci. U.S.A.* **77**, 4592–4596.

Chowdhry, V., Vaughan, R., & Westheimer, F. H. (1976). 2-Diazo-3,3,3- Trifluoropropionyl Chloride: Reagent for Photoaffinity Labeling. *Proc. Natl. Acad. Sci. U.S.A.* **73**, 1406–1408.

Chowdhry, V., & Westheimer, F. H. (1979). Photoaffinity Labeling of Biological Systems. *Annu. Rev. Biochem.* **48**, 293–325.

Coleman, R. (1973). Membrane-Bound Enzymes and Membrane Ultrastructure. *Biochim. Biophys. Acta* **300**, 1–30.

Cotmore, S. F., Furthmayr, H., & Marchesi, V. T. (1977). Immunochemical Evidence for the Transmembrane Orientation of Glycophorin A. Localization of Ferritin-Antibody Conjugates in Intact Cells. *J. Mol. Biol.* **113**, 539–553.

Cramer, J. A., Marchesi, V. T., & Armitage, I. M. (1980). Reversible Trifluoroacetic Acid-Induced Conformational Changes in Glycophorin A as Detected by Proton Nuclear Magnetic Resonance Spectroscopy. *Biochim. Biophys. Acta* **595**, 235–243.

Czernilofsky, A. P., Levinson, A. D., Varmus, H. E., Bishop, J. M., Tischer, E., & Goodman, H. M. (1980). Nucleotide Sequence of an Avian Sarcoma Virus Oncogene (*src*) and Proposed Amino Acid Sequence for Gene Product. *Nature (London)* **287**, 198–203.

Dailey, H. A., & Strittmatter, P. (1981). The Role of COOH-Terminal Anionic Residues in Binding Cytochrome b_5 to Phospholipid Vesicles and Biological Membranes. *J. Biol. Chem.* **256**, 1677–1680.

Demel, R. A., Wirtz, K. W. A., Kamp, H. H., Geurts van Kessel, W. S. M., & van Deenen, L. L. M. (1973). Phosphatidylcholine Exchange Protein from Beef Liver. *Nature (London) New Biol.* **246**, 102–105.

Demel, R. A., & de Kruijff, B. (1976). The Function of Sterols in Membranes. *Biochim. Biophys. Acta* **457**, 109–132.

Devaux, P. F., Moonen, P., Bienvenue, A., & Wirtz, K. W. A. (1977). Lipid-Protein Interaction in the Phosphatidylcholine Exchange Protein. *Proc. Natl. Acad. Sci. U.S.A.* **74**, 1807–1810.

Egmond, M. R., Williams, R. J. P., Welsh, E. J., & Rees, D. A. (1979). ¹H-Nuclear Magnetic Resonance Studies on Glycophorin and Its Carbohydrate-Containing Tryptic Peptides. *Eur. J. Biochem.* **97**, 73–83.

Enoch, H. G., Fleming, P. J., & Strittmatter, P. (1979). The Binding of Cytochrome b_5 to Phospholipid Vesicles and Biological Membranes: Effect of Orientation on Intermembrane Transfer and Digestion by Carboxypeptidase Y. *J. Biol. Chem.* **254**, 6483–6488.

Farley, R. A., Goldman, D. W., & Bayley, H. (1980). Identification of Regions of the Catalytic Subunit of (Na-K)-ATPase Embedded Within the Cell Membrane: Photochemical Labeling with [³H]Adamantane Diazirine. *J. Biol. Chem.* **255**, 860–864.

Fleming, P. J., Dailey, H. A. Corcoran, D., & Strittmatter, P. (1978). The Primary Structure of the Nonpolar Segment of Bovine Cytochrome b_5. *J. Biol. Chem.* **251**, 5369–5372.

Frank, G., Brunner, J., Hauser, H., Wacher, H., Semenza, G., & Zuber, H. (1978). The Hydrophobic Anchor of Small-Intestinal Sucrase-Isomaltase: N-Terminal Sequence of the Isomaltase Subunit. *FEBS Lett.* **96**, 183–188.

Furthmayr, H. (1977). Structural Analysis of a Membrane Glycoprotein: Glycophorin A. *J. Supramol. Struct.* **7**, 121–134.

Garoff, H., Frischauf, A.-M., Simons, K., Lebrach, H., & Delius, H. (1980). Nucleotide Sequence of cDNA Coding for Semliki Forest Virus Membrane Glycoprotein. *Nature (London)* **288**, 236–241.

Gazzotti, P., Bock, H.-G., & Fleischer, S. (1975). Interaction of D-β-Hydroxybutyrate Apodehydrogenase with Phospholipids. *J. Biol. Chem.* **250**, 5782–5790.

Gilles, A. M., Imhoff, J. M., & Keil, B. (1979). α-Clostripain-Chemical Characterization, Activity and Thiol Content of the Highly Active Form of Clostripain. *J. Biol. Chem.* **254**, 1462–1468.

Gitler, C., & Bercovici, T. (1980). Use of Lipophilic Photoactivatable Reagents to Identify the Lipid-Embedded Domains of Membrane Proteins. *Ann. New York Acad. Sci.* **346**, 199–211.

Goldman, D., Pober, J., White, J., & Bayley, H. (1979). Selective Labelling of the Hydrophobic Segments of Intrinsic Membrane Proteins with a Lipophilic Photogenerated Carbene. *Nature (London)* **280**, 841–843.

Grover, A. K., Slotboom, A. J., de Haas, G. H., & Hammes, G. G. (1975). Lipid Specificity of β-Hydroxybutyrate Dehydrogenase Activation. *J. Biol. Chem.* **250**, 31–38.

Gupta, C. M., Radhakrishnan, R., & Khorana, H. G. (1977). Glycerophospholipid Synthesis: Improved General Method and New Analogs Containing Photoactivable Groups. *Proc. Natl. Acad. Sci. U.S.A.* **74**, 4315–4319.

Gupta, C. M., Radhakrishnan, R., Gerber, G. E., Olsen, W. L., Quay, S. C., & Khorana, H. G. (1979a). Intermolecular Crosslinking of Fatty Acyl Chains in Phospholipids: Use of Photoactivable Carbene Precursors. *Proc. Natl. Acad. Sci. U.S.A.* **76**, 2595–2599.

Gupta, C. M., Costello, C. E., & Khorana, H. G. (1979b). Sites of Intermolecular Crosslinking of Fatty Acyl Chains in Phospholipids Carrying a Photoactivable Carbene Precursor. *Proc. Natl. Acad. Sci. U.S.A.* **76**, 3139–3143.

Hartman, F. C., & Wold, F. (1966). Bifunctional Reagents. Cross-Linking of Pancreatic Ribonuclease with a Diimide Ester. *J. Am. Chem. Soc.* **86**, 3890–3891.

Hoppe, J., & Sebald, W. (1980). Amino Acid Sequence of the Proteolipid Subunit of the Protein-Translocating ATPase Complex from the Thermophilic Bacterium PS-3. *Eur. J. Biochem.* **107**, 57–65.

Huang, K.-S., Bayley, H., & Khorana, H. G. (1980). Delipidation of Bacteriorhodopsin and Reconstitution with Exogenous Phospholipid. *Proc. Natl. Acad. Sci. U.S.A.* **77**, 323–327.

Huang, K.-S., & Law, J. H. (1981). Photoaffinity Labeling of *Crotalus atrox* Phospholipase A_2 by a Substrate Analogue. *Biochemistry* **20**, 181–187.

Kagawa, Y., & Racker, E. (1971). Partial Resolution of the Enzymes Catalyzing Oxidative Phosphorylation . XXV. Reconstitution of Vesicles Catalyzing $^{32}P_i$-Adenosine Triphosphate Exchange. *J. Biol. Chem.* **246**, 5477–5487.

Kahane, I., & Gitler, C. (1978). Red Cell Membrane Glycophorin Labeling from Within the Lipid Bilayer. *Science* **201**, 351–352.

Kamp, H. H., Wirtz, K. W. A., & van Deenen, L. L. M. (1975). Delipidation of the Phosphatidylcholine Exchange Protein from Beef Liver by Detergents. *Biochim. Biophys. Acta* **398**, 401–414.

Kamp, H. H., Wirtz, K. W. A., Baer, P. R., Slotboom, A. J., Rosenthal, A. F., Paltauf, F., & van Deenen, L. L. M. (1977). Specificity of the Phosphatidylcholine Exchange Protein from Bovine Liver. *Biochemistry* **16**, 1310–1316.

Karlish, S. J. D., Jorgensen, P. L., & Gitler, C. (1977). Identification of a Membrane-Embedded Segments of the Large Polypeptide Chain of (Na^+,K^+)ATPase. *Nature (London)* **269**, 715–717.

Khorana, H. G., Gerber, G. E., Herlihy, W. C., Gray, C. P., Anderegg, R. J., Nihei, K., & Biemann, K. (1979). The Amino Acid Sequence of Bacteriorhodopsin. *Proc. Natl. Acad. Sci. U.S.A.* **76**, 5046–5050.

Klip, A., & Gitler, C. (1974). Photoactive Covalent Labeling of Membrane Components from Within the Lipid Core. *Biochem. Biophys. Res. Commun.* **60**, 1155–1162.

Kyba, E. P., & Abramovitch, R. A. (1980). Photolysis of Alkyl Azides. Evidence for a Nonnitrene Mechanism. *J. Am. Chem. Soc.* **102**, 735–740.

Lutter, L. C., Ortanderl, F., & Fasold, H. (1974). The Use of a New Series of Cleavable Protein-Crosslinkers of the *Escherichia coli* Ribosome. *FEBS Lett.* **48,** 288–292.

Machida, K., & Ohnishi, S. I. (1978). A Spin-Label Study of Phosphatidylcholine Exchange Protein. Regulation of the Activity by Phosphatidylserine and Calcium Ion. *Biochim. Biophys. Acta* **507,** 156–164.

McRobbie, I. M., Meth-Cohn, O., & Suschitzky, H. (1976). Competitive Cyclisations of Singlet and Triplet Nitrenes. Part I. Cyclisation of 1-(2- Nitrenophenyl)Pyrazoles. *Tetrahedron Lett.* 925–928.

Nakashima, Y., & Konigsberg, W. (1974). Reinvestigation of a Region of the fd Bacteriophage Coat Protein Sequence. *J. Mol. Biol.* **88,** 598–600.

Ozols, J. (1974). Cytochrome b_5 from Microsomal Membranes of Equine, Bovine and Porcine Livers. Isolation and Properties of Preparations Containing the Membranous Segment. *Biochemistry* **13,** 426–434.

Ozols, J., Gerard, G., & Nobrega, F. G. (1976). Proteolytic Cleavage of Horse Liver Cytochrome b_5: Primary Structure of the Heme-Containing Moiety. *J. Biol. Chem.* **251,** 6767–6774.

Ozols, J., & Gerard, C. (1977a). Primary Structure of the Membranous Segment of Cytochrome b_5. *Proc. Natl. Acad. Sci. U.S.A.* **74,** 3725–3729.

Ozols, J., & Gerard, C. (1977b). Covalent Structure of the Membranous Segment of Horse Cytochrome b_5. *J. Biol. Chem.* **252,** 8549–8553.

Porter, A. G., Barber, C., Carey, N. H., Hallewell, R. A., Threlfall, G., & Emtage, J. S. (1979). Complete Nucleotide Sequence of an Influenza Virus Haemagglutinin Gene from Cloned DNA. *Nature (London)* **282,** 471–477.

Radhakrishnan, R., Gupta, C. M., Erni, B., Robson, R. J., Curatolo, W., Majumdar, A., Ross, A. H., Takagaki, Y., & Khorana, H. G. (1980). Phospholipids Containing Photoactivable Groups in Studies of Biological Membranes. *Ann. N.Y. Acad. Sci.* **346,** 165–198.

Radhakrishnan, R., Robson, R. J., Takagaki, Y., & Khorana, H. G. (1981). Synthesis of Modified Fatty Acids and Glycerophospholipid Analogs *Methods Enzymol.* **72D,** 408–433.

Rand, R. P., & Luzzati, V. (1968). X-Ray Diffraction Study in Water of Lipids Extracted from Human Erythrocytes. The Position of Cholesterol in the Lipid Lamellae. *Biophys. J.* **8,** 125–137.

Reiser, A., & Frazer, V. (1965). Ultra-violet Absorption Spectra of Aromatic Nitrenes and Dinitrenes. *Nature (London)* **208,** 682–683.

Reiser, A., Terry, G. C., & Willets, F. W. (1966). Observation of the Nitrene in the Flash-Photolysis of 1-Azidoanthracene. *Nature (London)* **211,** 410.

Reiser, A., Willets, F. W., Terry, G. C., Williams, V., & Marley, R. (1968). Photolysis of Aromatic Azides. 4. Lifetimes of Aromatic Nitrenes and Absolute Rates of Some of Their Reactions. *Trans. Faraday Soc.* **64,** 3265.

Rogers, J., Early, P., Carter, C., Calame, K., Bond, M., Hood, L., & Wall, R. (1980). Two mRNAs with Different 3' Ends Encode Membrane-Bound and Secreted Forms of Immunoglobulin μ Chain. *Cell* **20,** 303–312.

Rose, J. K., Welch, W. J., Sefton, B. M., Esch, F. S., & Ling, N. C. (1980). Vesicular Stomatitis Virus Glycoprotein Is Anchored in the Viral Membrane by a Hydrophobic Domain Near the COOH Terminus. *Proc. Natl. Acad. Sci. U.S.A.* **77,** 3884–3888.

Schmitz, E. (1962). Diaziridine, III. Die Einwirkung von Chloramin und Ammoniak auf Aldehyde. *Chem. Ber.* **95,** 688–691.

Schulte, T. H., & Marchesi, V. T. (1979). Conformation of Human Erythrocyte Glycophorin A and Its Constituent Peptides. *Biochemistry* **18,** 275–280.

Sebald, W., Wachter, E., & Tzagoloff, A. (1979). Identification of Amino Acid Substitutions in the Dicyclohexylcarbodiimide-Binding Subunit of the Mitochondrial ATPase Complex from Oligomycin-Resistant Mutants of *Saccharomyces cerevisiae*. *Eur. J. Biochem.* **100,** 599–607.

Sebald, W., & Wachter, E. (1980). Amino Acid Sequence of the Proteolipid Subunit of the ATP Synthetase from Spinach Cloroplasts. *FEBS Lett.* **122,** 307–311.

Sebald, W., Machleidt, W., & Wachter, E. (1980). N,N'-Dicyclohexylcarbodiimide Binds Specifically to a Single Glutamyl Residue of the Proteolipid Subunit of the Mitochondrial Adenosinetriphosphatases from *Neurospora crassa* and *Saccharomyces cerevisiae*. *Proc. Natl. Acad. Sci. U.S.A.* **77,** 785–789.

Selinger, Z., & Lapidot, Y. (1966). Synthesis of Fatty Acid Anhydrides by Reaction with Dicyclohexylcarbodiimide. *J. Lipid Res.* **7,** 174–175.

Sigrist, H., & Zahler, P. (1978). Characterization of Phenylisothiocyanate as a Hydrophobic Membrane Label. *FEBS Lett.* **95,** 116–120.

Sigrist-Nelson, K., Sigrist, H., Bercovici, T., & Gitler, C. (1977). Intrinsic Proteins of the Intestinal Microvillus Membrane: Iodonaphthylazide Labeling Studies. *Biochim. Biophys. Acta* **468,** 163–176.

Singh, A., Thornton, E. R., & Westheimer, F. H. (1962). The Photolysis of Diazoacetylchymotrypsin. *J. Biol. Chem.* **237,** PC3006–PC3007.

Skell, P. S., & Klebe, J. (1960). Structure and Properties of Propargylene. *J. Am. Chem. Soc.* **82,** 247–248.

Smith, R. A. G., & Knowles, J. R. (1975). The Preparation and Photolysis of 3Aryl-3H-Diazirines. *J. Chem. Soc., Perkin Trans. II* 686–694.

Snell, D. T., & Offord, R. E. (1972). The Amino Acid Sequence of the B-Protein of Bacteriophage ZJ-2. *Biochem. J.* **127,** 167.

Steck, T. L., & Dawson, G. (1974). Topographical Distribution of Complex Carbohydrates in the Erythrocyte Membranes. *J. Biol. Chem.* **249,** 2135–2142.

Steffens, G. J., & Buse, G. (1979). Studies on Cytochrome *c* Oxidase. IV. Primary Structure and Function of Subunit II. *Hoppe-Seyler's Z. Physiol. Chem.* **360,** 613–619.

Steglich, W., & Höfle, G. (1969). N,N-Dimethyl-4-Pyridinamine, A Very Effective Acylation Catalyst. *Angew. Chem. Int. Ed. Engl.* **8,** 981.

Takagaki, Y., Gupta, C. M., Nihei, K., Gerber, G. E., & Khorana, H. G. (1979). Phospholipid-Protein Interactions in Membranes: Investigation of the Tertiary Structure of the Membranous Component of Cytochrome b_5. Eleventh International Congress of Biochemistry, Toronto, Ont., Canada, July 8–13, 1979 (Abstr.), p. 381.

Takagaki, Y., Gerber, G. E., Nihei, K., & Khorana, H. G. (1980). Amino Acid Sequence of the Membranous Segment of Rabbit Liver Cytochrome b_5: Methodology for Separation of Hydrophobic Peptides. *J. Biol. Chem.* **255,** 1536–1541.

Takagaki, Y., & Radhakrishnan, R. (1981). Photolabeled Phospholipids in Studies of Protein-Lipid Interactions: A Study of Cytochrome b_5 in Model Membranes, 72nd Annual Metting ASBC, St. Louis, MO (Abstr.).

Tometsko, A. M., & Richards, F. M. (Eds.) (1980). Applications of Photochemistry in Probing Biological Targets. *Ann. New York Acad. Sci.* **346,** 1–502.

Tomita, M., Furthmayr, H., & Marchesi, V. T. (1978). Primary Structure of Human Erythrocyte Glycophorin A. Isolation and Characterization of Peptides and Complete Amino Acid Sequence. *Biochemistry* **17,** 4756–4770.

Turro, N. J. (1980). Structure and Dynamics of Important Reactive Intermediates Involved in Photobiological Systems. *Ann. New York Acad. Sci.* **346,** 1–17.

Van Wezenbeck, P. M. G. F., Hulsebos, T. J. M., & Schoenmakers, J. G. G. (1980). Nucleotide Sequence of the Filamentous Bacteriophage M13 DNA Genome: Comparison with Phage fd. *Gene* **11,** 129–148.

Vaughan, R. J., & Westheimer, F. H. (1969). A Method for Marking the Hydrophobic Binding Sites of Enzymes. An Insertion into the Methyl Group of an Alanine Residue of Trypsin. *J. Am. Chem. Soc.* **91,** 217–218.

Visser, L., Robinson, N. C., & Tanford, C. (1975). The Two-Domain Structure of Cytochrome b_5 in Deoxycholate Solution. *Biochemistry* **14,** 1194–1199.

Wachter, E., Schmid, R., Deckers, G., & Altendorf, K. (1980). Amino Acid Replacement in Dicyclohexylcarbodiimide-Reactive Proteins from Mutant Strains of *Escherichia coli* Defective in the Energy-Transducing ATPase Complex. *FEBS Lett.* **113,** 265–269.

Waxman, D. J., & Strominger, J. L. (1981). Primary Structure of the COOH-Terminal Membranous Segment of a Penicillin-Sensitive Enzyme Purified from Two Bacilli. *J. Biol. Chem.* **256,** 2067–2077.

Wirtz, K. W. A., Moonen, P., van Deenen, L. L. M., Radhakrishnan, R., & Khorana H. G. (1980). Identification of the Lipid Binding Site of the Phosphatidylcholine Exchange Protein with a Photosensitive Nitrene and Carbene Precursor of Phosphatidylcholine. *Ann. New York Acad. Sci.* **348,** 244–255.

Wold, F. (1972). Bifunctional Reagents. *Methods Enzymol.* **25,** 623–651.

Woodworth, R. C., & Skell, P. S. (1959). Methylene, CH_2. Stereospecific Reaction with *cis*- and *trans*-2-Butene. *J. Am. Chem. Soc.* **81,** 3383–3386.

Zupancic, J. J., & Schuster, G. B. (1980). Chemistry of Fluoroenylidene. Direct Observation of, and Kinetic Measurements on, a Singlet and a Triplet Carbene at Room Temperature. *J. Am. Chem. Soc.* **102,** 5958–5960.

Five

Interactions Between Proteins and Amphiphiles

JACQUELINE A. REYNOLDS

*Department of Physiology and Whitehead Medical Research
Institute, Duke University Medical Center, Durham, North Carolina
U.S.A.*

ABBREVIATIONS

CMC, critical micelle concentration
HDL, high density lipoprotein
HLB, hydrophilic-lipophilic balance

LDL, low density lipoprotein
SDS, sodium dodecylsulfate
$\bar{V}$, partial specific volume
$\bar{v}$, number of moles of ligand bound per mole of protein, also referred to as r.

I. PROPERTIES OF AMPHIPHILIC COMPOUNDS

Amphiphiles are molecules that contain spatially separated hydrophobic and hydrophilic regions and possess unique physical properties reflecting this multidomain nature. Relevant literature and the theoretical basis for the properties of these compounds are discussed in detail in Tanford (1980), so exhaustive references are not provided in this chapter. Rather, I summarize briefly the solution physical chemistry of amphiphilic molecules as it relates to the interactions between them and different types of proteins.

A. Self-Association and Critical Micelle Concentration

All amphiphiles have a finite monomer solubility in water, frequently called the critical micelle concentration (CMC). Since one domain of an amphiphile is hydrophobic (i.e., tends to be excluded from contact with water), the self-association that occurs at total concentrations greater than the CMC is thermodynamically driven by the requirement to remove this hydrophobic region from solvent contact. Infinite growth of the association product (phase separation) is prevented by an unfavorable free energy (opposing force) arising from head group interactions. If the head group is ionic, this positive contribution to the free energy is satisfactorily predicted by ionic repulsion, but the repulsive term for uncharged head groups is less readily defined from first principles.

Amphiphiles that contain a normal alkyl or alkenyl chain attached to a polar head group generally form micelle structures that are well-approximated as oblate ellipsoids (Fig. 1a). Naturally occurring lipids that contain two fatty acyl chains per polar head group tend to form extended bilayer structures in order to minimize the curvature at the polar surface, hence to minimize the contact of the hydrophobic interior with water. An unusual type of amphiphile is found in the bile salts, which have hydrophobic faces and are thought to form self-association products similar to that in Fig. 1b (Small, 1971). Although some disagreement exists as to the exact structure of these micelles, there is no question that bile salts do not form an oblate ellipsoid as shown in Fig. 1a.

Tables 1 and 2 present data on the CMC, average aggregation number, and partial specific volume of a number of common commercial detergents and a few phospholipids. More complete tabulations are found in Steele et al. (1978) and Helenius et al. (1979). The average aggregation number and the CMC of ionic detergents (Table 1) are a function of ionic strength—the aggregation number increasing and the CMC decreasing as the ionic strength is raised. Nonionic de-

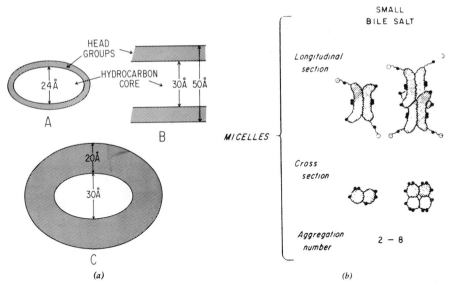

Figure 1. Left: Approximate dimensions of disk-shaped micelles in comparison with a planar bilayer. *A*, Sodium dodecylsulfate (SDS) micelle at ionic strength 0.1; *B*, Bilayer formed by egg yolk phosphatidylcholine; *C*, Lubrol WX micelle. The thickness of the hydrophobic core is determined solely by the alkyl chain length. The shaded areas reflect the distances to which the head groups extend from the core surface. The head groups themselves do not fill all the space within these regions, space between head groups being occupied by solvent. Reproduced by permission of C. Tanford and J. A. Reynolds and Elsevier Scientific Publishing Co., Amsterdam (Tanford & Reynolds, 1976). Right: Proposed structures for the primary and secondary bile salt micelles. Reproduced with permission of A. Helenius, K. Simons, and Elsevier Scientific Publishing Co., Amsterdam (Helenius & Simons, 1972).

tergents, on the other hand, are relatively insensitive to ionic strength but do have a tendency to secondary association to a larger particle under some conditions. Secondary aggregation is particularly severe when the polar head group of polyoxyethylene detergents is small relative to the hydrophobic region (e.g., $C_{12}E_8$), and these particular amphiphiles may present severe experimental problems for that reason.

It is frequently not appreciated that biological lipids also have a finite free monomer concentration in equilibrium with the self-associated bilayers (Table 2). Although these monomer concentrations are extremely low relative to those of detergents, the concept of a CMC for lipids is extremely important in subsequent discussions of interactions between lipids and proteins and of reconstitution procedures.

In addition to the experimental parameters, CMC and average aggregation number, detergent chemists frequently refer to an HLB (hydrophilic-lipophilic balance) number when describing nonionic detergents. This terminology refers specifically to the *ratio* of the size of the hydrophilic and hydrophobic domains. It is this ratio

TABLE 1 Properties of Some Common Commercial Detergents

Detergent[a]	CMC $(M \times 10^3)$	Average Aggregation Number	Partial Specific Volume $\overline{V}$ (cm^3/g)	Ionic Strength
$C_{12}OSO_3^-Na^+$	8.13	62	0.864	0
	2.30	84		0.05
	0.92	103		0.20
$C_{12}NMe_3^+Br^-$	14.6	61	0.97	0
	5.71	72		0.05
	2.54	83		0.25
$C_8\phi E_{9.6}$ (Triton X-100)	0.3	140	0.908	
$C_9\phi E_{10}$ (Triton N-101)	0.097	100	0.922	
$C_9\phi E_{15}$	0.087	52	0.965	
$C_{12}E_8$	0.087	120	0.973	
$C_{16}E_{20}$ (Brij)	0.0039	70	0.919	
C_{12} Sorbitan E_{20} (Tween 20)	0.059		0.869	
C_{18} Sorbitan E_{20} (Tween 80)	0.012	60	0.896	
C_8 Glucoside	25.0			
Deoxycholate	2.0	1.7	0.778	0 (pH 9)
	1.6	2.2		0.01
	0.91	22.0		0.15
Cholate	10.0	2	0.771	0 (pH 8–9)
	5.0	2.8		0.01
	3.0	4.8		0.15

Data from Helenius et al. (1979) and Steele et al. (1978).

[a]C_n indicates the number of carbon atoms in the hydrophobic region; E_m indicates the number of ($-O-CH_2-CH_2$) groups in the hydrophilic region.

TABLE 2 Properties of Diacyl Phospholipids

Phospholipid[a]	CMC $(M \times 10^3)$	Partial Specific Volume, $\overline{V}$ (cm^3/g)
diC$_7$PC	1.4	0.865
diC$_8$PC	0.27	0.888
diC$_{10}$PC	0.005	0.927
diC$_{16}$PC	4.7×10^{-7}	0.976
Egg yolk PC	Not determined	0.981

From Tanford (1980).
[a]diC$_n$PC refers to the fully saturated diacyl phosphatidylcholine containing n carbon atoms per fatty acyl chain.

that governs the tendency to secondary micellar association mentioned above. However, interactions between detergents and proteins or lipid bilayers are primarily related to the absolute size of the hydrophilic and hydrophobic domains, so that HLB numbers are likely to be of little value in predicting the applicability of a particular detergent to a specific biological problem.

B. Monolayer Formation

One thermodynamic process for removing the hydrophobic region of amphiphiles from contact with water is the formation of monolayers at interfaces. Investigations of monolayer properties are carried out at either air-water or hydrocarbon-water interfaces. An amphiphile at an air-water interface has its hydrophobic domain in contact with the air and its polar head group dissolved in the aqueous medium; pressure versus area measurements in this type of system reflect both head group interactions and the van der Waals attraction between the hydrocarbon tails. On the other hand, amphiphiles at hydrocarbon-water interfaces have their hydrophobic domains in an environment similar to that in micelles and bilayers, and pressure versus area curves in this case reflect only the head group interactions over a wide range of surface areas.

Monolayers of amphiphilic compounds at an air-water interface are often used to study protein-amphiphile interactions. However, molecular interpretations of these data are difficult since proteins themselves tend to "surface denature" and, further, the hydrocarbon milieu is not truly analogous to that in a micelle or bilayer. A more serious problem arises for the biological scientist who is normally interested in amphiphiles that have charged head groups and a finite monomer solubility in water. No rigorous theory has yet been derived for charged, soluble monolayers, and the classical insoluble monolayer thermodynamic treatment is inapplicable to these systems (Mingins et al., 1975; Aveyard & Haydon, 1973).

C. Thermodynamics and Kinetics of Amphiphile Self-Association

The thermodynamic equilibrium involved in micellization (or the formation of a lipid bilayer) can be satisfactorily, if not rigorously, described by the following equation (see Tanford, 1980, for a detailed theoretical approach):

$$mL \rightleftharpoons L_m \tag{1}$$

$$K = \frac{[L_m]}{[L]^m}$$

where L represents the concentration of amphiphile in solution, m is the average aggregation number, and L_m is the concentration of micellar amphiphile. It is apparent that if m is large, the concentration of monomer $[L]$ in solution will increase very little above the CMC; that is, when $[L_m]$ becomes a significant fraction of the total concentration. This relationship is important in considering the interaction between amphiphilic compounds and proteins in the following sections, since all such interactions are competitive with the formation of pure self-association products as defined by Eq. 1.

The association constant K in Eq. 1 is the quotient of two rate constants, k_1 (forward) and k_2 (reverse). As the CMC decreases, the reverse rate constant becomes smaller, resulting in slower rates of dissociation of the monomer from the micelles. The practical consequences of this phenomenon are the slow removal of detergents with low CMCs from protein-detergent or protein-lipid-detergent complexes and nonequilibrium binding data due to the slow approach to equilibrium. Conversely, amphiphiles with high CMCs have relatively rapid rates of dissociation from their own micelles and from protein-amphiphile complexes. Experimental design must be based on these considerations. For example, exchange of a bound amphiphile with a high CMC for one with a low CMC (e.g., a diacylphospholipid) from a protein-amphiphile complex may lead to exposure of a membrane protein to water and subsequent denaturation if the ligand with the more rapid dissociation rate is removed before association with the second (lower CMC) ligand can occur.

II. THERMODYNAMIC TREATMENT OF INTERACTIONS

The interaction of a solvent component with a protein is defined in general terms as the difference in solvent composition in the immediate vicinity of the protein relative to that in the bulk solvent mixture. Rigorously this interaction is the perturbation of the chemical potential of a solvent component by the addition of protein (i.e., the preferential interaction of a ligand with a protein is equal to $(\partial g_3/\partial g_2)_{T,P,\mu_3}$ where g_3 is grams of ligand 3 associated with the protein per grams of water, and μ_3 is the chemical potential of ligand 3). Note that the chemical potential of ligand 3, temperature (T), and pressure (P) are held constant. The equilibrium condition for interaction requires that $\mu_{3(bound)} = \mu_{3(free)}$. Preferential interaction can be positive as in the case of amphiphile-protein interactions or negative as in the exclusion of glycerol from the region immediately surrounding a protein molecule. The thermodynamic definition of preferential interaction applies to all perturbations of the chemical potential of the solvent component, both weak and strong. However, in this chapter we are primarily concerned with interactions in which $(\partial g_3/\partial g_2)_{T,P,\mu_3}$ is positive and of significant magnitude. For example, the collision of a gas molecule with a vessel wall clearly perturbs the chemical potential of the gas molecule, but the magnitude of this interaction is very small compared to the binding of oxygen to hemoglobin.

Another way of looking at the relative strength and weakness of preferential interaction is to consider the relative residence time of a solvent component in the vicinity of a protein. If the residence time is short (i.e., reverse dissociation rate is rapid), the interaction is weak. Our hypothetical gas molecule colliding with a vessel wall is an example of short residence time, while hemoglobin binding of oxygen represents a long residence time.

It is customary in biological sciences to measure the number of grams of ligand bound per gram of protein or number of moles of ligand bound per mol of protein $(\bar{v})$. (Some investigators use "r" to express this ratio.) Experimentally one normally varies the amount of ligand added to a system and determines $\bar{v}$ as a function of free or unbound ligand concentration. The data are conventionally reported in the form of a graph of $\bar{v}$ versus the logarithm of the unbound ligand concentration. Although it is not rigorously correct to neglect activity coefficients of the solvent components, most binding measurements are made at sufficiently low solute concentrations that significant errors are not introduced by using concentration rather than activity in the appropriate equations.

The mathematical relationship between $\bar{v}$ and the system variables is derived from equations that express the correct thermodynamic equilibria. Table 3 gives appropriate equations for several possible binding models that are not all inclusive but serve to demonstrate several common ligand binding modes.

The most familiar type of binding expression is given by equation 1 in Table 3, in which the protein is assumed to contain a set of independent, identical binding sites for the ligand being investigated. Other types of binding equilibrium are also presented in Table 3 and typical plots of $\bar{v}$ versus free ligand concentration that correspond to these equations are shown in Fig. 2. If there is no change in the state of association of the protein as the result of ligand binding, $\bar{v}$ does not depend on protein concentration (equations 1–4, Table 3). Conversely, when ligand binding induces an alteration in the state of association of the protein, a complicated dependence on total concentration of protein ensues (equations 5 and 6, Table 3). In these cases $[P]$, the free protein concentration, appears in the mathematical expressions, and usually it is not possible to measure this variable directly. If the ligand itself undergoes self-association as do amphiphilic molecules, this phenomenon must be taken into account in writing the thermodynamic equations. An example is provided by equation 7 in Table 3. Note that experimentally a measure of free ligand concentration in this system will be the sum of both monomeric and micellar ligands.

TABLE 3 Thermodynamic Binding Equations[a]

Equilibrium Equations	$\bar{v}$ as a Function of Solution Variables
	No Protein Concentration Dependence

1. $\displaystyle\sum_{i=0}^{n-1}(PL_i + L \rightleftharpoons PL_{i+1})$

$$\bar{v} = \frac{nK[L]}{1+K[L]}$$

2. $P + L \rightleftharpoons PL \qquad (K_1)$
$PL \rightleftharpoons P^*L \qquad (K_2)$
$P \rightleftharpoons P^* \qquad (K_3)$

$$\bar{v} = \frac{(K_1 + K_1K_2)\,[L]}{(1+K_3)+(K_1+K_1K_2)\,[L]}$$

3. $\displaystyle\sum_{i=0}^{n-1}(PL_i + L \rightleftharpoons PL_{i+1}) \qquad (K_1)$

$\displaystyle\sum_{j=0}^{m-1}(P^*L_j + L \rightleftharpoons P^*L_{j+1}) \qquad (K_2)$

$P \rightleftharpoons P^* \qquad (K_3)$

$$\bar{v} = \frac{[L]}{1+[P_T^*]/[P_T]}\left(\frac{nK_1}{1+K_1[L]} + \frac{mK_2[P_T^*]/[P_T]}{1+K_2[L]}\right)$$

$$\frac{[P_T^*]}{[P_T]} = K_3\frac{(1+K_2[L])^m}{(1+K_1[L])^n}$$

4. $P + nL \rightleftharpoons PL_n$

$$\bar{v} = \frac{nK[L]^n}{1+K[L]^n}$$

Protein Concentration Dependence

5. $nP \rightleftharpoons P_n$ (K₁)

$$\bar{\nu} = \frac{K_2[L]}{1 + K_2[L] + nK_1[P]^{n-1}}$$

   $P + L \rightleftharpoons PL$ (K₂)

$$[P_T] = [P] + K_1[P]^n + nK_1K_2[P]^n[L]$$

6. $nP \rightleftharpoons P_n$ (K₁)

$$\bar{\nu} = \frac{K_1K_2[L]}{[P]^{n-1} + K_1 + nK_1K_2[L]}$$

   $P_n + L \rightleftharpoons P_nL$ (K₂)

$$[P_T] = [P] + K_1[P]^n + nK_1K_2[P]^n[L]$$

Ligand Self-Association

7. $P + L \rightleftharpoons PL$ (K₁)

$$\bar{\nu} = \frac{K_1[L]}{1 + K_1[L]}$$

   $mL \rightleftharpoons L_m$ (K₂)

$$[L_T] = [L] + K_1[P][L] + K_2[L]^m$$

Linearized Forms of Equation 1

(Scatchard, 1949)

(Klotz, 1953)

$$\frac{\bar{\nu}}{[L]} = nK - \bar{\nu}K$$

$$\frac{1}{\bar{\nu}} = \frac{1}{n} + \frac{1}{nK[L]}$$

$\bar{\nu}$ = mol of ligand bound per mol of protein; $[P_T]$ = total protein concentration; $[P]$ = free protein concentration; $[L_T]$ = total ligand concentration; $[L]$ = unbound ligand concentration; K_i = equilibrium association constant; $[P_T^*]$ and $[P^*]$ refer to conformationally altered states.

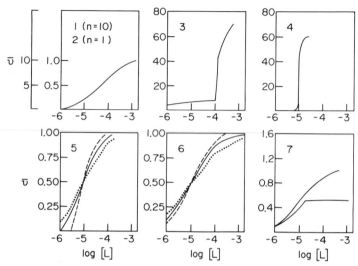

Figure 2. Binding isotherms corresponding to the equations in Table 3. The numbers on each graph refer to the corresponding equations of Table 3. Curves 1 and 2 (upper left), calculated for $n = 10$, $k = 2 \times 10^4$ M^{-1} corresponding to equation 1 and for $n = 1$, $(1 + K_3)/(K_1 + K_1 K_2) = 5 \times 10^{-5}$, corresponding to equation 2, can be superimposed when scaled to the same plateau height (n). Curve 3, $n = 8$, $K_1 = 1.2 \times 10^6$ M^{-1}, $m = 80$, $K_2 = 8 \times 10^3$ M^{-1}, $K_3 = 5 \times 10^{-7}$ M^{-1}. Curve 4, $n = 60$, $K = 10^{300}$ M^{-1}. Curve 5, $n = 2$ for all three isotherms and $[P_T] = 10^{-5}$ M: dashed curve, $K_1 = 10^{10}$ M^{-1}, $K_2 = 3.15 \times 10^7$ M^{-1}; solid curve, $K_1 = 10^6$ M^{-1}, $K_2 = 3.7 \times 10^5$ M^{-1}; dotted curve, $K_1 = 10^3$ M^{-1}, $K_2 = 10^5$ M^{-1}. Curve 6, $n = 2$ and $[P_T] = 10^{-5}$M for all three isotherms: dashed curve, $K_1 = 10^{10}$ M^{-1}, $K_2 = 10^5$ M^{-1}; solid curve, $K_1 = 10^6$ M^{-1}, $K_2 = 1.37 \times 10^5$ M^{-1}; dotted curve, $K_1 = 10^3$M^{-1}, $K_2 = 1.02 \times 10^7$ M^{-1}. Equation 7, $K_1 = 10^5$, $K_2 = 0$ (upper curve), $K_1 = 10^5$, $K_2 = 10^{343}$, $m = 70$ (lower curve).

The equation (equation 1, Table 3) describing simple independent site interaction can be algebraically manipulated into a number of linear forms, two of which are shown in the bottom portion of Table 3. The "Scatchard" plot (Scatchard, 1949) and the double reciprocal plot (Klotz, 1953) are commonly used to test for validity of the independent site model. If the data points do not deviate from linearity, n and K can be obtained in a direct manner. An excellent discussion of the problems inherent in linearization of this type of equation is given by P. J. F. Henderson (1978). Summarizing briefly, the "Scatchard" plot contains the dependent variable $\bar{v}$ on both the x and y axes, and therefore, the *usual* method of least-squares fit cannot be applied. On the other hand, taking reciprocals of data containing finite errors leads to a greater distortion of experimental errors and will actually conceal a poor fit between the data and a straight line. In this day of modern computers it is difficult to see the advantage of linearization, since a weighted least-squares fit to the hyperbolic equation (equation 1, Table 3) is readily carried out.

Examination of the binding isotherms in Fig. 2 demonstrates clearly that phenomena such as "positive" and "negative" cooperativity can arise from multiple models. Binding measurements alone will not distinguish among these, and additional types of experiments must be carried out in order to eliminate incorrect thermodynamic models.

Direct methods for measuring ligand-protein interaction are given in the Appendix. Indirect experimental procedures such as optical perturbations of the protein in the presence of ligand or alterations in some measurable property of the ligand due to added protein need to be approached with caution. Implicit in the use of these indirect procedures is the assumption that the experimental parameter being followed is a linear function of the perturbation of the chemical potential of the ligand, an assumption that must be substantiated by an independent method.

III.   INTERACTION BETWEEN AMPHIPHILES AND WATER-SOLUBLE PROTEINS

A.   Sets of Independent and Identical Sites

Very few water-soluble proteins in their native state contain binding sites for amphiphilic ligands. The now classical example of one that does is serum albumin, which interacts with a wide variety of amphiphiles but most strongly with those containing normal alkyl chains and anionic head groups. Several classes of amphiphilic binding sites have been identified on this protein—one high affinity site that is probably the *in vivo* site for fatty acid transport, and a second group of 8–10 sites of lower affinity that are often treated as mathematically identical but are distinguishable by differing affinities for anionic or cationic head groups. Many studies of binding to serum albumin have led to a postulated model for these binding sites that envisions a hydrophobic cleft of limited size with at least one lysine residue involved in head group interaction (see Tanford, 1980, for a brief summary of the experimental support for this model, and Chapter 2, Volume 1, for a differing view). Equation 1, Table 3, and Fig. 2 (upper left) are the thermodynamic descriptions of this type of binding.

B.   Cooperative Binding and Protein Conformational Changes

Protein chemists have known for many years that the interaction of water-soluble proteins with anionic detergents containing *n*-alkyl chains longer than 10 carbon atoms leads to conformational changes and loss of native structure. More recently, it has been observed that cationic detergents have the same effect (Nozaki et al., 1974). Detergent induced denaturation does not lead to a random coil protein structure, and all binding of this type takes place below the CMC. The fact that this interaction competes effectively with micelle formation indicates that a significant contribution to the free energy is made by favorable head group–protein interactions. Nonionic detergents are not capable of binding in this mode to water-soluble proteins, presumably because the free energy of micelle formation is more favorable than the free energy of binding.

One practical development that has arisen from studies of these systems has been the use of sodium dodecylsulfate (SDS) polyacrylamide gel electrophoresis. Since most water-soluble proteins that have been investigated bind equal amounts of this detergent (on a gram per gram basis) and assume a hydrodynamic size that is a unique function of the molecular weight, it is possible to estimate the molecular weights of such proteins by observing their mobility in SDS polyacrylamide gels (as first suggested by Weber & Osborn, 1969). Unfortunately, the experimental demonstration of a constant charge to mass ratio and a hydrodynamic size related to the molecular weight has been obtained only in phosphate buffer at pH 7.5 (see Nielsen & Reynolds, 1978). Thus, there is no sound basis for the use of this technique to estimate molecular weights in other buffer systems, and since detergent binding has been demonstrated to be a function of both pH and ionic strength for a number of water-soluble proteins, the use of pH or ionic strength gradient systems is risky. These problems, of course, do not detract from the use of this method as a separation technique or a comparative procedure for cataloguing protein components under identical electrophoretic conditions.

IV. INTERACTIONS BETWEEN AMPHIPHILES AND PROTEINS NORMALLY ASSOCIATED WITH LIPIDS *IN VIVO*

Two quite different complexes between proteins and lipids are found in living systems—cellular membranes and water-soluble lipoproteins. (In this section the discussion of the latter is restricted to human serum lipoproteins.) Despite the fact that the molar ratios of lipid to protein vary over the same range in both these types of complex, the former are insoluble permeability barriers and the latter are water-soluble, lipid-transport particles. Clearly, the protein structures and the organization of lipids and proteins must be different in the two classes, and it has been the aim of recent research to elucidate these differences.

In summarizing amphiphile-protein interactions in this section I emphasize general, as opposed to specific, concepts and also point out the areas of greatest divergence among investigators in this field. Einstein (1950) described science as "the attempt to make the chaotic diversity of our sense experience correspond to a logically uniform system of thought." As will be apparent to readers of this volume, all scientists involved in membrane and lipoprotein research have not come to the same "logically uniform system of thought," but it is from attempts to reconcile divergent interpretations of observations that general principles finally emerge. Thus, one hopes that exposure of differing views to discussion and criticism will lead to a reduction in the current state of "chaos."

A. Serum Lipoproteins

The circulating serum lipoproteins are complex particles containing protein and varying ratios of lipids (Osborne & Brewer, 1977, Chapter 6, Volume 1). Several classes of serum lipoproteins are separable on the basis of particle density, and it is clear from numerous analytical studies that specific types of protein are required to interact with and to solubilize specific mixtures of lipids (Table 4).

The two major polypeptides from human serum high density lipoprotein (HDL)

TABLE 4   Molar Composition of Human Serum Lipoproteins[a]

Lipoprotein	Apoproteins[b]	Composition (mol lipid/mol lipoprotein particle)[c]			
		PL	FC	CE	TG
HDL$_2$[d]	4 AI	137	56	89	18
	2 AII				
HDL$_3$[d]	2 AI	43	13	24	7
	1 AII				
LDL$_2$[e]	2 apo B	621	475	1,311	281

[a]The molar composition is calculated from the lipid analysis of Scanu (1971) and the weight percent protein of Osborne and Brewer (1977). The apoprotein composition of HDL$_2$ and HDL$_3$ was determined by Friedberg and Reynolds (1976). Assuming that the sum of AI and AII represent 95% of the total protein in the particle, and using the weight percent protein and lipid, one can calculate the molar composition without reference to the experimentally determined molecular weight of the holo particle. These latter numbers contain large uncertainties.
[b]1 mol AI contains 28,000 g protein, 1 mol AII contains 17,400 g protein, and 1 mol apo B contains 250,000 g protein.
[c]PL, phospholipid; FC, free cholesterol; CE, cholesterylester; TG, triglyceride.
[d]High density lipoproteins also contain variable amounts of C peptides. These proteins and cholesterol exchange among lipoprotein particles intravascularly; hence the compositions are averages.
[e]Since LDL$_2$ is an intravascular catabolic product of VLDL, the lipid compositions are averages obtained from fasting, normal donors.

are referred to as AI and AII. Both have been sequenced—AI has a molecular weight of 28,400 and AII is a disulfide bonded dimer of 17,400 (two 8,700 polypeptide chains). (Complete sequences are given in Chapter 6, Volume 1.) Predictions of secondary structure from the primary sequences of both proteins has led to postulated amphipathic helices that possess hydrophilic and hydrophobic faces (Segrest et al., 1974). The extension of these predicted protein structures to postulates regarding modes of interaction of amphipathic helices with lipids is on a much less firm foundation. For example, it has been suggested that reconstituted apo HDL–phospholipid particles consist of a truncated phospholipid bilayer surrounded by the apoprotein amphipathic helices (see, e.g., Andrews et al., 1976). The means by which this proposed bilayer could incorporate triglycerides and cholesteryl esters that do not partition significantly into pure phospholipid bilayers (Loomis et al., 1974) is not clear.

Investigations of the amphiphile binding properties of AI and AII have led to the following thermodynamic descriptions of interactions with single-tail amphiphilic molecules (Reynolds, 1980):

$$2(AI) \rightleftharpoons (AI)_2 \tag{2a}$$

$$\sum_{i=0}^{3} \left\{ (AI)L_i + L \rightleftharpoons (AI)L_{i+1} \right\} \tag{2b}$$

$$(AI)L_4 \rightleftharpoons (AI)*L_4 \tag{2c}$$

$$(AI)*L_4 + mL \rightleftharpoons (AI)*L_{m + 4} \tag{2d}$$

where L is the single-tail amphiphilic ligand. AI contains three or four independent and identical binding sites in the monomeric state for n-alkyl detergents (Eq. 2b), the occupancy of which leads to a conformational change (Eq. 2c) and concomitant cooperative binding of a large number of amphiphilic molecules (Eq. 2d). The independent site binding described by Eq. 2b is primarily hydrophobic, with little or no contribution from interactions with the polar head group of the ligand. Saturation binding levels (Eq. 2d) of a large number of these detergents show that the maximum moles of ligand bound per mole of protein decreases as the chain length increases. This behavior is in direct contrast to that of pure detergent self-association in which the average aggregation number increases with increasing chain length. Knowing the volume of the hydrophobic domain of the bound amphiphile, it is possible to calculate the total hydrophobic volume occupied by each bound ligand. AI interacts with a constant hydrophobic volume of n-alkyl chains of approximately $25,000$ cm^3 per mol of protein, as is shown in Table 5.

AII exhibits binding properties similar to those of AI.

$$2(AII) \rightleftharpoons (AII)_2 \tag{3a}$$

$$\sum_{i = 0}^{j} \left\{ (AII)_2 L_i + L \rightleftharpoons (AII)_2 L_{i + 1} \right\} \tag{3b}$$

$$(AII)_2 L_j \rightleftharpoons (AII)_2^* L_j \tag{3c}$$

$$(AII)_2^* L_j + mL \rightleftharpoons 2(AII)*L_{(m + j)/2} \tag{3d}$$

where (AII) refers to the disulfide bonded dimer of 17,400 molecular weight and $(AII)_2$ is 34,800 molecular weight. $(AII)_2$ contains a set of independent and identical hydrophobic binding sites (Eq. 3b), the occupancy of which leads to a conformational change (Eq. 3c). Subsequent cooperative binding of large numbers of n-alkyl detergent molecules is observed together with the dissociation to (AII) (Eq. 3d). As with AI, saturation binding levels decrease with increasing chain length and the maximum hydrophobic volume bound by AII is approximately 16,000 cm^3/ 17,400 g AII. The absolute number of independent sites represented by Eq. 3b and the association constants cannot be determined unambiguously since reactions described by Eqs. 3c and 3d set in prior to saturation of these sites (see, e.g., Fig. 2, top center). For example, the binding of sodium dodecylsulfate and tetradecyl trimethyl ammonium chloride to the independent sites on $(AII)_2$ are equally well fit by $n = 4$, $K = 3 \times 10^4$ M^{-1} and $n = 10$, $K = 8 \times 10^3$ M^{-1}.

The conformational change that is induced in both AI and AII by amphiphilic ligand binding can be followed by a number of different optical measurements. Table 6 shows the increase in negative ellipticity at 208 nm observed at saturation

TABLE 5   Amphiphile Binding Parameters for AI and AII

Apoprotein	Ligand	Independent Site, $\Delta G°$ (kcal/mol)[b]	Maximum Cooperative Binding (mol/mol protein)	Maximum[a] Hydrophobic Volume (cm^3)
AI	hexane	-5.7		
	octane	-7.5		
	$C_{12}OSO_3^-$	-8.2	138	26,000
	$C_{14}NME_3^+$	-8.2	117	25,900
	$C_{16}PC$	-9.0	97 ± 10	23,700
	$C_{16}E_{20}$	nd	80 ± 10	21,500
	$diC_{10}PC$	0	95 ± 10	28,100
AII	$C_{12}E_8$	nd	85 ± 10	16,000
	$C_{12}OSO_3^+$	-7.5 to -7.8	84 ± 7	16,000
	$C_{14}NME_3^+$	-7.5 to -7.8	74 ± 7	16,300
	$C_{12}E_{20}$	nd	49 ± 7	13,200
	$diC_{10}PC$	nd	62 ± 10	18,300
	egg yolk PC	nd	27 ± 3	15,000

Data from Reynolds (1980b).

[a]The maximum hydrophobic volume is subject to the uncertainty of whether the carbon atom nearest the polar head group is counted. The values given here are averages in which this carbon atom is given a value of 0.5 relative to the others. In addition there is a $\pm 10\%$ uncertainty due to the errors in maximal binding levels.

[b]nd, not determined.

TABLE 6   Spectral Properties of Protein-Amphiphile Complexes

Ligand	$\theta_{208}/\theta_{208(no\ ligand)}$ AI	AII
None[a]	1.0	1.0
$C_{12}OSO_3^-$	0.97	1.53
$C_{12}E_8$	nd	1.47
$C_{14}NMe_3^+$	0.91	1.44
$C_{16}E_{20}$	nd	1.68
$C_{16\&18}E_{17}$	1.06	nd
$C_{16}PC$	1.16	nd
$diC_{10}PC$	1.16	1.65
EY-PC	nd	1.59

[a]In the absence of ligands AI is monomeric (28,400 g/mol) and AII is dimeric (34,800 g/mol) at the concentrations employed.

levels of binding of a number of ligands. Relatively small changes accompany the packaging of large hydrophobic volumes by AI, but much larger alterations in presumed secondary structure are seen with AII.

The binding of didecanoylphosphatidylcholine (di $C_{10}PC$) to AI and AII has been studied (Reynolds et al., 1977; Reynolds, 1980b). This diacylphospholipid does not interact with AI until the three to four independent sites for single-tail amphiphiles (Eq. 2b) have been occupied. The binding equilibria are consistent with the following thermodynamic equations:

$$\sum_{i=0}^{3} \left\{ (AI)B_i + B \rightleftharpoons (AI)^*B_{i+1} \right\} \tag{4a}$$

$$2(AI)^*B_4 + 190 \ PL \rightleftharpoons (AI)^*_2B_8PL_{190} \tag{4b}$$

where B represents lysolipid and/or fatty acid and PL is diacylphospholipid. AII on the other hand interacts cooperatively with didecanoylphosphatidylcholine at or near the CMC ($5 \times 10^{-6}M$) in the absence of single-tail amphiphiles (Reynolds, 1980b). Saturation levels of binding to both proteins are given in Table 5 and are seen to correspond to the maximum hydrophobic volume of single-chain amphiphilic ligands. Both AI and AII are self-associated in the presence of didecanoylphosphatidylcholine, in contrast to the monomeric state observed at saturation binding of the simple detergent ligands; that is, $(AI)_2$ and $(AII)_2$ are formed in the presence of the phospholipid ligand.

A number of investigators have reported attempts to saturate AI and AII with other diacylphospholipids, and it is with these data that a wide divergence of results and interpretations arises. Two independent laboratories (Nichols et al., 1974; Assmann & Brewer, 1974) report no interaction between AI and diacylphospholipid in the absence of single-tail amphiphiles, a result explained by Reynolds et al. (1977) as a failure to fill the independent sites on AI that induce the conformational change. Table 5 demonstrates an apparent maximum free energy of binding to these sites of -9.0 kcal/mol, far lower than the free energy of association of most diacyl phospholipids. Thus, the interaction with protein in its nonliganded state cannot compete effectively with lipid-lipid interactions.

Other workers using primarily dimyristoylphosphatidylcholine (di $C_{14}PC$) (CMC $= 10^{-10}$ M, calculated from the measured CMCs of other lipids shown in Table 2) report formation of AI-lipid complexes of widely varying molar ratios (from 40 mol/AI to 254 mol/AI). In all cases the products were heterogeneous with respect to lipid binding, and thermodynamic equilibrium was not demonstrated by a rigorous method. Often no indication of the purity of the phospholipid is included, so it is difficult to rule out the presence of lysolipid and/or fatty acid. One systematic study of the binding of dimyristoylphosphatidylcholine to AI has been carried out (Swaney, 1980) in which the protein concentration was sufficiently low so that over 85% of non-liganded AI should have been in monomeric form. The interaction was measured as a function of increasing lipid to protein ratios and the results are shown in Table 7, together with comparable binding data obtained by Reynolds et al.

TABLE 7 Interaction of Diacyl Phospholipid with AI

(1)	Initial Concentrations		Binding Ratios Observed				
	AI (mol/liter × 10^5)	diC_{14}PC (mol/liter × 10^4)	mol/ 28,400 g (AI)	mol/ 28,400 g Dimer (AI)_2	mol/ 28,400 g Trimer (AI)_3	Free AI	Free Lipid (mol/liter × 10^4)
	1.81	2.92		80		yes	0
	1.79	7.20		116		yes	0
	1.72	17.4		140		yes	0
	2.09	42.2		172	280	0	0
	1.465	59.0		195	400	0	yes
	1.33	80.4		200	372	0	yes
(2)		diC_{10}PC (mol/liter × 10^4)					
	2.0	1.45	6.8				0.092
	1.84	1.76	4.0				1.0
	0.88	6.02	20.0				4.26
	0.88	9.54		28			7.08
	0.82	19.4		39			16.2
	0.88	15.1		58			10.0
	0.39	19.4		75			16.5
	0.39	38.8		87			35.4
	0.59	97.0		93			91.5

(1) Swaney (1980)
(2) Reynolds et al. (1977)

(1977) for the binding of didecanoylphosphatidylcholine to AI. The complexes formed in both cases are homogeneous with respect to binding.

Swaney found two types of complex separable by gel filtration chromatography and assumed to be (AI)_2 and (AI)_3 based on their elution positions. Reversibility was not demonstrated in this study, and the lipid was not analyzed for hydrolytic products or contaminants. Non-liganded (AI), (AI)_2-phospholipid, and (AI)_3-phospholipid were separable during the time required for elution, indicating that this is not a rapidly equilibrating system. These results illustrate dramatically the problem encountered in studying the interaction between long chain phospholipids and the apoproteins from HDL. The low CMC of dimyristoylphosphatidylcholine results in a slow "off time" of the lipid from any complex once formed, and, therefore, it is experimentally difficult to determine the true equilibrium state of the protein-lipid interaction product.

In light of the foregoing discussion, experimentalists in this field are faced with the following two questions. (1) Is a single-tail amphiphile such as lysolipid or fatty acid (or *in vivo* possibly bile salt) required to bind to the independent sites of AI prior to saturation with diacylphospholipid? (2) What is the true equilibrium state of AI saturated with diacylphospholipid regardless of the mechanism by which binding occurs? Support for an affirmative answer to the first question has been presented by Nichols et al. (1974), Assmann and Brewer (1974), and Reynolds et al. (1977). In the data in which complexes have presumably been formed in the absence of single-tail amphiphiles, there is no quantitative information as to the maximum level of single-tail contaminants that might have been present.

The most obvious explanation of the variable stoichiometry and protein state of association observed by numerous investigators is that the complexes formed are not equilibrium states. Consistent with this view are the studies of Pownall and his co-workers (e.g., Pownall et al., 1978), who demonstrated a slow kinetic interaction between AI and diacylphospholipids under the conditions of their experiments, Scanu and co-workers (see Chapter 6, Volume 1), who have shown variable stoichiometry in reconstitution experiments as a function of initial protein concentration, and Swaney (1980), who demonstrated that a rapid interconversion of species does not occur with dimyristoylphosphatidylcholine as the ligand. It is apparent, therefore, that additional experimental work is necessary in order to reconcile the variable results reported for diacyl phospholipid interaction with AI and to determine the true equilibrium state of the protein-lipid complexes. Complete characterization of a kinetically stable intermediate may provide some structural information; to understand an assembly mechanism and relate it to an *in vivo* process, however, it is necessary that all *equilibrium* states along a proposed pathway be well defined.

Many fewer studies of AII-phospholipid interactions have been reported (Chapter 6, Volume 1), but as with AI the results are varied and the complexes poorly characterized. Additional studies are clearly needed.

Some of the unique characteristics of AI and AII in terms of amphiphile interactions are listed here for comparison with the properties of membrane-bound proteins discussed in Section IV.B. These characteristics rest on firm thermodynamic studies and are not subject to the uncertainties enumerated above.

1. Apoproteins AI and AII contain a set of independent, identical hydrophobic binding sites, the occupancy of which leads to a conformation change and concomitant cooperative binding of large numbers of amphiphilic ligands (for a summary see Reynolds, 1980b).

2. Amphiphilic ligands are synergistic in their interactions with AI and AII; that is, the binding of one type of amphiphile can facilitate the binding of another by filling the set of independent sites and inducing the required conformational change (Stone & Reynolds, 1975; Nichols et al., 1974).

3. AI and AII interact with a constant, specific hydrophobic volume of n-alkyl chains when the ligand is a single-tail amphiphile or didecanoylphosphatidylcholine. (Note that the data in Table 7 for binding between dimyristoylphosphatidylcholine and AI are not consistent with this observation. Hence either this complex is a nonequilibrium state or the mode of binding of the longer chain diacylphospholipids is different.)

4. The structure of the final complex formed between amphiphilic ligands and AI or AII is governed (directed) by the protein, not by the normal association state of the amphiphile. AI and AII do not bind the equivalent of detergent micelles, nor do they form an equilibrium complex with preformed lipid bilayers.

The major polypeptide constituent of LDL, apolipoprotein B (apo B), is insoluble in aqueous medium in the absence of bound amphiphiles, and therefore, interaction studies must be carried out as competition experiments (one amphiphilic ligand is substituted for another that is already bound to the protein). Apo B has been obtained in homogeneous form as a complex with several detergents by substituting the single-tail amphiphiles for the naturally occurring lipids. In all systems thus far investigated apo B is dimeric in the detergent-protein complex (510,000 g protein per particle), and as has been previously pointed out (Tanford & Reynolds, 1979), this is the same state of association as that found *in vivo*. This property is in direct contrast to the observed behavior of AI and AII from HDL, which are monomeric in the presence of saturating levels of detergents and are self-associated only in the presence of bound phospholipids. There is clearly strong protein-protein interaction in the apo B–amphiphilic ligand complex.

Much confusion has existed over the polypeptide chain molecular weight of apo B, primarily due to the difficulty of working with this protein in the absence of bound amphiphiles and the susceptibility of the polypeptide to proteolytic cleavage. A rigorous determination of the molecular weight has been carried out in two separate laboratories (Smith et al., 1972; Steele & Reynolds, 1979), and the 510,000 g of apo B per mol of LDL represents two 255,000 molecular weight polypeptides.

Table 8 summarizes the binding data that have been obtained for apo B. In the anionic detergent sodium dodecylsulfate, the secondary structure at maximum binding is altered from that found in holo LDL. However, at concentrations of this detergent below the CMC and in the presence of nonionic detergents, the secondary structure as well as the antigenic determinants are preserved.

A complete binding isotherm for the interaction of apo B with an amphiphilic ligand has been obtained only with sodium dodecylsulfate (Steele & Reynolds, 1979). The midpoint of the observed cooperative binding coincides with the CMC of the ligand, suggesting the formation of a protein-detergent mixed micellar system. This behavior is reminiscent of that reported for intrinsic membrane proteins (Section IV.B) and differs from the interaction of AI and AII with detergents. In the latter case binding occurs below the CMC of the amphiphile.

Extensive hydrodynamic and morphological studies of apo B : $C_{12}E_8$ have demonstrated that this complex is an asymmetric, flexible rod with a strong tendency to lateral aggregation at neutral pH (Zampighi et al., 1981). Titration of the protein to more alkaline pH or addition of small amounts of charged amphiphiles produces a complex containing 500,000 g of protein per mol which is homogeneous and monodisperse (see Fig. 3).

It is apparent from the few studies in the literature and from Fig. 3 that apo B complexed with single-tail amphiphiles does not form a spherical particle such as

TABLE 8 Interaction of Amphiphiles with Apo B

Ligand	Maximum Binding (mol/500,000 g)	Circular Dichroism	Immunological Determinants	References
$C_{12}OSO_3^-$	$2,880 \pm 100$	Intact ($<$ CMC)	Intact ($<$ CMC)	Steele & Reynolds (1979)
$C_{12}E_8$	930 ± 100	Intact	Intact	Watt & Reynolds (1980)
$C_{16}PC$	600 ± 100	nd[b]	nd	Steele & Reynolds (1979)
Triton X-100	403 ± 54	nd	Intact	Helenius & Simons (1972)
DOC$^-$	771 ± 133	nd	Intact	Helenius & Simons (1972)
Egg yolk lecithin	300^a	Intact	Intact	Watt & Reynolds (1981)

[a]There has not yet been a direct demonstration that this value is a saturation binding number.
[b]nd, not determined.

212

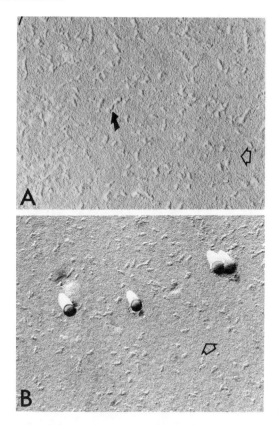

Figure 3. Apo B in $C_{12}E_8$, pH 10; (A) Shadowed with platinum and carbon at 12°; arrows indicate direction of shadowing. Magnification: 49,000 ×. (B) Shadowed at 30°. Magnification: 49,000 ×. From Zampighi et. al. (1981).

is found *in vivo* when the ligands are naturally occurring lipids. Thus, the tertiary structure of these detergent-protein particles is quite different from that of holo LDL, even though both secondary and quaternary structures are maintained. We must then assume that the binding of an amphiphile with two hydrophobic regions such as a diacylphospholipid induces a more symmetric complex without significant alteration in polypeptide chain order (helix or β-pleated sheet).

Egg yolk lecithin has been substituted for bound detergent on apo B and an electron micrograph of the product is shown in Fig. 4 (Watt & Reynolds, 1981). Particularly notable is the large number of symmetric particles seen in this field when compared with Fig. 3. Apo B does not interact with phospholipid vesicles or liposomes in this system (i.e., a stable thermodynamic complex is not formed between preformed bilayers and apo B), but rather packages a relatively small number of phospholipid molecules in water-soluble form.

The summary of saturation level binding for amphiphilic ligands to apo B shown in Table 8 demonstrates that this protein interacts with much larger numbers of

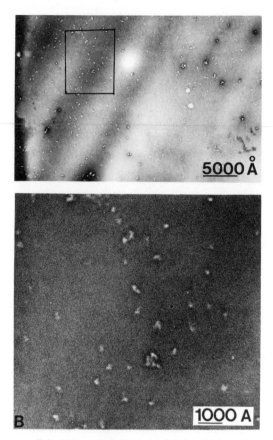

Figure 4. Apo B in egg yolk lecithin, negatively stained with uranyl acetate. (A) at 21,000 × and (B) 85,000 ×. (See also Watt & Reynolds, 1981.)

ligand molecules than do AI and AII and that there is no relationship between the normal state of association of the pure amphiphile and the maximum binding levels. As with AI and AII, it is the protein itself, not the preferred state of self-association of the amphiphile, that dictates the final structure of the complex.

B. Membrane-Bound Proteins

Diacylphospholipids self-associate in bilayer form in an aqueous medium, and it is this structure that forms the permeability barrier separating the cytoplasm of the cell from its surrounding medium and the intracellular compartments from one another. Intrinsic membrane proteins are multidomain in character, containing regions that associate with the hydrophobic milieu of the bilayer and others that are water soluble. In this section I consider only this type of membrane protein, not those that interact only with the bilayer surface. It is convenient to consider the binding of intrinsic membrane proteins to amphiphilic ligands from the standpoint of the separate domains, keeping in mind that interaction with one part of the protein may affect interaction with the other.

1. The Hydrophilic Domains

Very few water-soluble proteins have binding sites for individual amphiphilic ligands (Section III.A); thus one does not expect association between the water-soluble domain of an intrinsic membrane protein and nonionic detergents or ionic detergents with low CMCs. However, it is anticipated that the so-called denaturing detergents (Section III.B) will interact with these domains in much the same manner as with totally water-soluble proteins. The concentration of free detergent required for this type of binding may be much higher in the case of membrane proteins, a phenomenon that is best considered in light of a simple thermodynamic expression of the equilibria involved (Wyman, 1964). If a protein has two accessible states, which I shall call N and D, and both states can bind ligand, the equilibrium between the two states is governed by the following equation:

$$\frac{d \ln K}{da_L} = \bar{\nu}_{D,L} - \bar{\nu}_{N,L} \tag{5}$$

where K is the equilibrium constant for the reaction $N \rightleftharpoons D$, a_L is the activity of the ligand, $\bar{\nu}_{D,L}$ is mol ligand per mole of D, and $\bar{\nu}_{N,L}$ is moles ligand bound per mole of N. If $\bar{\nu}_{N,L}$ is greater than $\bar{\nu}_{D,L}$, the protein will be stabilized in state N. Thus if a membrane protein binds more detergent in its native state, N, than in the denatured state, D, the former will predominate. Native cytochrome b_5, for example, binds approximately one micelle of sodium dodecylsulfate to its hydrophobic region and requires very high free concentration of the detergent to denature the water-soluble domain (Robinson & Tanford, 1975).

The hydrophilic domain of a membrane protein may in some cases exhibit binding specificity for head groups of amphiphilic ligands; however, some caution must be exercised in interpretation of enzymatic activity data when detergents are substituted for the naturally occurring membrane lipids. It is often assumed without experimental demonstration that a membrane protein requires a specific lipid head group for activity (or even that the entire lipid molecule is required) when a function is lost as the result of detergent replacement. The most commonly used detergents for solubilization of membrane proteins are polyoxyethylene derivatives, which have very large head group regions relative to those of the phospholipids (Fig. 1). The polarizability of the head group of these detergents is very different from that of water, and it is not difficult to envision a perturbation of the water-soluble domain of a membrane protein with an active site embedded in this milieu.

A sample calculation of the weight percent water and oxyethylene in the solvent permeated mantle of a $C_{16}E_{20}$ micelle illustrates the problem. In this case the volume occupied by the polyoxyethylene head groups contains 67% water by weight and

TABLE 9 Detergent Interaction with Delipidated Membrane Proteins

Protein	Detergent	Ionic Strength	Maximum Binding	Aggre-gation Number (Detergent)	Midpoint of Cooper-ative Binding (mol/liter)	CMC (mol/liter) Detergent	State of Protein Association	Functionally Active	References
Glycophorin A[a]	$C_{12}OSO_3^-$	0.026	53 ±10	115	6.3×10^{-3}	3×10^{-3}	monomeric		Grefrath (1974); Mimms et al. (1980)
	DOC^-	0.26			1.6×10^{-3}	8.2×10^{-4}	monomeric		
		0.03	68		5×10^{-3}	4×10^{-3}	monomeric		
	$C_{14}NMe_3^+$	0.30			3×10^{-3}	1.8×10^{-3}	monomeric		
		0.027			4×10^{-3}	4×10^{-3}			
Cytochrome b_5	$C_{12}OSO_3^-$	0.01	45	65	2.5×10^{-3}	6.5×10^{-3}	monomeric		Robinson & Tanford (1975)
		0.10	51	85	8×10^{-4}	1.3×10^{-3}	monomeric		
	DOC^-	0.01	20	4	5×10^{-3}	5×10^{-3}	monomeric		
		0.10	30	13	3×10^{-3}	3×10^{-3}	monomeric		
	Triton X-100		100	140	6.3×10^{-4}	3×10^{-4}	monomeric		
fd Coat Protein	$C_{12}OSO_3^-$	0.10	60	85	3×10^{-3}	3×10^{-3}	dimeric		Makino et al. (1975)
	DOC^-	0.10	16	13			dimeric		
Thy-1	DOC^-	0.01	7	4	5×10^{-3}	5×10^{-3}	monomeric	yes	Kuchel et al. (1978)
	Brij 96						monomeric	yes	

Bacterio-rhodopsin	Triton X-100		177	140	monomeric	yes	Reynolds & Stoeckenius (1977)
Rhodopsin[b]	$C_{12}E_8$		240	120	monomeric	no	McCaslin & Tanford (1981)
	Cholate	0.15	11	10	trimeric	yes	
Acetylcholine receptor	Brij 58		40	75	monomeric	yes	Reynolds & Karlin (1978)
			18	75	dimeric	yes	
Ca^{2+}-ATPase	DOC⁻	0.02	56	6	monomeric	no	Le Maire et al. (1976)
		0.11	89	13	monomeric	no	
	$C_{12}E_8$		264.	120	monomer-dimer	yes	Dean & Tanford (1978)
Dopamine hydroxylase	Brij 58	$C_{12}E_8$			tetrameric	yes dimer-tetramer	Albanesi (1980) no
	DOC⁻					no	
	Cholate					no	
	Triton X-100					yes	

[a]Determination of maximum binding by Grefrath (1974) has subsequently shown to have been carried out with detergent containing 35% $C_{14}OSO_3^-$. The value in this table was determined with homogeneous $C_{12}OSO_3^-$.
[b]Activity refers to the ability to bleach reversibly.

217

33% oxyethylene, a composition far different from that of the dilute aqueous solutions found *in vivo*.

2. The Hydrophobic Domains

The hydrophobic domain of some intrinsic membrane proteins binds single-tail amphiphiles in a manner reminiscent of mixed micelle formation. Unfortunately data addressed to this type of interaction are sparse, and interpretation is complicated by the general insolubility of intrinsic membrane proteins in aqueous solution in the absence of bound amphiphiles. Table 9 presents data on the interaction of amphiphiles with a number of intrinsic membrane proteins that have been totally delipidated. The first four proteins in the table are the only ones for which complete binding isotherms have been obtained. Glycophorin A (human red cell membrane), cytochrome b_5 (liver endoplasmic reticulum), and Thy-1 (antigenic cell surface protein from thymocytes) are soluble in the absence of amphiphilic ligands, forming large aggregates that dissociate to monomers at saturation binding levels. This phenomenon complicates the analysis of binding isotherms, since the interaction is a function of both protein and ligand concentration (Section II). Nevertheless, it is apparent qualitatively that these proteins form complexes with the detergent investigated in a cooperative mode near the CMC of the pure ligand. The total detergent bound in the case of ionic ligands varies as a function of ionic strength in the same manner as does the aggregation number of the pure detergent.

Many other intrinsic membrane proteins have been solubilized, delipidated, and purified in detergents. Some of these proteins are listed in Table 9 together with the maximum binding, state of protein association, and the effect of the detergent on a measurable function. Important observations emerge from an inspection of these data.

1. Most solubilizations and delipidations have been carried out with detergents containing short hydrophobic tails (e.g., $C_{12}E_8$) or small hydrophobic regions (bile salts). This point is discussed further in Section V, but there is an increasing concern that amphiphilic ligands of these types may disrupt the oligomeric structure of some proteins.
2. Proteins whose function is primarily binding of a water-soluble ligand (Thy-1, acetylcholine receptor) appear to retain activity in delipidated form in a wide variety of detergents, perhaps reflecting the insensitivity of the hydrophilic domain to the nature of the ligand bound to the hydrophobic portion.

Contrasting the amphiphile binding properties of intrinsic membrane proteins with those discussed previously for serum lipoproteins, one observes that in the former case mixed micellar systems are formed such that the properties of the pure amphiphile appear to govern the properties of the association product. The apoproteins from HDL and LDL, on the other hand, package naturally occurring lipids and amphiphiles in structures directed by the protein itself.

V. MIXED DETERGENT-LIPID SYSTEMS

The addition (and removal) of detergent to a lipid-protein bilayer can be adequately described by the following set of equations:

$$
\text{Pr-PL (bilayer)} \underset{+D}{\rightleftharpoons} \text{Pr-PL-D (bilayer)} \underset{+D}{\rightleftharpoons} \text{Pr-PL-D (bilayer fragment)}
$$

$$
+ \text{ D} \qquad\qquad\qquad\qquad (6)
$$

$$
D_m + \text{PL-D (micelle)} + \text{Pr-D (micelle)} \underset{+D}{\rightleftharpoons} \text{Pr-PL-D (mixed micelle)}
$$

Where Pr denotes the membrane protein and PL the phospholipid. In the first step detergent (D) partitions into the bilayer. As the detergent concentration increases in the lipid milieu, defects are formed due to nonideal mixing, curvature at the surface increases, and fragmentation ensues. In the final step, detergent is in such a large excess that protein (Pr) and lipid (PL) are dispersed in individual detergent micelles.

In principle, this process is completely reversible, and attempts at reconstitution are based on the premise that a mixture of lipid-detergent and protein-detergent mixed micelles will reform protein-lipid bilayers on removal of the single-tail amphiphile. In practice, there is a very real kinetic problem involved in reversing the last step of Eq. 6. Since lipid and membrane proteins have very low aqueous solubilities, these species do not equilibrate rapidly between the two types of mixed micelle (i.e., PL-D and Pr-D). Removal of the detergent from this system can lead to a kinetically stable mixture of lipid bilayers and aggregated protein. It is clearly preferable to reconstitute from an intermediate state (Pr-PL-D, mixed micelle) in which lipid and protein are already together in the same particle.

The thermodynamic parameters governing the partition of detergent into lipid-protein bilayers are dependent on the type and concentration of each of the three components. We do not know the equilibrium constants for any of the above reaction steps in any system and hence cannot predict the concentrations of specific detergents required for solubilization (for a more complete discussion of this problem, see Reynolds, 1980a). Furthermore, in multicomponent systems such as these, the CMC of the detergent is not equal to that found in the pure amphiphile-water mixture (e.g., Tanford, 1980) but is always at lower concentration. In the absence of direct partition measurements, it is not possible to relate the CMC of the pure detergent to its potential for interaction with protein-lipid bilayers, since we do not know the effects of (1) nonideal mixing of hydrophobic moieties of differing size and shape, (2) the magnitude of the interaction (repulsive or attractive) between the different head groups, or (3) possible direct detergent–membrane protein interactions.

One general principle that appears to be emerging from many experimental studies is that detergents containing hydrophobic regions that are much smaller than those of the native lipids solubilize biological membranes at lower total concentrations than do those that are more closely matched to the hydrocarbon region of the bilayer. This probably is a result of the nonideal mixing described above, in that defects in the bilayer and increased surface curvature occur at lower concentrations of these amphiphiles (carbon chain shorter than 14) relative to those with longer hydrocarbon chains. Unfortunately, these detergents also disrupt oligomeric protein structures in a number of systems that have been investigated: rhodopsin (McCaslin & Tanford, 1981), (Na^+, K^+)-ATPase (Freytag & Reynolds, unpublished observation); Ca^{2+}-ATPase (le Maire et al., 1976; Dean & Tanford, 1978). Because the bile salts with small hydrophobic regions and the short n-alkyl chain detergents have relatively high CMCs many investigators have chosen them for "reconstitution" systems. The relative ease of removal obviously needs to be balanced against the possibility of irreversible disruption of native oligomeric structure of the membrane proteins.

VI. CONCLUSIONS

In the introduction to *Equilibrium in Solutions* George Scatchard (1976) points out that "the study of solutions may be divided into three parts: (1) the study of equilibrium; (2) the study of the approach to equilibrium; and (3) the study of transport problems". This chapter has attempted to emphasize the first of these principles—the study of equilibrium. This is not to say that valuable information is not obtained from kinetics; rather, it restates the old axiom that the testing of models requires a precise knowledge of the equilibrium states encountered along a proposed reaction pathway. The interactions between amphiphilic molecules and proteins or between two types of amphiphile are an important aspect of our studies of real biological systems, but reliable thermodynamic data are often difficult to obtain even when the need for them is recognized. There are many gaps in our understanding of these interactions and the forces that govern them, and one hopes that in the next few years more scientists will be stimulated to provide the equilibrium studies that are so sorely needed.

ACKNOWLEDGMENTS

I am particularly indebted to a number of colleagues who through discussions and their own scientific writing have clarified problems of common interest: Drs. Charles Tanford, Sergei Timasheff, Jacinto Steinhardt, and Joseph Foster. A special debt is owed to Dr. George D. Halsey for first directing me toward interactions between macromolecules and solvent components. This work was partially supported by grant HL 22570 from the National Institutes of Health.

APPENDIX: EXPERIMENTAL METHODS FOR MEASUREMENT OF INTERACTIONS

1. EQUILIBRIUM DIALYSIS

In equilibrium dialysis the protein is restricted to a compartment by the use of a semi-permeable membrane that allows free passage of all other solvent components but prevents the movement of the macromolecule. At equilibrium the unbound ligand is at constant chemical potential on both sides of the membrane. It is common practice to measure the free ligand concentration in the protein-free compartment and the total ligand concentration in the presence of protein. The difference in these two numbers then gives total moles of bound ligand. The protein concentration should be kept sufficiently low so that the concentration of solvent components expressed as grams/cm^3 or moles/cm^3 is approximately equal to grams (or moles)/gram H$_2$O (i.e., the excluded volume occupied by the protein is negligible). The ionic strength must also be sufficiently high so that the Donan effect is suppressed when the ligand and/or protein is charged at the pH being used in the experiment. Equilibrium dialysis techniques are applicable when all solvent components and the ligand of interest can freely pass the semi-permeable membrane. In most cases this restricts the investigator to free amphiphile concentrations below the CMC. However, if the protein moiety is large, it may be possible to choose a membrane having a pore size large enough to ensure that small micelles are also permeable.

2. GEL FILTRATION CHROMATOGRAPHY

Gel filtration chromatography is applicable to ligands below and above their CMCs. The protein is pre-equilibrated with an excess of ligand in an appropriate buffer and eluted from a gel filtration column with a solvent containing the ligand at a fixed concentration (free or unbound concentration). The excess ligand that elutes with the protein is determined, together with the protein concentration, providing a direct measurement of $\bar{\nu}$ at a fixed unbound ligand concentration. If one is working above the CMC, the protein-amphiphile complex must be separable from pure amphiphile micelles (or association products) on the support material used. If the amphiphile has a low CMC, serious adsorption problems can occur and pre-saturation of the column may be required. Good experimental practice requires a demonstration of close to 100% recovery of added protein and ligand from the column.

Sucrose density gradients are often used in the same manner as gel filtration chromatography to measure binding. In this case, the ligand is distributed at a fixed, free concentration in the gradient, and protein and ligand are added. At equilibrium, protein concentration and bound and free ligand concentrations are determined after appropriate collection procedures. Implicit in analysis of data obtained in this manner is that the presence of high concentrations of another solvent component, namely sucrose, does not perturb ligand-protein interactions.

3. ANALYTICAL ULTRACENTRIFUGATION

Analytical ultracentrifugation is also applicable above and below the CMC of the ligand if a rotor speed can be chosen such that the amphiphile itself does not sediment appreciably. The accuracy obtained is less than that from the previously described procedures, but the amounts of material required are small and a specific chemical analysis for the ligand is not required.

At equilibrium in the centrifuge one measures directly from a plot of ln protein concentration vs. the square of the distance from the center of rotation $M_p(1 - \phi'\rho)$, where M_p is the molecular weight of the protein and $(1 - \phi'\rho)$ is the buoyant density factor:

$$(1 - \phi'\rho) = (1 - \bar{v}_p\rho) + \sum_i \delta_i(1 - \bar{v}_i\rho)$$

where $\bar{v}_p$ is the partial specific volume of the protein, δ_i is the grams of component i per gram of protein, $\bar{v}_i$ is the partial specific volume of component i, ϕ' is the effective partial specific volume, and ρ is the density of the solvent. Thus, if all partial specific volumes and the molecular weight of the protein are known, δ_i is directly obtained in a system containing only one bound species. Note that bound water does not enter into the equation above since $\bar{v}_{H_2O} = 1/\rho$. If the system contains a high concentration of another solvent component (such as sucrose, ethylene glycol, CsCl, etc.), the buoyant density factor due to H_2O does not cancel and this term must be known to obtain δ_i.

The use of this method requires an analytical ultracentrifuge equipped with a photoelectric scanner so that the protein moiety can be measured alone. Rayleigh interference optics will simultaneously measure all species present, including unbound ligand that is not sedimenting with the protein.

4. INDIRECT METHODS

Spectrophotometric techniques that measure the perturbation of some optical property of the protein and/or ligand as a function of interaction are classified as indirect procedures and must be calibrated against a rigorous thermodynamic technique. It cannot be assumed that such perturbations are a linear function of $\bar{v}$ without prior demonstration.

REFERENCES

Albanesi, J. (1980). Structural and Kinetic Studies of Dopamine Hydroxylase Purified from Membranes of Bovine Adrenal Chromaffin Granules, Ph.D. thesis, Duke University, Durham, NC.

Andrews, A. L., Atkinson, D., Barratt, M. D., Finer, E. G., Hauser, H., Henry, R., Leslie, R. B., Owens, N. L., Phillips, M. C., & Robertson, R. N. (1976). Interaction of Apoprotein from Porcine HDL with Dimyristoyl Lecithin. *Eur. J. Biochem.* **64**, 549–563.

Assmann, G., & Brewer, H. D. (1974). Lipid Protein Interactions in HDL. *Proc. Natl. Acad. Sci. U.S.A.* **71**, 989–993.

Aveyard, R., & Haydon, D. A. (1973). *An Introduction to the Principles of Surface Chemistry,* Cambridge University Press, Cambridge, England.

Dean, W. L., & Tanford, C. (1978). Properties of a Delipidated, Detergent-Activated Ca^{2+}ATPase. *Biochemistry* **17**, 1683–1690.

Einstein, A. (1950). *Out of My Later Years.* Philosophical Library, New York.

Friedberg, S. J., & Reynolds, J. A. (1976). The Molar Ratio of the Two Major Polypeptide Components of Human High Density Lipoprotein. *J. Biol. Chem.* **251**, 4005–4009.

Grefrath, S. P. (1974). Biological Membranes: Compositional and Structural Analysis. Ph.D. thesis, Duke University, Durham, NC.

Helenius, A., & Simons, K. (1972). The Binding of Detergents to Lipophilic and Hydrophilic Proteins. *J. Biol. Chem.* **247**, 3656–3661.

Helenius, A., McCaslin, D. R., Fries, E., & Tanford, C. (1979). Properties of Detergents. *Methods Enzymol.* **56**, 734–749.

Henderson, P. J. F. (1978). Statistical Analysis of Enzyme Kinetic Data. In *Techniques in Protein and Enzyme Biochemistry*, Vol. BI/II, p. 113, Elsevier/North Holland, Amsterdam.

Klotz, I. M. (1953). In *The Proteins* (H. Neurath & K. Bailey, Eds.), Vol. I, Academic Press, New York.

Kuchel, P. W., Campbell, D. G., Barclay, A. N., & Williams, A. F. (1978). Molecular Weights of the Thy-1 Glycoproteins from Rat Thymus and Brain in the Presence and Absence of Deoxycholate. *Biochem. J.* **169**, 411–417.

Le Maire, M., Jorgensen, K. E., Roigaard-Petersen, H., & Moller, J. V. (1976). Properties of Deoxycholate Solubilized Sarcoplasmic Reticulum Ca^{2+} ATPase. *Biochemistry* **15**, 5805–5812.

Loomis, C. R., Janiak, M. J., Small, D. M., & Shipley, G. G. (1974). The Binary Phase Diagram of Lecithin and Cholesteryl Linolenate. *J. Mol. Biol.* **86**, 309–324.

Makino, S., Woolford, J. L., Tanford, C., & Webster, R. E. (1975). Interaction of Deoxycholate with the Coat Protein of Bacteriophage fl. *J. Biol. Chem.* **250**, 4327–4332.

McCaslin, D. R., & Tanford, C. (1981). Different States of Aggregation of Unbleached and Bleached Rhodopsin in Two Different Detergents. *Biochemistry* **20**, 5212–5221.

Mimms, L. T. (1980). Physical Properties of Native and Deglycosylated Glycophorin A: Analysis of Glycophorin A Reconstituted Vesicles. Ph.D. thesis, Duke University, Durham, NC.

Mingins, J., Taylor, J. A. G., Owens, N. F., & Brooks, J. H. (1975). Surface Equation of State for Very Dilute Charged Monolayers at Aqueous Interfaces. *Adv. Chem.* **144**, 28–43.

Nichols, A. V., Forte, T., Gong, E., Blanche, P., & Verdery, R. B. (1974). Effect of Lysophosphatidylcholine on Interaction between Phosphatidylcholine and Activator Protein of Lecithin : Cholesterol Acyltransferase. *Scand. J. Clin. Lab. Invest.* **33**, 147–156.

Nielsen, T. B., & Reynolds, J. A. (1978). Measurements of Molecular Weight by Gel Electrophoresis. *Methods Enzymol.* **48**, 3–10.

Nozaki, Y., Reynolds, J. A., & Tanford, C. (1974). The Interaction of a Cationic Detergent with Bovine Serum Albumin and Other Proteins. *J. Biol. Chem.* **249**, 4452–4459.

Osborne, J. G., & Brewer, H. B. (1977). The Plasma Lipoproteins. Adv. Prot. Chem. **31**, 253–338.

Pownall, H. J., Massey, J. B., Kusserow, S. K., & Gotto, A. M. (1978). Kinetics of Lipid-Protein Interactions: Interactions of Apolipoprotein AI from HDL with Phosphatidylcholines. *Biochemistry* **17**, 1183–1188.

Reynolds, J. A. (1980a). Solubilization and Characterization of Membrane Proteins. In *Techniques for Membrane Receptor Characterization and Purification* (S. Jacobs & P. Cuatrecasas, Eds.), Chapman & Hall, Ltd., London.

Reynolds, J. A. (1980b). Binding Studies with Apolipoproteins. *Ann. N.Y. Acad. Sci.* **348**, 174–186.

Reynolds, J. A., and Karlin, A. (1978). Molecular Weight in Detergent Solution of Acetylcholine Receptor from *Torpedo californica*. *Biochemistry* **17**, 2035–2038.

Reynolds, J. A., and Stoeckenius, W. (1977). Molecular Weight of Bacteriorhodopsin Solubilized in Triton X-100. *Proc. Natl. Acad. Sci. U.S.A.* **74**, 2803–2804.

Reynolds, J. A., Stone, W. L., and Tanford, C. (1977). Interaction of L-Didecanoyl Phosphatidylcholine with the AI Polypeptide of HDL. *Proc. Natl. Acad. Sci. U.S.A.* **74**, 3796–3799.

Robinson, N. C., & Tanford, C. (1975). The Binding of Deoxycholate, Triton X-100, Sodium Dodecylsulfate, and Phosphatidylcholine Vesicles to Cytochrome b_5. *Biochemistry* **14**, 369–377.

Scanu, A. M. (1971). Plasma Lipoproteins. In Biochemical Society Symposium No. 33 (R.M.S. Smellie, Ed.), pp. 29–45. Academic Press, New York.

Scatchard, G. (1949). *Ann. N.Y. Acad. Sci.* **51**, 660–672.

Scatchard, G. (1976). *Equilibrium in Solutions.* Harvard University Press, Cambridge, MA.

Segrest, J. P., Jackson, R. L., Morrisett, J. D., & Gotto, A. M. (1974). A Molecular Theory of Lipid Protein Interactions in the Plasma Lipoproteins. *FEBS Lett.* **38**, 247–253.

Small, D. M. (1971). The Physical Chemistry of Cholanic Acids. In *The Bile Acids* (P. P. Nair & D. Kritchevsky, Eds.) Vol. 1, pp. 249–356, Plenum Press, New York.

Smith, R., Dawson, J. R., & Tanford, C. (1972). The Size and Number of Polypeptide Chains in Human Serum Low Density Lipoprotein. *J. Biol. Chem.* **247**, 3376–3381.

Steele, J. C. H., & Reynolds, J. A. (1979). Characterization of Apolipoprotein B Polypeptide of Human Plasma LDL in Detergent and Denaturant Solutions. *J. Biol. Chem.* **254**, 1633–1638.

Steele, J. C. H., Tanford, C., & Reynolds, J. A. (1978). Determination of Partial Specific Volumes for Lipid-Associated Proteins. *Methods Enzymol.* **48**, 11–23.

Steinhardt, J., & Reynolds, J. A. (1969). *Multiple Equilibria in Proteins,* pp. 10–32, Academic Press, New York.

Stone, W. L., & Reynolds, J. A. (1975). Hydrophobic Interactions of the apo-Gln-I Polypeptide Component of Human HDL. *J. Biol. Chem.* **250**, 3584–3587.

Swaney, J. B. (1980). Properties of Lipid-Apolipoprotein Association Products. *J. Biol. Chem.* **255**, 877–881.

Tanford, C. (1961). *Physical Chemistry of Macromolecules,* pp. 526–585, Wiley, New York.

Tanford, C. (1980). *The Hydrophobic Effect,* 2nd ed., Wiley, New York.

Tanford, C., & Reynolds, J. A. (1979). Structure and Assembly of Human Serum Lipoproteins. In *The Chemistry and Physiology of the Human Plasma Proteins* (D. H. Bind, Ed.), pp. 111–126, Pergamon Press, New York.

Watt, R. M., & Reynolds, J. A. (1980). Solubilization and Characterization of Apolipoprotein B from LDL in *n*-Dodecyl Octaethyleneglycol Monoether. *Biochemistry* **19**, 1593–1598.

Watt, R. M., & Reynolds, J. A. (1981). Interaction of Egg Yolk Lecithin with Apolipoprotein B from LDL. *Biochemistry* **20**, 3897–3901.

Weber, K., & Osborn, M. (1969). The Reliability of Molecular Weight Determinations by Dodecyl-sulfate-Polyacrylamide Gel Electrophoresis. *J. Biol. Chem.* **244**, 4406–4412.

Wyman, J. (1964). Linked Functions and Reciprocal Effects in Hemoglobin: A Second Look. *Adv. Protein Chem.* **19**, 224–286.

Zampighi, G., Reynolds, J. A., & Watt, R. M. (1980). Characterization of Apolipoprotein B from LDL. An Electron Microscopic Study, *J. Cell Biol.* **87**, 555–561.

Six

Equilibrium Constants and Number of Binding Sites for Lipid-Protein Interactions in Membranes

O. HAYES GRIFFITH, JAAKKO R. BROTHERUS, AND PATRICIA C. JOST

Institute of Molecular Biology and Department of Chemistry, University of Oregon, Eugene, Oregon U.S.A. (O.H.G. & P.C.J.)

Department of Medical Chemistry, University of Helsinki, Helsinki, Finland (J.R.B.)

I. INTRODUCTION

Integral membrane proteins embedded in the bilayer are in contact with lipids, which in turn are in equilibrium with the two-dimensional fluid phospholipid bilayer. To characterize the distribution of specific lipids and other hydrophobic molecules between bulk lipid and the hydrophobic protein surface, it is useful to describe the system by means of a multiple binding equilibria model. To do this, it is necessary to recast the classical equilibrium treatment into a form appropriate for the equilibria occurring between lipids and proteins within the membrane continuum. When treating water-soluble proteins or detergent binding, water is the reference solvent and quantities are expressed as moles of solute per liter of aqueous solution (see Chapter 5). In the intact bilayer the lipid is the reference solvent and aqueous concentrations do not enter into the equations explicitly. The effective concentration of membrane proteins in the lipid phase is orders of magnitude higher than the analogous case of concentrations of water-soluble proteins in the aqueous phase. In membranes, protein-associated lipid represents a significant fraction of the total solvent lipid. In addition, different quantities are measurable for membranes (see Chapters 2 and 5). For these two reasons the appropriate binding equations differ from the classical multiple equilibria binding equations even though the underlying thermodynamics is the same. This chapter outlines a treatment for lipid-protein interactions* that can provide the number of lipid binding sites, binding constants relative to the solvent lipid, and derived thermodynamic properties in detergent-free membranes. Computer-generated binding curves and experimental data are included to illustrate specific cases.

II. MULTIPLE EQUILIBRIA BINDING EQUATIONS

The starting point is an exchange reaction between a solute lipid (L^*) and a solvent lipid (L) occupying a binding or contact site on the hydrophobic surface of a membrane protein (P). The reaction for the hypothetical case where the protein has only one site is

$$L^* + PL \rightleftharpoons L + PL^*$$

and the relative binding constant is defined by the mass action equation $K = [L] [PL^*] / [L^*] [PL]$, where the concentrations are expressed as mole fractions of lipids in the membrane. For the more realistic protein with N binding sites there is a set of equilibrium equations of the form

$$L^* + PL_{N-i}L_i^* \rightleftharpoons L + PL_{N-i-1}L_{i+1}^*$$

However, we assume that the N sites are independent so that it is not necessary to consider combinations and permutations of sites occupied by L and L^*. Instead, the equilibrium equation for each site is stated separately

$$L^* + BL \rightleftharpoons L + BL^* \tag{1}$$

where BL and BL^* refer to binding sites occupied by L and L^*, respectively. The equilibrium constant K is initially expressed in mole fraction units but, since there are an equal number of exchanging species, K reduces to

$$K = \frac{L_f L_b^*}{L_f^* L_b} \tag{2}$$

where L_f and L_f^* are moles of free solvent lipid and solute lipid, respectively, in the fluid bilayer. The corresponding quantities L_b and L_b^* are moles of solvent lipid and solute lipid, respectively, bound to the protein sites, B.

*The treatment presented here is adapted from Griffith and Jost (1979) and Brotherus et al. (1981).

We consider first the case of N identical sites. There are three mass balance equations:

$$\text{binding sites: } L_b + L_b^* = N \cdot P \tag{3}$$

$$\text{solute lipid: } L_f^* + L_b^* = L_t^* \tag{4}$$

$$\text{solvent lipid: } L_f + L_b = L_t \tag{5}$$

where P is the total protein, L_t^* the total solute lipid, and L_t the total solvent lipid, all in moles. Equation 3 states that there are no unoccupied sites, all sites being occupied by solvent or solute lipid. Equations 4 and 5 describe the conservation of solute and solvent lipids, respectively. The experimentally measurable quantities are $x^* = L_t^*/P$, the total solute lipid per protein; $x = L_t/P$, the total solvent lipid per protein; and $y = L_f^*/L_b^*$, the bilayer to bound ratio of the solute lipid. Lipid phosphorus and protein analyses, combined with the known amount of L_t^* added, determine x and x^*. The variable y is more troublesome, but it can be estimated, for example, from electron spin resonance (ESR) spectral components when lipid spin labels are used (see Chapter 2).

From Eq. 2 and the definition of y, $Ky = L_f/L_b$. From the mass balance equations (Eqs. 3–5), $L_f = L_t - N \cdot P - L_b^*$, $L_b = N \cdot P - L_b^*$ and $L_b^* = L_t^*(1 - y)^{-1}$. Combining these equations to eliminate L_f, L_b, and L_b^* yields

$$\frac{x^*}{1 + y} + \frac{x}{1 + Ky} = N \tag{6}$$

Equation 6 is a general expression for lipid-protein associations applicable for any solute lipid concentration and one class of binding sites. If the amount of solute lipid present is small (i.e., $x^* \ll x$), the first term in Eq. 6 can be discarded, yielding Approximation I:

$$\text{Approximation I:} \quad y = \frac{x}{NK} - \frac{1}{K} \tag{7}$$

According to Eq. 7 a plot of the observed ratio of bilayer to bound solute lipid (y) versus the lipid to protein ratio (x) will yield a straight line with y-intercept of $-1/K$, x-intercept of N, and slope of $1/NK$.

A second linear approximation to Eq. 6 applicable at somewhat higher levels of x^* is:

$$\text{Approximation II:} \quad y = \frac{x + x^*}{NK} - \frac{1}{K} \tag{8}$$

Plots of the exact solution and the two approximations are compared in Fig. 1.

A more general case involves n_1 sites of binding constant K_1, n_2 sites of binding constant K_2, etc. For m classes of binding sites Eqs. 2 and 3 become a set of m

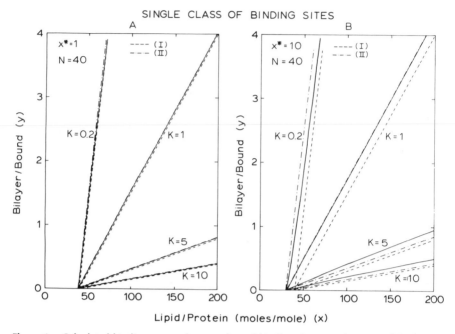

Figure 1. Calculated binding curves for one class of binding sites as a function of the solvent lipid to protein ratio (x). Solid lines are plots of the exact Eq. 6, and I and II refer to the approximate solutions (Eqs. 7 and 8 respectively). For this example the total number of binding sites (N) is 40 and the solute to protein ratios (x*) are 1 (A) and 10 (B). The binding constant (K) is varied between 0.2 and 10.

equations, and together with Eqs. 4 and 5, there are a set of $2m + 2$ independent equations from which $2m + 2$ variables ($L_{b_i}^*$, L_{b_i}, L_f^*, L_f) can be solved in terms of the parameters of the system L_t^*, L_t, P, N, n_i, K_i), where $L_{b_i}^*$ and L_{b_i} are the moles of solute lipid and solvent lipid, respectively, bound to the protein sites B_i; n_i is the number of binding sites of class i per protein and K_i is the relative binding constant of a site in class i. N is still the total number of the binding sites per protein. Solving for $\bar{v} = (L_t^* - L_f^*)/P$ in terms of L_f^* and L_f gives

$$\bar{v} = \sum_{i=1}^{m} \frac{n_i K_i (L_f^*/L_f)}{1 + K_i(L_f^*/L_f)} \tag{9}$$

The quantity $\bar{v}$ is the ratio of bound solute to protein, also referred to as r.

In applications such as spin labeling, it is advantageous to write Eq. 9 in terms of the variables y, x, and x^*. In order to do this, we note that the total amount of bilayer in the system is $L_t + L_t^* - N \cdot P$ so that the free solvent lipid in the bilayer L_f is $(L_t + L_t^* - N \cdot P) - L_f^*$, where $N \cdot P$ is the total number of binding (or contact) sites on the hydrophobic surface of the proteins. Substituting into Eq. 9 and using the definitions given above for y, x, and x^* yields the equation

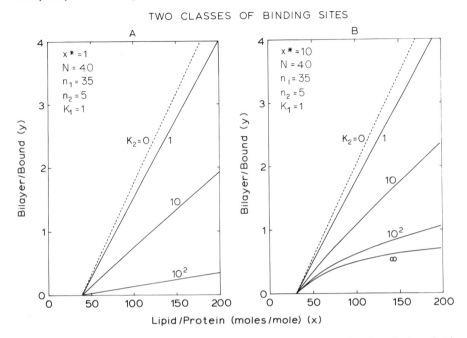

Figure 2. Calculated binding curves for two classes of binding sites as a function of solvent lipid to protein ratio (x), with $n_1 = 35$ non-selective sites $(K_1 = 1)$ and $n_2 = 5$ selective sites. The total number of sites (N) is 40, and the solute to protein ratios (x) are 1 (A) and 10 (B). The dashed line $(K_2 = 0)$ denotes the excluded site case, where the selective sites discriminate *against* the solute lipid.

$$\sum_{i=1}^{m} \frac{n_i K_i y}{(1 + K_i y)(1 + y)^{-1} x^* + x - N} = 1 \tag{10}$$

Other possible choices for the dependent variable, besides $\bar{\nu}$ and y, include the ratio of free solute to total solute (L_f^*/L_t^*), and the ratio of bound solute to total solute $(L_t^* - L_f^*)/L_t^*$. Figures 2 and 3 show plots of Eq. 10 for the special case of a protein that has two classes of binding sites: one class of nonselective (contact) sites with $K_1 = 1$ and another class of selective sites with different binding affinities. The presence of high affinity binding sites causes a characteristic curvature at higher solute levels $(x^* \geq n_2)$, which provides a way of detecting the presence of high affinity sites.

For many labeling experiments the concentration of solute is so low that the terms containing x^* may be neglected and Eq 10 becomes

$$\sum_{i=1}^{m} \frac{n_i K_i y}{x - N} = \frac{N K_{av} y}{x - N} = 1 \tag{11}$$

where K_{av} is the weighted average of the binding constants of the different classes of sites:

TWO CLASSES OF BINDING SITES

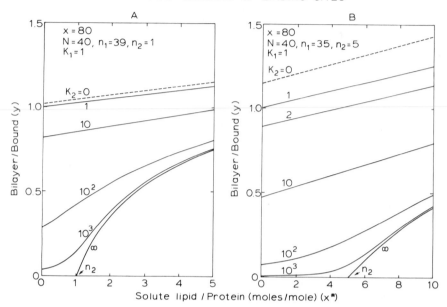

Solute lipid / Protein (moles/mole) (x^*)

Figure 3. Calculated binding curves for the two-class model as a function of the solute lipid to protein ratio x^*, with 1 specific site (A) and 5 specific sites (B); the total number of sites (N) in both plots is 40. The solvent lipid (x) is 80 molecules per protein. As in Fig. 2, $K_1 = 1$ for the non-specific sites. The increase of y with increasing x^*, even for the case of $K_1 = 1$, is caused by the extension of the free bilayer by the added solute. [For plots of the case in which the total lipid $(x + x^*)$ is held constant see Fig. 4 of Brotherus et al. (1981).]

$$K_{av} = \frac{1}{N} \sum_{i=1}^{m} n_i K_i \tag{12}$$

Equation 11 is identical to Approximation I, with the exception that the single-site binding constant K is replaced by the average binding constant K_{av}. This is an interesting result, since it shows that, without knowing the precise solute to protein ratio x^* (other than that it is small) or the number of classes of binding sites, a plot of y versus x gives the average binding constant and the total number of binding sites. Similarly, Approximation II is applicable by replacing K by K_{av}.

The most familiar use of binding treatments in biochemistry is for dilute aqueous systems (see Chapter 5). The double reciprocal plot, $1/\bar{\nu} = 1/K_a n[L] + 1/n$, and the Scatchard equation, $\bar{\nu}/[L] = nK_a - \bar{\nu}K_a$ are used to determine the number of binding sites n and the association constant K_a from the experimental values of $\bar{\nu}$, the bound ligand to protein ratio, and [L], the concentration of free ligand (or $[L_f^*]$ in the present notation). These equations follow from Eq. 9. The relationship between this familiar treatment and the membrane case is evident if L_t (the total solvent lipid) is greatly increased so that the concentrations of protein and lipid solute become very dilute in the lipid solution. Then L_f^*/L_f (free ligand lipid to free solvent lipid ratio) in Eq. 9 approaches the concentration $[L_f^*]$ expressed as moles of free solute (ligand) lipid per liter of lipid in the bilayer times a constant [for water as the solvent the constant is $(55 \text{ M})^{-1}$ and is conventionally incorporated into the binding constant].

In the membrane case, the double reciprocal plot is $1/\bar{\nu}$ versus $(L_f^*/L_f)^{-1}$ and the Scatchard-like plot is $\bar{\nu}(L_f^*/L_f)^{-1}$ versus $\bar{\nu}$. Here the free solvent lipid L_f is a variable because of the high concentration of protein and lipid ligand (solute) and cannot be incorporated into the binding constant. Figure 4 shows the behavior of the double reciprocal plots and Scatchard plots over the relevant range of the variables for membranes. These plots could be used to differentiate between a single class of sites and multiple classes of sites and to determine the equilibrium constants, providing that reliable methods of determining L_f^*/L_f are developed. This can be accomplished, in principle, by using pairs of samples in which the solvent lipid is labeled in one sample and the solute lipid labeled in the other. Another approach would be to measure the free components of the solvent and solute lipids simultaneously, but this has proved to be an illusive goal. At present, the most accessible experimental quantities are x, x^*, and y, and therefore it is not a simple matter to use the plots in Fig. 4, since L_f^*/L_f is also a function of N, the total number of binding sites. This is easily seen from the formulas relating $\bar{\nu}$ and L_f^*/L_f to the quantities measured in spin labeling: $\bar{\nu} = x^* (y + 1)^{-1}$ and $L_f^*/L_f = x^*y [x^* + (y + 1) (x - N)]^{-1}$. The plots in Figs. 1–3 give a direct estimate of N and are more useful in current lipid-protein binding studies.

III. EXPERIMENTAL APPLICATIONS

Experimental examples of the three possible cases where the average relative binding affinity of the solute lipid is equal to, greater than, or smaller than one are shown in Figs. 5–7. In all three cases the solvent lipid is one of several unsaturated phosphatidylcholines at 25°C (well above their transition temperatures), the samples are suspended in aqueous buffer, and the range of lipid to protein values (x) covers the range estimated for native membranes (Jost & Griffith, 1980).

The theory predicts that at low solute lipid levels (Approximations I and II) the plot of the spectroscopic parameter (y) versus the compositional parameter (x) will be linear. Figure 5 shows equilibria binding data for the mitochondrial protein cytochrome oxidase, and the plot is linear within experimental error. It is remarkable that the curves remain linear for such concentrated solutions ($x = 100$ corresponds to about 10^{-2} M solute lipid in the lipid bilayer solution). The ratio of the activity coefficients evidently remains close to unity because of the structural similarity of the solute and solvent. The linearity also argues that non-specific protein aggregation is not an important factor. Linear plots are also observed in Figs. 6 and 7, and this confirms that both the theory and the spectroscopic parameters are a good starting point for quantitating lipid-protein binding equilibria in membranes.

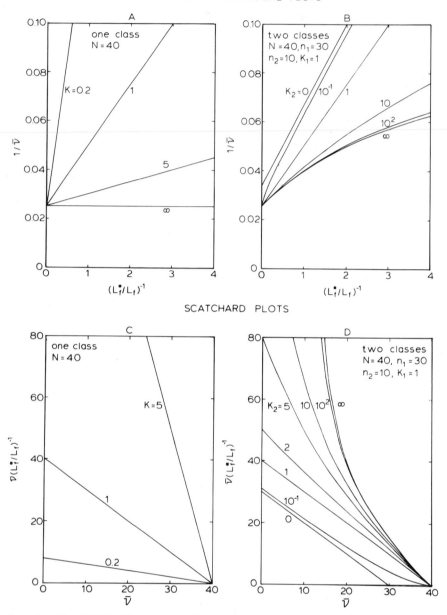

Figure 4. (A) and (B) Double reciprocal plots and (C) and (D) Scatchard-like plots comparing the model for one class of sites (left) and two classes of sites (right), based on Eq. 9. The plots are for the membrane case where lipids, rather than water, form the solvent for the equilibrium binding. As in the case of the water-soluble proteins, where water is treated explicitly, the one class of sites case gives linear plots and the two classes of sites case introduces nonlinearities. The total number of binding sites (N) is 40. For two classes of sites, the number of non-selective sites (n_1) is 30, with $K_1 = 1$, and the selective binding constant (K_2) is varied for the 10 selective sites (n_2).

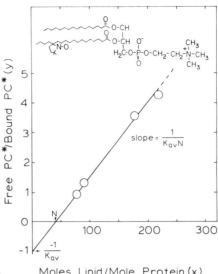

Figure 5. Experimental binding data at 25°C for beef heart cytochrome oxidase where both the solvent and the solute are phosphatidylcholines. The solvent is 1,2-dioleoyl-*sn*-3-phosphatidylcholine and the solute is the phospholipid spin label PC* (14-proxylphosphatidylcholine) shown at the top of the figure. In these samples x* ≪ x, and the solvent lipid is well above the gel→liquid crystalline phase transition temperature. From Brotherus et al. (1981).

Most labeling techniques will introduce at least some perturbation that could shift the measured equilibria. Any such shift can be determined by using labeled and unlabeled lipid with the same polar head group and looking for deviations from $K_{av} = 1$. In the case of spin labeling, for example, the presence of the nitroxide moiety could reduce the binding for steric reasons, or any hydrogen bonding between the nitroxide and the protein surface could increase the binding. Steric effects could be large, especially in the case of some of the aromatic fluorescent lipid probes, but in any case, once the shift in binding has been measured, such labels can still be used after appropriate corrections. In the case of spin-labeled lipids with the nitroxide group near the distal end of the acyl chain, any perturbation of the equilibrium is negligible, as shown in Fig. 5, where $K_{av} \approx 1$. Technically, in this system there is a small correction for approximately one to two molecules of tightly bound diphosphatidylglycerol that were not replaced by the solvent in the reconstitution procedure. This corresponds to two to four phosphatidylcholine sites, and thus from Eq. 12, the maximum effect is $N \cdot K_{av} = 40(1.0) = n_1K_1 + n_2K_2 = (36)K_1 + 4(0)$, so that $K_1 = 40/36 = 1.1$, which cannot be distinguished from 1.0 within experimental error.

The total number of lipid binding sites N given by the x-axis intercept in Fig. 5 is $N \sim 40$. This includes all contact sites on the irregular hydrophobic protein surface that restrict the motion of the spin label; and does not imply either specific or long-lived interactions, only that the exchange between the bulk bilayer and the protein surface is slow on the ESR time scale—that is, two spectral components are resolvable (Jost et al., 1977). N also includes any specific sites. The data of Fig. 5 are consistent with data on the yeast cytochrome oxidase reconstituted in a saturated lipid, dimyristoylphosphatidylcholine at 32°C (see Fig. 31 Chapter 2) and the early data using fatty acid spin labels (Jost et al., 1973; Griffith & Jost, 1979).

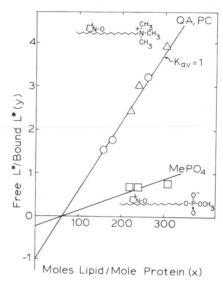

Figure 6. The effect of the solute lipid head group on binding to the shark (Na⁺,K⁺)-ATPase. The squares indicate the alkylmethylphosphate lipid label (MePO₄) and the triangles the analogous alkyl quaternary amine (QA). The solvent lipid in both cases is 1-palmitoyl-2-oleoyl-*sn*-3-phosphatidylcholine (POPC).

Experimental procedure: the *Squalus acanthias* enzyme, purified by the method of Dixon and Hokin (1978), was delipidated by gel filtration on Sepharose 6B (Ottolenghi, 1975) and reconstituted at various lipid to protein ratios with POPC in the buffered detergent used for delipidation. Each reconstitution was divided into two aliquots before labeling, so that the lipid to protein ratio was identical for each pair of QA- and MePO₄-labeled samples. In these experiments the labeling level was $x^* = 0.5$ per $\alpha_2\beta_2$ dimer, assuming a molecular weight of 314,000, so that $x^* \ll x$, and the data were recorded from samples at 25°C. Labeling and data analysis are the same as described in Brotherus et al. (1980). One reconstitution at a low lipid level ($x \sim 90$) gave anomalously high values of y (QA, $y \sim 1.9$; MePO₄, $y \sim 0.5$). At low lipid levels sample heterogeneity is often encountered (see pp. 71–72), and for clarity the data for this sample have been omitted. For comparison the PC points (circles), replotted from Brotherus et al. (1981), are for the same phospholipid spin label lipid as shown in Fig. 5 incorporated into the same (Na⁺,K⁺)-ATPase, which was lipid-substituted with 1,2-dioleoylphosphatidylcholine (DOPC) as the solvent lipid. These data are for the samples at 25°C. The solid lines are theoretical lines for $N = 60$, $K_{av} = 1$ (QA, PC), and $K_{av} = 5$ (MePO₄)

The data in Fig. 6 illustrate an example of K_{av} greater than one. In this experiment two single-tail spin labels differing only in the polar head groups were used. The structures are shown as inserts in Fig. 6. The phosphate label carries a net negative charge over a wide pH range and the quaternary amine label is positively charged. The hydrocarbon chain is identical in both lipids, factoring out any possible effect of the spin label moiety in the ratio of binding constants. Small amounts of the spin labels (0.5 label per protein) were diffused into aliquots of sodium- and potassium-activated adenosinetriphosphatase (Na⁺,K⁺-ATPase) reconstituted with phosphatidylcholine. From an earlier study the total number of binding sites was estimated to be $N \approx 60$ (Brotherus et al., 1980), and data for the phosphatidylcholine spin label (circles) are from Brotherus et al. (1981).

Despite the point scatter of the preliminary data in Fig. 6, it is clear that the binding affinities of the two labels are significantly different. The two lines are drawn using $N = 60$, $K_{av} = 1$ and $K_{av} = 5$. The correspondence of these lines with the experimental points at higher lipid to protein ratios shows that the negatively charged phosphate spin label has an average binding constant (K_{av}) approximately five times larger than the analogous positively charged amine. This corresponds to an average free energy difference of about -1 kcal/mol in the binding affinities. However, these averages are over all 60 sites, and the values would be much larger if there were only a few specific sites involved. In the limit of only one specific site for the negatively charged lipid among 59 non-specific sites, Eq. 12 gives $N \cdot K_{av} = 60(5) = 59(1) + 1(K_2)$. This gives $K_2 \approx 240$, corresponding to a ΔG of about -3 kcal/mol.

The observation that the quaternary amine spin label has approximately the same equilibrium distribution as the solvent lipid ($K_{av} \approx 1$) suggests that the latter interpretation is correct and that there is only a small number of specific sites. Otherwise, as in Fig. 1, the slope of the quaternary amine line would be increased above the $K = 1$ line approximately to the same extent as the slope of the phosphate falls below (assuming that sites attracting negative charges tend to repel positively charged lipids). This is not the case. For the case of one specific site for the phosphate from which the quaternary amine would be excluded, Eq. 12 gives $N \cdot K_{av} = 60(K_{av}) = 59(1) + 1(0)$, so that $K_{av} = 59/60 = 0.98$, which is consistent with the experimentally observed line ($K_{av} \approx 1$) for the quaternary amine label. Even five totally excluded sites would have little effect on the slope (see the $K_2 = 0$ line in Fig. 2). To separate out the various classes of binding sites contributing to K_{av} will require additional experiments varying the labeling level (e.g.,

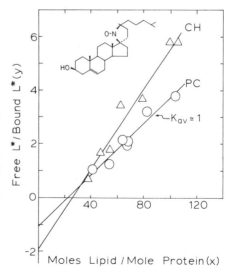

Figure 7. Experimental solute lipid binding curves, comparing the interaction of cholesterol and phosphatidylcholine labels with rabbit sarcoplasmic reticulum Ca^{2+}-ATPase reconstituted in egg yolk phosphatidylcholine (solvent lipid). The cholesterol (CH) label is shown, and the PC label is the same as that of Fig. 5. The labeling levels are $x^* = 0.5$/ATPase, assuming a molecular weight of 115,000. The solid lines are independent least-squares fits of $N = 23$, $K_{av} = 0.91$ (for PC), and $N = 25$, $K_{av} = 0.54$ (for CH). From Silvius et al. (1982).

Figs. 2 and 3). For other interesting cases in which selectivity in lipid binding is reported see Fig. 16 in Chapter 2, Cable and Powell (1980) and Knowles et al. (1981).

An average binding constant of less than one is shown in Fig. 7. In this example, taken from Silvius et al. (1982), sarcoplasmic reticulum Ca^{2+}-ATPase was reconstituted with egg yolk phosphatidylcholine as the solvent lipid. A reference line is provided by the phosphatidylcholine spin label as the solute (circles), where the average relative binding constant is very nearly one. In contrast, when the solute is the cholesterol spin label (inset in Fig. 7), the line is rotated, indicating a lower average binding constant, K_{av} on the order of 0.6. Either the cholesterol analogue is excluded from many sites on the protein or the average binding at all sites is reduced relative to the phospholipid.

It is instructive to compare the lipid-protein binding in membranes with binding of substrates to enzymes in aqueous solutions. We choose as the basis for comparison a 90% saturation of the specific binding sites. For an aqueous substrate concentration of 10^{-4} M, a binding constant of approximately 10^5 M^{-1} (5.5×10^6 in mole fraction units) is required. For a membrane protein present in a bilayer composed of 10 mol % of a specific lipid (L^*) it takes a binding constant of only approximately 100 in mole fraction units to achieve the 90% occupancy. Therefore, as a consequence of the concentrated nature of the system, relatively small binding constants may play a significant role in the composition of lipids in contact with membrane proteins.

In summary, lipid-protein interactions in membranes can be formulated as an exchange reaction in which lipid at the protein surface exchanges with lipid in the bilayer. The equations presented here, combined with experimental measurements of the lipid at the lipid-protein interface, are applicable to the intact membrane. At low solute concentrations, the limiting behavior of the system is described by linear approximations (I and II) that yield the total number of lipid binding or contact sites, N, and the average binding constant, K_{av}, relative to the solvent lipid. Since the general equations (Eqs. 9 and 10) apply at all solute concentrations, it is possible to analyze experiments involving saturation of sites to determine the number of selective binding sites. At low solute levels the experimental data show linear plots as predicted by theory and demonstrate that the protein can influence its immediate lipid composition.

Note added in proof: Recently fluorescence quenching by spin-labeled phospholipids has been used as a method of determining relative binding constants for lipid-protein associations. The effect of acyl chain characteristics (chain length, double bonds, and degree of unsaturation) on the binding of phosphatidylcholines to the sarcoplasmic reticulum Ca^{2+}-ATPase have been examined. Acyl chain properties had no effect on the relative apparent binding affinities of phosphatidylcholines for the enzyme [E. London and G. W. Feigenson (1981) *Biochemistry* 20, 1939–1948; M. Caffrey and G. W. Feigenson (1981) *Biochemistry* 20, 1949–1961]. The fluorescence quenching method is used at higher lipid to protein ratios and over a wider range of spin label concentrations. It does not provide the number of binding sites nor is it sensitive to a small number (10%) of binding sites with a higher affinity. Thus the two approaches are complementary, and the results obtained thus far are consistent.

ACKNOWLEDGMENTS

We thank J. Craig Baumeister for analysis of the (Na^+,K^+)-ATPase data, William Stewart for technical assistance, and Dr. Lowell E. Hokin and Mary Lochner for the purified (Na^+,K^+)-ATPase protein. We thank Drs. John F. W. Keana and John R. Silvius for useful discussions. This work was supported by grant GM 25698 from the U.S. Public Health Service.

REFERENCES

Brotherus, J. R., Jost, P. C., Griffith, O. H., Keana, J. F. W., & Hokin, L. E. (1980). Charge Selectivity at the Lipid-Protein Interface of Membranous Na,K-ATPase, *Proc. Natl. Acad. Sci. U.S.A.* **77**, 272–276.

Brotherus, J. R., Griffith, O. H., Brotherus, M. O., Jost, P. C., Silvius, J. R., & Hokin, L. E. (1981). Lipid-Protein Multiple Equilibria Binding in Membranes. *Biochemistry* **20**, 5261–5267.

Cable, M. B., & Powell, G. L. (1980). Spin-Labelled Cardiolipin: Preferential Segregation in the Boundary Layer of Cytochrome *c* Oxidase. *Biochemistry* **19**, 5679–5686.

Dixon, J. F., & Hokin, L. E. (1978). A Simple Procedure for the Preparation of Highly Purified (Sodium + Potassium)-Adenosinetriphosphatase from the Rectal Salt Gland of *Squalus acanthias* and the Electric Organ of *Electrophorus electricus*. *Anal. Biochem.* **86**, 378–385.

Griffith, O. H., & Jost, P. C. (1979). The Lipid-Protein Interface in Cytochrome Oxidase. In *Cytochrome Oxidase* (B. Chance, T. E. King, K. Okunuki, & Y. Orii, Eds.), pp. 207–218, Elsevier/North Holland Biomedical Press, Amsterdam.

Jost, P. C., & Griffith, O. H. (1978). The Spin Labeling Technique. In *Methods of Enzymology*, Vol. 49, *Enzyme Structure*, Part G (C. H. W. Hirs & S. N. Timasheff, Eds.), pp. 369–418, Academic Press, New York.

Jost, P. C., & Griffith, O. H. (1980). The Lipid-Protein Interface in Biological Membranes. *Ann. N.Y. Acad. Sci.* **348**, 391–407.

Jost, P. C., Griffith, O. H., Capaldi, R. A., & Vanderkooi, G. (1973). Evidence for Boundary Lipid in Membranes. *Proc. Natl. Acad. Sci. U.S.A.* **70**, 480–484.

Jost, P. C., Nadakavukaren, K. K., & Griffith, O. H. (1977). Phosphatidylcholine Exchange Between the Boundary Lipid and Bilayer Domains in Cytochrome Oxidase-Containing Membranes. *Biochemistry* **16**, 3110–3114.

Knowles, P. F., Watts, A., & Marsh, D. (1981). Spin-Label Studies of Head-Group Specificity in the Interaction of Phospholipids with Yeast Cytochrome Oxidase. *Biochemistry* (in press).

Ottolenghi, P. (1975). The Reversible Delipidation of a Solubilized Sodium plus Potassium Ion-Dependent Adenosine Triphosphatase from the Salt Gland of the Spiny Dogfish. *Biochem. J.* **151**, 61–66.

Silvius, J. R., Jost, P. C., & Griffith, O. H. (1982). Differential Association of Phosphatidylcholine and Cholesterol with the Sarcoplasmic Reticulum Ca^{2+}-ATPase. Manuscript in preparation.

Seven

Thermotropic Phase Transitions of Pure Lipids in Model Membranes and Their Modification by Membrane Proteins

JOHN R. SILVIUS

Institute of Molecular Biology, University of Oregon, Eugene, Oregon U.S.A.

ABBREVIATIONS USED IN THE TABLES

ANS	1-anilino-8-naphthalene sulfonate
2AP, 16AP	2- or 16-anthroylpalmitic acid
12AS	12-anthroylstearic acid
6dM-PC	phosphatidylcholine substituted at the glycerol 2-position with a 6-doxylmyristoyl chain
DNS-PE	5-dimethylaminonaphthylsulfonyl phosphatidylethanolamine
doxyl	1-oxyl-2,2-dimethyloxazolidinyl
DPH	1,6-diphenyl-1,3,5-hexatriene
5dP-PC	phosphatidylcholine substituted at the glycerol 2-position with a 5-doxylpalmitoyl chain
DSC	differential scanning calorimetry
5dS, 12dS, 16dS	1-oxyl-2,2-dimethyloxazolidine derivatives of 5-, 12-, or 16-ketostearic acid ("doxyl" stearic acids)
5dS-PC, 12dS-PC, 16dS-PC	phosphatidylcholines substituted at the glycerol 2-position with 5-, 12,-, or 16-doxylstearoyl chains
DTA	differential thermal analysis
DTBN	di-*tert*-butyl nitroxide
ESR	electron spin resonance
FI	fluorescence intensity
FP	fluorescence polarization
GLC	gas-liquid chromatography
ΔH	enthalpy of transition (kcal/mol)
HSDSC	high sensitivity differential scanning calorimetry
LS	90° light scattering
NMR	nuclear magnetic resonance
NPN	*N*-phenylnaphthylamine
ParA	parinaric acid
2-ParA-PC	phosphatidylcholine substituted at the 2-position with a parinaroyl chain
T_c	phospholipid gel → liquid crystalline phase transition temperature (°C)
ΔT_c	difference in the T_c values of two phospholipids (°C)
T_p	phospholipid pretransition temperature (°C)
TEMPO	2,2,6,6-tetramethylpiperidine-1-oxyl
TLC	thin layer chromatography
Trp	tryptophan (protein intrinsic fluorescence chromophore)
X-ray	X-ray diffraction

The abbreviations used to denote lipid structures are described in the notes for Tables 1–3.

I. INTRODUCTION

Since the demonstration in the 1960s of thermotropic phase transitions in both pure hydrated lipid systems and biological membranes [for reviews, see (1–4)], variation of the experimental temperature, and thereby of the lipid physical state, has been a frequently used tool in studies of model and biological membranes. Although the significance of lipid phase transitions in the normal functioning of biological membranes remains to be fully clarified, it is undeniably true that temperature is a most useful experimental variable in studies of the thermodynamics of lipid-lipid, lipid-protein, and lipid-ion interactions, and in studies on the effects of the lipid physical state on the functions of a variety of lipid-protein systems. The thermodynamic information derived from studies of the former type, while of considerable interest in its own right, can also greatly aid the design and interpretation of experiments of the latter type.

In the tables of this chapter, I have summarized the more frequently reported types of thermodynamic parameters describing the thermotropic behavior of model membrane systems consisting of a single lipid, two lipid species, or a pure lipid plus protein. The compilation attempts to be as comprehensive as possible with respect to the variety of systems listed, but no effort has been made to include all reported data for some of the more commonly studied lipid systems. Wherever possible, data obtained for a given system using more than one technique to monitor the thermotropic behavior are included in the tables. Among the references to this chapter are listed several papers that generally discuss the technical and/or theoretical considerations relevant to the use of some of these techniques, including scanning calorimetric methods (3, 175, 176), fluorescence spectroscopy (177, 178), dilatometry (179), and electron spin resonance (10, 31, 180), to detect lipid phase transitions and lateral phase separations. Until recently, the lamellar phases formed by membrane lipids were considered to be most important in terms of biological function, but in recent years evidence has been obtained that nonlamellar phases exist in some biological membranes and in certain lipid mixtures similar to those found in membranes (5–7). However, because many of the methods most frequently used to monitor lipid phase transitions (e.g., calorimetric methods) do not reliably detect thermotropic lamellar $\rightarrow$ nonlamellar phase transitions, relatively few determinations of such transitions have been made for pure synthetic lipids, and they are not included in this compilation [for a review of this area, see (8)].

Pure synthetic phospholipids exhibit two types of lamellar phase transition. The first, which is common to almost all phospholipids with acyl or alkyl chains of roughly 12 or more carbons, is the familiar gel $\rightarrow$ liquid crystalline phase transition. This transition is generally fairly sharp [cooperative units are typically of the order of hundreds of molecules (25)] and energetic ($\Delta H \gtrsim 5$ kcal/mol for most phospholipids with acyl or alkyl chains of 14 or more carbons) and has been shown to be a true first-order transition in the case of dipalmitoyl phosphatidylcholine (9). The transition involves a large disordering of the acyl chain packing from the quasi-crystalline arrangement of the gel state to the relatively fluid and disordered arrangement characteristic of the liquid crystalline state (1, 2, 10). The second common type of lamellar phase transition, generally called a "pretransition," has an enthalpy lower than that of the main transition but a comparable cooperative unit size (25), and it appears to involve a rearrangement of the quasi-crystalline packing of the lipid molecules in the gel state without a major disordering of the lipid hydrocarbon chains (11–13). Such pretransitions, once thought to be a unique feature of saturated phosphatidylcholine phase behavior, have now been demonstrated for alkali cation salts of saturated phosphatidylglycerols, phosphatidylserines, and doubly ionized phosphatidic acids (see Table 2). Since disaturated phospholipids do not comprise the major lipid component in most biological membranes, the biological significance of the pretransition is questionable. However, the high sensitivity of the pretransition to perturbations of the lipid matrix [e.g., by low levels of contaminants such as fluorescent probes (14)] provides a valuable indicator of subtle influences on the lipid behavior that may be exerted by, for example, a drug or a protein. Reported values for the gel → liquid crystalline phase transition temperatures and enthalpies of a variety of hydrated phospholipids and phospholipid analogues are listed in Table 1, and the corresponding data for pretransitions are listed in Table 2.

To understand the thermotropic behavior of lipids in biological membranes, it is not sufficient to consider the behavior of pure isolated lipid species alone, for biological membranes not only are highly heterogeneous in their phospholipid composition but also contain a variety of nonphospholipid components, such as proteins and in many cases sterols, which interact with the phospholipids and can influence their behavior. Tables 3 and 4 present data pertinent to two aspects of this problem.

Table 3 summarizes reported data on the phase behavior of various binary mixtures of phospholipids. Such systems can be treated by the thermodynamic formalisms appropriate to two-component systems (water, the third component, being kept in great excess), and from their phase diagrams a variety of features of the lipid-lipid interaction, such as the presence of gel or liquid crystalline-phase immiscibility or the magnitudes of the excess enthalpies of mixing of the two lipid components, can be determined (15, 16). Because the reported results of studies of binary lipid mixtures have been presented in many different ways, the entries in Table 3 are given in a simple common form, as described in the notes for the table, and must not be regarded as complete descriptions of the sometimes quite sophisticated analyses presented in the papers cited.

Table 4 summarizes data pertaining to the thermotropic behavior of another important class of model membranes, namely, reconstituted lipid-protein systems. The theoretical analysis of the behavior of such systems is complicated by the fact that proteins, which have much larger molecular dimensions than do phospholipids, cannot be considered to be simply dilute contaminants in the lipid phase whose perturbations of the lipid phase behavior are understandable solely in terms of colligative properties (for examples of the current state of theoretical modeling of lipid-protein interactions, see refs. 17–19). The analysis of the behavior of such systems is also complicated by practical problems such as the difficulty of obtaining a true two-component membrane (again assuming water, the third component, to be in great excess) free of detergents, lipids, or contaminating proteins that may have been associated with the protein of interest, the need for careful reconstitution of protein and lipid to avoid major compositional heterogeneity in the sample, and the sometimes unappreciated influence of the size of a lipid or lipid-protein vesicle (which may vary with composition) on its phase behavior. These theoretical and practical difficulties are sizable and deserve careful attention in the design and interpretation of studies of this type, but they are gradually being overcome by progressively more sophisticated studies, which constitute a rapidly developing and most interesting aspect of membrane research.

II. DATA DESCRIBING THE THERMOTROPIC PHASE BEHAVIOR OF PURE PHOSPHOLIPID AND SPHINGOLIPID SYSTEMS

A. Notes on Tables 1–3

It is well established that certain features of lipid systems other than composition can influence their thermotropic behavior. To allow comparison of the various entries for a particular compound in these tables, each entry is defined as fully as possible with respect to the experimental conditions which may have appreciable effects on the measured values. In Tables 1–3, all values listed were obtained with large, unsonicated multilamellar vesicles unless otherwise indicated. It has been found that small unilamellar vesicles with high radii of curvature ($\sim$ 100–150 Å) exhibit a phase transition that is broader and occurs at a lower temperature than does the gel $\rightarrow$ liquid crystalline phase transition in multilamellar vesicles (20, 21). Results that are obtained in the presence of ethylene glycol are clearly indicated as such; recent studies (22, 44) indicate that this antifreezing agent can significantly alter both the temperature and the enthalpy of the phase transition for certain lipids. In Table 2, because phospholipid pretransitions can be strongly affected by the presence of impurities (23, 24) and by the rate at which the temperature is varied through the pretransition range (14), both the sample purities and the temperature scan rates are listed where these data are available.

Entries in Tables 1–3 are grouped according to lipid head group structure and are ordered within groups on the basis of acyl or alkyl chain structure. Except where otherwise noted, all glycerolipids listed are L-α-glycerolipids (i.e., 1,2-dia-cyl- or 1,2-di-O-alkyl-sn-glycerol derivatives). In the latter sections of Table 1 and in Tables 2 and 3, head group structures are abbreviated as follows: PA, phosphatidic acid; PC, phosphatidylcholine; PE, phosphatidylethanolamine; PG, phosphatidyl-glycerol (the head group glycerol is usually racemic); PS, phosphatidylserine (the head group serine is usually the L-isomer). In all three tables, lipid acyl chains are represented by the following sequence of numbers and symbols:

(*total carbon number*) : (*number of double bonds,* or cp for a cyclopropane ring) (*chain configuration symbol:* c for *cis-* double bond or cyclopropane ring, t for a *trans* double bond or ring, i for an isobranched chain, or ai for an anteisobranched chain) (*superscript number(s)* giving the position(s) of any double bonds or cyclopropane rings)

A palmitoyl (hexadecanoyl) chain is thus represented as 16:0 and an oleoyl (*cis*-9-octadecenoyl) chain as $18:1c\Delta^9$. In Tables 1 and 2, lipid acyl substituents are given as (A/B), where A is the acyl chain found at the glycerol 1-position and B is that found at the 2-position. In Table 3, lipids with like acyl chains, which constitute nearly all the entries, are abbreviated as di-(A) PC, etc. for economy of space.

In Table 3, the numbers ΔT_c after the lipid names represent the differences of the T_c's of the two components of the mixtures. Miscibilities are classified as follows:

Perfect	Ideal mixing and no broadening of the phase transition in the mixture.
H, High	Mixing ideal or very nearly so; phase transitions of mixtures are markedly broader than those of the pure components.
M, Moderate	Phase diagram deviates markedly from ideal behavior, but the solidus curve is not flat over any significant range of compositions.
L, Low	Solidus curve in phase diagram is essentially flat over some range of compositions encompassing a region 20–50% wide over the X_B-axis.
VL, Very low	Solidus curve is essentially flat over most of the range of compositions (along $> 50\%$ of the X_B-axis).

(A flat portion of the solidus curve in the phase diagram gives evidence of solid-solid immiscibility of the components). When full phase diagrams are not available, miscibilities are estimated from the phase behavior of mixtures at molar ratios close to 1 : 1, where calorimetric data can often give good indications of solid state immiscibility or of compound formation.

The transition and pretransition temperatures T_c and T_p in Tables 1 and 2, as well as the transition temperature differences ΔT_c in Table 3, are given in degrees Celsius. In most cases, the listed T_c and T_p values represent the midpoints of the phase transitions observed upon heating the indicated samples. However, in differential scanning calorimetric studies using relatively rapid temperature scan rates (5 °C/min or more), it is more common to determine T_c by extrapolating the rising phase of the transition peak to the baseline; the intersection temperature is then reported as T_c. Throughout the table, I have attempted to list the transition temperature values that the cited authors judge to be the best estimates of the true thermodynamic transition temperatures. All transition enthalpies ΔH are given in units of kilocalories per mol.

A list of the abbreviations used in Tables 1–3 precedes the Introduction.

B. Table 1. Phospholipid Major Endothermic Phase Transition Parameters (in Excess Water)

Species	T_c	ΔH	Method	Reference
Saturated n-Acyl PC's				
12:0/12:0	∿0		DSC	1
	-1.8	1.7	HSDSC	25
	0	4.3	DSC	26
13:0/13:0	13.5		DTA	27
14:0/14:0	23	6.7	DSC	1
	23	6	DSC	12
	23.9+0.1	5.0+0.2	HSDSC	14
	23.8+0.25		FP(DPH)	14
	23.9	5.4	HSDSC	25
	23	6.8	DSC	26
	24.0		DTA	27
	24.0		Dilatometry	28
	24.4+0.5		FP(DPH)	29
	23.59	5.03	HSDSC	30
	23.2		ESR(TEMPO)	31
	22.33		ESR(TEMPO)	32,71
	22.42		ESR(5dS-PC)	32,71
15:0/15:0	34.2		DTA	27
	33.7		Dilatometry	28
16:0/16:0	41	8.7	DSC	1
	42.0	9	DSC	12
	41.3+0.1	8.6+0.3	HSDSC	14
	40.6+0.25		FP(DPH)	14
	41.4	8.7	HSDSC	25
	41.5	8.6	DSC	26
	41.5		DTA	27
	41.4		Dilatometry	28
	41.1+0.5		FP(DPH)	29
	41.05	6.43	HSDSC	30
	40.5		ESR(TEMPO)	31
	40.17		ESR(TEMPO)	32,71
	39.26		ESR(5dS-PC)	32,71
	41.4	8.2	HSDSC	33
	42.8		DTA	34
	40.5	8.5	DSC	35
		8.6	DSC	36
	42.4+0.2	7.7	DSC	37
17:0/17:0	48.8		DTA	27
	48.5		Dilatometry	28
	47.8+0.4	9.2	DSC	38
18:0/18:0	58	10.7	DSC	1
	54.5+0.1	10.2+0.4	HSDSC	14
	53.7+0.1		FP(DPH)	14
	54.9	10.6	HSDSC	25
	54.8		DTA	27
	54.1		Dilatometry	28
	54.15	7.89	HSDSC	30

Table 1 *(Continued)*

Species	T_c	ΔH	Method	Reference
18:0/18:0	54.0		ESR(TEMPO)	31
	56		DTA	34
		11.9	DSC	36
	55		DTA	39
19:0/19:0	60.9		DTA	27
20:0/20:0	66		DSC	172
22:0/22:0	75	14.9	DSC	1
	75		DSC	40

Unsaturated n-Acyl PC's

Species	T_c	ΔH	Method	Reference
$16:1c\Delta^9/16:1c\Delta^9$	-36[a]	9.1	DSC	26
	-35.5		DTA	41
$17:1c\Delta^9/17:1c\Delta^9$	-27.6 ± 0.5		DSC	38
$18:1c\Delta^2/18:1c\Delta^2$	41	9.6	DTA/DSC	42
$18:1c\Delta^3/18:1c\Delta^3$	35	8.7	DTA/DSC	42
$18:1c\Delta^4/18:1c\Delta^4$	23	8.2	DTA/DSC	42
$18:1c\Delta^5/18:1c\Delta^5$	11	7.8	DTA/DSC	42
$18:1c\Delta^6/18:1c\Delta^6$	1	7.8	DTA/DSC	42
$18:1c\Delta^7/18:1c\Delta^7$	-8	7.6	DTA/DSC	42
$18:1c\Delta^8/18:1c\Delta^8$	-13	7.5	DTA/DSC	42
$18:1c\Delta^9/18:1c\Delta^9$	-22	7.6	DSC	1
	-14[a]	11.2	DSC	26
		11.2	DSC	36
	-15.8		DTA	41
	-21	7.7	DTA/DSC	42
	-20		DSC	43
	-23		DSC	44
$18:1c\Delta^{10}/18:1c\Delta^{10}$	-21	7.6	DTA/DSC	42
$18:1c\Delta^{11}/18:1c\Delta^{11}$	-19.5		DTA	41
	-19	7.8	DTA/DSC	42
$18:1c\Delta^{12}/18:1c\Delta^{12}$	-8	7.9	DTA/DSC	42
$18:1c\Delta^{13}/18:1c\Delta^{13}$	1	8.2	DTA/DSC	42
$18:1c\Delta^{14}/18:1c\Delta^{14}$	7	8.6	DTA/DSC	42
$18:1c\Delta^{15}/18:1c\Delta^{15}$	24	8.9	DTA/DSC	42
$18:1c\Delta^{16}/18:1c\Delta^{16}$	35	9.6	DTA/DSC	42
$18:1c\Delta^{17}/18:1c\Delta^{17}$	45		DTA	42
$16:1t\Delta^9/16:1t\Delta^9$	-4.0		DTA	41
$18:1t\Delta^9/18:1t\Delta^9$	9.5	7.3	DSC	26
		10.0	DSC	36
	12.9		DTA	41
	13		DSC	43
	12		ESR(TEMPO)	45
$18:1t\Delta^{11}/18:1t\Delta^{11}$	13.2		DTA	41

Mixed-Acyl PC's

Species	T_c	ΔH	Method	Reference
$14:0/16:0$[c]	35.05, 35.08 [b]	6.92	HSDSC	30
	35.3	7.9	DSC	46
$16:0/14:0$[c]	27.28, 27.54	5.16	HSDSC	30

Table 1 *(Continued)*

Species	T_c	ΔH	Method	References
16:0/14:0[c]	27.2	6.5	DSC	46
14:0/18:0[c]	38.55, 38.63 [b]	6.93	HSDSC	30
18:0/14:0[c]	29.37, 29.62 [b]	5.20	HSDSC	30
16:0/18:0[c]	48.98	8.33	HSDSC	30
	47.4		DSC	46
18:0/16:0[c]	43.87	7.26	HSDSC	30
	44.0		DSC	46
16:0/18:1cΔ^9	-5	8.0	DSC	36
	3		DSC	43
	~4		HSDSC	47
	-3	8.1	DSC	48
	-1.5		Raman	49
18:1cΔ^9/16:0	-10.3		DSC	171
18:0/18:1cΔ^6	31		DTA	42
18:0/18:1cΔ^9	13		DSC	43
	6.3	5.3	DSC	48
18:0/18:1cΔ^{12}	19		DTA	42
18:0/18:1cΔ^{16}	44		DTA	42
18:1cΔ^9/18:0		6.7	DSC	36
	~11	6.7	DSC	50
18:0/18:1tΔ^9	33		DSC	1
16:0/22:6cccccc	-30to-25		FP(ParA)	51

Isoacyl PC's

12:0i/12:0i	-18.5		DTA	52
13:0i/13:0i	-9.5		DTA	52
14:0i/14:0i	6.5		DTA	52
15:0i/15:0i	6.5		DTA	52
16:0i/16:0i	22		DTA	52
17:0i/17:0i	27		DTA	52
18:0i/18:0i	36.5		DTA	52

Anteisoacyl PC's

(-)13:0ai/(-)13:0ai	None >-60		DTA	53
(+)13:0ai/(+)13:0ai	None >-60		DTA	53
(+)14:0ai/(+)14:0ai	-33.7		DTA	53
(-)15:0ai/(-)15:0ai	-16.5		DTA	53
(+)15:0ai/(+)15:0ai	-14.8		DTA	53
(+)16:0ai/(+)16:0ai	-3.0		DTA	53
(-)17:0ai/(-)17:0ai	7.6		DTA	53
(+)17:0ai/(+)17:0ai	8.0		DTA	53
(+)18:0ai/(+)18:0ai	18.7		DTA	53
(-)19:0ai/(-)19:0ai	26.4		DTA	53
(+)19:0ai/(+)19:0ai	26.8		DTA	53

Cyclopropane Acyl or C^{10}-Derivatized Acyl PC's

| di-(+)17:cp,cΔ^9 | -19.9 | | DTA | 41 |

Table 1 *(Continued)*

Species	T_c	ΔH	Method	Reference
di-($\pm$)17:cp,tΔ^9	-0.3		DTA	41
di-($\pm$)19:cp,cΔ^9	-0.5		DTA	41
di-($\pm$)19:cp,cΔ^{11}	-3.5		DTA	41
di-($\pm$)19:cp,tΔ^9	16.3		DTA	41
di-($\pm$)19:cp,tΔ^{11}	14.0 d		DTA	41
di-($\pm$)-10-Me-18:0	<5 d		FP(DPH)	54
di-($\pm$)-10-Br-18:0	<5 d		FP(DPH)	54

Phosphatidylethanolamines

Species	T_c	ΔH	Method	Reference
12:0/12:0	29.0	4.0	DSC	26
14:0/14:0	49.5	5.8	HSDSC	25
	47.5	6.4	DSC	26
16:0/16:0	60	8.5	DSC	26
	63		ESR(TEMPO)	30
	63.5	8.8	DSC	35
	66		Raman	55
18:0/18:0	71		DSC	56
16:1cΔ^9/16:1cΔ^9	-33.5	4.3	DSC	26
18:1cΔ^9/18:1cΔ^9	-16	4.7	DSC	5
	-16	4.5	DSC	26
18:1tΔ^9/18:1tΔ^9	35	7.0	DSC	26
	38.5	4.3	DSC	57
	37.5		ESR(DTBN)	57
	36.5		ESR(TEMPO)	57
	37.5	6.3	HSDSC	58
16:0/18:1cΔ^9	20		FI(NPN)	59
18:0/18:1cΔ^9	30.4		LS	60

Phosphatidic Acids

Species	T_c	ΔH	Method	Reference
12:0/12:0				
-H$_2$(pH<3)	36.5		DSC	61
-NaH (pH 5.5)	33		DSC	61
-Na$_2$ (pH>10)	0		DSC	61
-Ca (pH 6)	None <100		DSC	62
14:0/14:0				
-H$_2$ (pH<3)	54		DSC	61
	45		FI(NPN)	63
-NaH (pH 5.5-7.4)	50	6.8	DSC	61
	~52		FI(NPN)	63
	~52		FI(NPN)	64
	48.8		DSC	65
(100mM NaCl)	49	7.1	DSC	66
(dist. water)		5.6	DSC	66
-Na$_2$ (pH>11)	28		FI(NPN)	63
	27		FI(NPN)	64
-Mg (pH 6)	60		FI(NPN)	64
-Ca (pH 6)	None <100		DSC	62
16:0/16:0				
-H$_2$(pH<3)	63		FI(NPN)	63
-NaH (pH 6)	71		FI(NPN)	63

Table 1 *(Continued)*

Species	T_c	ΔH	Method	Reference
16:0/16:0				
-NaH (pH 6.5)	67+1	5.2	DSC	37
(pH 9.1)	58$\overline{+1}$	2.9	DSC	37
-Na$_2$ (pH >11)	46$^-$		FI(NPN)	63

Phosphatidylglycerols

Species	T_c	ΔH	Method	Reference
12:0/12:0				
-H	30		DTA	39
-Na	∿4		DTA	39
	0		DSC	67
	0	4.5	DSC	68
-1/2 Mg	65	10	DSC	69
-1/2 Ca	67		DSC	67
	75	10	DSC	69
14:0/14:0				
-H (pH<2)	47		DTA	39
	42		DSC	61
	41.7		ESR(TEMPO)	70
	42.2		DSC	71
-Na	23.7	6.94	HSDSC	39
	23	6.8	DSC	61
	23.3		ESR(TEMPO)	70
	23.5		DSC	71
	23		FP(DPH)	72
-K	23.4		DSC	71
-1/2 Mg	44&68 or 68^e		DTA	73
	$\overline{71}$	12.4	DSC	74
-1/2 Ca	80-90	15.2	DSC	61
15:0/15:0				
-Na	33.5		HSDSC	173
16:0/16:0				
-H	64.5		DTA	39
	61		DSC	61
	57		ESR(TEMPO)	70
	56.4	6.4	DSC	71
-Na	41.0+0.2	7.9+0.4	DSC	37
	41$\overline{.5}$	8.$\overline{9}$	DSC	39
	41.5	8.8	DSC	61
	41.5		DSC	65
	39.5	9.1	DSC	75
(pH 7,.5M NaCl)	42.5		DSC	37
-K	40.2		ESR(TEMPO)	70
	40.5	9.3	DSC	71
-1/2 Mg	53.3		DSC	37
	59 or 76^f		DTA	73
-1/2 Ca	80-90	19	DSC	61
18:0/18:0				
-H	75		DTA	39
-Na	54.5	10.5	DSC	39

Table 1 *(Continued)*

Species	T_c	ΔH	Method	Reference
$18{:}1c\Delta^9/18{:}1c\Delta^9$				
-Na	-18		DTA	39
$18{:}1t\Delta^9/18{:}1t\Delta^9$				
-Na	10	7.1	DSC	75
Phosphatidylserines				
$14{:}0/14{:}0$				
-H (pH<3)	55		DSC	61
-Na (pH 6)	37	8.4	DSC	61
(pH 7.4)	39		DTA	76
(pH 7.0)	36	7.0+0.5	DSC,ESR(TEMPO)	77
(pH 9.5)	37	8.3	DSC	75
-Ca	None <100		DSC	61
$16{:}0/16{:}0$				
-H (pH<3.2)	69	9.0	DSC	78
	72		Turbidity	78
-Na (pH 7)	55		Turbidity	78
	50.8		ESR(TEMPO)	79
	53	9.0+0.5	DSC,ESR(TEMPO)	77
(pH 9)	55		Turbidity	78
-K (pH 7)	55		Turbidity	78
-3/2 Mg [g]	69		Turbidity	78
(excess Mg^{2+})	None <85		DSC	78
-1/2 Ca [g]	69		Turbidity	78
(excess Ca^{2+})	None <85		DSC	78
$18{:}1c\Delta^9/18{:}1c\Delta^9$				
-Na (pH 7.0)	-11	8.8+0.5	DSC,^{31}P-NMR	77
(pH 9.5)	-10	7.5	DSC	75
$18{:}1t\Delta^9/18{:}1t\Delta^9$				
-Na (pH 7.5)	22.5		DSC	80
Sphingolipids[h]				
DL-erythro-N-16:0-	40.8	6.7	HSDSC	33
sphingosinephos-	44	8.3	DSC	40
phorylcholine	41.3	6.8	HSDSC	47
DL-erythro-N-18:0-	52.8	17.9	HSDSC	47
sphingosinephos-	46 or 57[i]	20	HSDSC	81
phorylcholine				
DL-erythro-N-24:0-	48.8	15.3	HSDSC	33
sphingosinephos-	48.6	15.3	HSDSC	47
phorylcholine				
DL-erythro-N-16:0-	47.8	9.4	HSDSC	47
dihydrosphingo-				
sinephosphoryl-				
choline				
N-16:0-galactosyl-	82+2		^{2}H-NMR	82
ceramide				

Table 1 *(Continued)*

Species	T_c	ΔH	-	Method	References
Diphosphatidylglycerols (cardiolipins)					
di-(12:0/12:0)·(NH$_4$)$_2$	25.2	8.7		DSC	83
di-(14:0/14:0)·(NH$_4$)$_2$	39.8	12.5		DSC	83
di-(16:0/16:0)·(NH$_4$)$_2$	57.8	17.7		DSC	83
di-(16:0/16:0)·Na$_2$	39.5	8.9		DSC	83
di-(16:0/16:0)·K$_2$	54.3	12.4		DSC	83
di-(16:0/16:0)·Ca	88.3	24.2		DSC	83
di-(16:0-16:0)·Mg	70.8	30.5		DSC	83
3,3'-Phosphatidyldiacylglycerols					
di-(12:0/12:0)·Na	52.1	10.2		DSC	83
di-(14:0/14:0)·Na	61.2	17.7		DSC	83
di-(16:0/16:0)·Na	73.3	22.2		DSC	83
di-(16:0/16:0)·K	74.8	19.6		DSC	83
di-(16:0/16:0)$_2$·Ca	52.9			DSC	83
di-(16:0/16:0)$_2$·Mg	62.1	23.2		DSC	83
Ether Phospholipids (1,2-Dialkyl-sn-glycerol-3-phosphate derivatives)					
Ditetradecyl PA					
-NaH (pH 7)	61.5	4.7		DSC	84
-Na$_2$ (pH 14)	31.5	5.4		DSC	84
Ditetradecyl PG					
-Na (pH 4.6)	26			DSC	85
-1/2 Ca (pH 4.6)	81 or 57[j]	9.0 or 5.0[j]		DSC	85
Dihexadecyl PC	43.4	9.4		DSC	35
Dihexadecyl PE	68	7.6		DSC	35
Dihexadecyl-N-methyl PE	62	9.1		DSC	35
Dihexadecyl-N,N-dimethyl PE	51	9.5		DSC	35
Dihexadecyl PA					
-H$_2$	62			FI(NPN)	63
	61	13.9		DSC	66
(pH 3.5)	74	2.5		DSC	66
-NaH (pH 7)	75	5.6		DSC	66
-Na$_2$	49	5.2		DSC	66
	49			FI(NPN)	63
Dihexadecylglycero-phosphoryl:					
-propanolamine	62.5			DSC	86
-N-methylpro-panolamine	58.5			DSC	86
-N,N-dimethyl-propanolamine	49.0			DSC	86
-N,N,N-trimethyl-propanolammonium	43.0			DSC	86

Table 1 *(Continued)*

Species	T_c	ΔH	Method	Reference
Miscellaneous Phospholipids				
DL-α-16:0/16:0 PC	41.3[k]		DSC	87
rac-1,2-di-O-hexadecyl-glycerol-3-phosphoryl:				
-ethanol	47.0		DSC	88
-ethanediol	50.5		DSC	88
-(2'-dimethylsulfo-nium) ethanol	47.8		DSC	88
-ethanolamine	67.5		DSC	88
-N-methylethanol-amine	61.5		DSC	88
-N,N-dimethyl-ethanolamine	51		DSC	88
-choline	42.5		DSC	88
	44.7		FI(trans-ParA)	89
-3'-aminopropanol	62.5		DSC	88
-(3'-trimethylammo-nium)propanol	43.0		DSC	88
-(3'-triethylammo-nium)propanol	47.3		DSC	88
DL-α-18:0/18:0 PC	56.3		DSC	88
rac-1,2-dioctadecanoyl-glycerol-3-phosphoryl-(2'-trimethylphospho-nium)ethanol	55.3		DSC	88
1,3-di-16:0-glycerol-2-phosphorylcholine	37-38	9.4	DSC	90
1,3-di-14:0-glycerol-2-phosphate				
-H_2	38		DSC	63
-NaH	42		DSC	63
-Na_2	23		DSC	63
(16:0/16:0)-N-methyl PE	58	8.6	DSC	35
(16:0/16:0)-N,N-dimethyl PE	48	10.0	DSC	35
	49.0		FI(cis-ParA)	91
	50.0		FI(trans-ParA)	91
1,2-di-16:0-rac-glycerol-3-phosphoryl-1'-rac-glycerol				
-Na	40.9		DSC	67
-K	40.5		DSC	67
-NH_4	43.3		DSC	67
-H	60.9		DSC	67
-1/2 Mg	53.3		DSC	67
1,2-di-16:0-sn-glycerol-3-phosphoryl(3'-trime-thylammonium)propanol	41	3.7	DSC	92
2,3-dihexadecyl-sn-gly-cerol-1-(hydroxyphos-phinylpropyl)trimethyl-ammonium hydroxide	45	8.3	DSC	93

253

Table 1 *(Continued)*

Species	T_c	ΔH	Method	Reference
rac-1-16:0-2-18:1cΔ^9 glycerol-3-phospho- rylcholine	11.0		FI(<u>trans</u>-ParA)	89
rac-1-18:1cΔ^9-2-O- hexadecylglycerol- 3-phosphorylcholine	-3		DSC	36
rac-1-O-hexadecyl- 2-18:1cΔ^9-glycerol- 3-phosphorylcholine	9.4		FI(<u>trans</u>-ParA)	89
rac-1-18:1cΔ^9-2-C- hexadecylpropane- diol-3-phosphoryl- choline	0		DSC	36
rac-1-O-hexadecyl-2- O-octadec-<u>cis</u>-9'- enylglycerol-3-phos- phorylcholine	17.0		FI(<u>trans</u>-ParA)	89
rac-1,2-di-O-octadec- <u>cis</u>-9'-enylglycerol- 3-phosphorylcholine	-21±4		Raman	94
rac-2-hexadecoxy-3- octadecoxypropyl(2'- trimethylammonium- ethyl)phosphinate	65	13.2	DSC	36
1,2-diphytanoyl PC	None from -120 to +80		DSC	95

Phospholipids Used as Spectroscopic Probes

Species	T_c	ΔH	Method	Reference
di-(d$_{27}$-14:0) PC	∿24		Raman	96
di-(d$_{31}$-16:0) PC	37.5		DTA	34
	37.5		Raman	97
	37.5		DSC	98
	36-37		Raman	99
di-(d$_{35}$-18:0) PC	50		DTA	34
(16',16',16'-d$_3$-16:0)/ 16:1cΔ^9) PC	∿-15		^{2}H-NMR	100
di-(4',4'-difluoro- 14:0) PC	28.5	11.7	HSDSC	101
di-(8',8'-difluoro- 14:0) PC	16.1	11.1	HSDSC	101
di-(12',12'-difluoro- 14:0) PC	19.7	9.4	HSDSC	101
di-(7',7'-difluoro- 16:0) PC	31.5		HSDSC	102
di-(7',7'-difluoro- 18:0) PC	∿45		HSDSC	102
di-(12',12'-difluoro- 18:0) PC	48.8		HSDSC	102
(16:0/5'-doxyl-18:0) PC	29		HSDSC	103
(16:0/12'-doxyl-18:0) PC	<0		HSDSC	103
(16:0/16'-doxyl-18:0) PC	34		HSDSC	103

Footnotes for Table 1

[a]Measured in the presence of 50% ethylene glycol, which has been reported to increase both the transition enthalpy and the transition temperature for unsaturated phosphatidylcholines (see Section II.A.)

[b]Multiple transition peaks observed; the numbers given refer to the midpoints of the separate transitions.

[c]The mixed-acyl phospholipids invariably contain some of the reverse isomer (e.g., 1-palmitoyl-2-stearoyl PC in a preparation of the 1-stearoyl-2-palmitoyl species). The listed T_c values and ΔH values were determined by extrapolation to the pure lipid properties from measurements of the transition characteristics in a series of preparations with varying contents of the reverse isomer.

[d]The lipid was dispersed in the form of small unilamellar vesicles by sonication, which has been shown to depress the temperature and sharpness of the transition for at least some phospholipids (see Section II.A).

[e]A double transition endotherm is observed when the sample is cooled from above 70°C before the heating scan, while a single transition at 68°C is observed when the sample has been previously cooled from 55°C.

[f]The 76°C transition is seen in samples equilibrated for several days at 20°C before the heating scan; after the first heating scan, a 59°C endotherm is seen in subsequent scans.

[g]The listed lipid/ion ratios represent the ratios of the total molar quantities of these components present in the sample, and it cannot be concluded a priori that a sample consists purely of a lipid/ion complex whose stoichiometry is given by the listed lipid/ion molar ratio.

[h]Synthetic sphingosine derivatives can contain significant amounts of the corresponding dihydrosphingosine derivatives (usually 5–15%). For data regarding the purities of these preparations, the reader is referred to the original papers.

[i]A 46°C transition is seen when the sample is cooled from above 57°C to room temperature and a heating scan is run shortly thereafter. When the sample is left at room temperature for several hours, a 57°C endotherm is seen on subsequent heating; the listed ΔH value is that for this latter endotherm.

[j]The first listed transition for this compound is seen when it is rapidly cooled from 100°C and subsequently heated. The second is seen when the sample is incubated at 70°C for more than 2 h and then rapidly cooled to 0°C and subjected to a heating scan.

[k]The D- and L-isomers of dipalmitoyl PC have been found to be ideally miscible in all proportions (Upreti and Jain, J. Membrane Biol., in press).

C. Table 2. Phospholipid Pretransition Properties

Species	Purity	T_p
(14:0/14:0) PC	>98%(TLC)	11
	Pure(TLC)	14.1+0.3
	Pure(TLC)	12.4
	NG	14.2
	>99%(TLC,GLC)	15.8
	NG	13-14
	>98%(TLC)	9.3-14
	Pure(TLC)	14.4
	Pure(TLC)	11
	Pure(TLC)	10
	Pure(TLC)	8
	NG	16
(15:0/15:0) PC	>99%(TLC,GLC)	24.1
	NG	22-23
(16:0/16:0) PC	Pure(TLC)	34.5
	Pure(TLC)	35.2+0.3
	Pure(TLC)	29.8
	Pure(TLC)	31.9
	NG	35.3
	>99%(TLC,GLC)	35.6
	NG	33-35
	>98%(TLC)	25.2-33.9
	Pure(TLC)	34.8
	∿98%(TLC)	39.5
	Pure(TLC)	34
	NG	37+1
	Pure(TLC)	32
	Pure(TLC)	∿30
	NG	34
(17:0/17:0) PC	>99%(TLC,GLC)	42.2
	NG	41-43
	NG	42
(18:0/18:0) PC	Pure(TLC)	48.5+0.3
	Pure(TLC)	45.6
	NG	51.5
	>99%(TLC,GLC)	51.8
	NG	48-50
	98%(TLC)	42.5-47.9
	Pure(TLC)	50.4
	NG	53.5
	Pure(TLC)	48
	Pure(TLC)	49
	Pure(TLC)	47
(19:0/19:0) PC	>99%(TLC,GLC)	60.9

ΔH (kcal mole^{-1})	Scan Rate (deg min^{-1})	Method	Reference
∿1.0	5	DSC	12
1.0±0.5	0.25	HSDSC	14
	0.3	FP(DPH)	14
1.0	0.5	HSDSC	25
	3	DTA	27
	0.017	Dilatometry	28
		FP(DPH)	29
0.83	0.5	HSDSC	30
	<2	FP(trans-ParA)	91
	<2	FP(cis-ParA)	91
	<2	FP($\overline{2}$-Par PC)	91
	8	DSC	104
	3	DTA	27
	0.017	Dilatometry	28
∿2	5	DSC	12
1.5±0.5	0.25	HSDSC	14
	0^a	FP(DPH)	14
	0.3	FP(DPH)	14
1.8	0.5	HSDSC	25
	3	DTA	27
	0.017	Dilatometry	28
		FP(DPH)	29
0.92	0.5	HSDSC	30
	10	DTA	34
	10	DSC	35
1.2±0.2	5	DSC	37
	<2	FP(cis- or trans-ParA)	91
	<2	FP(2-Par PC)	91
1.6	8	DSC	104
	3	DTA	27
	0.017	Dilatometry	28
	5	DSC	38
1.7±0.5	0.25	HSDSC	14
	0.3	FP(DPH)	14
1.8	0.5	HSDSC	25
	3	DTA	27
	0.017	Dilatometry	28
		FP(DPH)	29
0.89	0.5	HSDSC	30
	10	DTA	34
	<2	FP(trans-ParA)	91
	<2	FP(cis-ParA)	91
	<2	FP($\overline{2}$-Par PC)	91
	3	DTA	27

Table 2 *(Continued)*

Species	Purity	T_p
(20:0/20:0) PC	Pure(TLC)	62.9
(14:0/16:0) PC	b 94%[c]	22.76±0.17 25
(16:0/14:0) PC	b 92%[c]	10.77±0.17 12
(18:0/14:0) PC	b	20.03±0.16
(16:0/18:0) PC	b 81%[c]	39.91±0.18 38
(18:0/16:0) PC	b 96%[c]	30.78±0.18 32
di-(d_{31}-16:0) PC	>98% deuteration NG	33.0 28
di-(d_{35}-18:0) PC	∿98% deuteration	48
1,2-dihexadecyl PC	Pure(TLC)	32
1,3-di-16:0-glycerol-2-phosphorylcholine	NG	37-38
1,2-dihexadecyl PE (1M NaCl, pH>13)	Pure(TLC)	∿39
(14:0/14:0) PG·Na	>98%(TLC,GLC) Pure(TLC)	∿17 ∿10
(14:0/14:0) PG·K	Pure(TLC) 97-98%(GLC)	13.1 13.1
(16:0/16:0) PG·Na	>98%(TLC,GLC) Pure(TLC) NG	∿35 39 35±1
(16:0/16:0) PG·K	Pure(TLC)	32.8 36.0
1,2-dihexadecyl PG·Na (pH 5.8)	>98%(TLC)	∿36
1,2-ditetradecyl PA·Na$_2$ (pH 14, 2.6 M KCl)	Pure(TLC)	23.5
(14:0/14:0) PS·Na$_2$ (pH 9.5)	Pure(TLC)	∿17

ΔH (kcal mole^{-1})	Scan Rate (deg min^{-1})	Method	Reference
	1.25	DSC	172
.22+.04	0.5	HSDSC	30
	5	DSC	46
.28+.04	0.5	HSDSC	30
	5	DSC	46
1.95+.12	0.5	HSDSC	30
.36+.08	0.5	HSDSC	30
	5	DSC	46
.52+.08	0.5	HSDSC	30
	5	DSC	46
	10	DTA	34
		HSDSC	98
	10	DTA	34
	10	DSC	35
1.7-4.2	NG	DSC	90
1.4	2.5	DSC	174
	10	DTA	39
	5	DSC	61
	1.8	ESR(TEMPO)	70
		DSC	71
	10	DTA	39
	5	DSC	65
0.5+0.2	5	DSC	37
	1.8	ESR(TEMPO)	70
0.84		DSC	71
	20	DSC	67
0.24	2.5	DSC	84
	5	DSC	75

Footnotes for Table 2

[a] T_p was determined at a series of scan rates and the zero-rate value determined by extrapolation.

[b] Because mixed-acyl phospholipids invariably contain variable but significant amounts of the reverse isomer, the reported T_p and ΔH_p values were obtained from phase diagrams constructed from a series of samples containing varying amounts of the reverse isomer.

[c] The reported pretransition temperatures and enthalpies were determined for the preparations of these compounds that were found to be the most nearly free of contaminating reverse isomer. "Purity" in this case indicates how much of the preparations used were the correct isomer.

D. Table 3. Miscibility of Phospholipid and Sphingolipid Pairs

Pair (ΔT_c)	Miscibility	Method	Reference
di-12:0 PC/di-16:0 PC (43)	L	DSC	16
di-12:0 PC/di-18:0 PC (57)	VL	HSDSC	25
di-14:0 PC/di-16:0 PC (17)	H	DSC	16,104
	H	HSDSC	25
	H	ESR(TEMPO)	31
	H	FP(DPH)	105
di-14:0 PC/di-18:0 PC (31)	L	DSC	16
	M	HSDSC	25
	M	ESR(TEMPO)	31
	L-M	FP(DPH)	105
	M	Dilatometry	106
di-14:0 PC/di-20:0 PC (42)	VL	Dilatometry	106
di-16:0 PC/di-18:0 PC (14)	H	ESR(TEMPO)	31
di-18:1cΔ^9 PC/di-14:0 PC (40)	VL	DSC	1
di-18:1cΔ^9 PC/di-16:0 PC (57)	VL	DSC	1
	VL	DTA	41
di-18:1cΔ^9 PC/di-18:0 PC (71)	VL	DSC	1,36
di-18:1cΔ^9 PC/di-22:0 PC (91)	VL	DSC	1
di-18:1tΔ^9 PC/di-14:0 PC (11)	M	DSC	16
	L	ESR(TEMPO)	45
di-18:1tΔ^9 PC/di-16:0 PC (28)	L	DSC	16
	M	DTA	41
	M	ESR(TEMPO)	45
di-18:1tΔ^9 PC/di-18:0 PC (42)	L	ESR(TEMPO)	45
di-19:cp,cΔ^9 PC/di-16:0 PC (42)	VL	DTA	41
di-19:cp,tΔ^9 PC/di-16:0 PC (25)	M	DTA	41
(16:0/18:1cΔ^9)PC/di-16:0 PC (44)	VL	DSC	48
(18:0/18:1cΔ^9)PC/di-16:0 PC (35)	VL	DSC	48
(16:0/22:6cccccc)PC/di-16:0 PC (∼70)	VL	FP(ParA)	51
di-14:0 PC/di-(4',4'-difluoro-14:0) PC (5)	L(Compound formation)	HSDSC	101
di-14:0 PC/di-(8',8'-difluoro-14:0) PC (8)	L-M	HSDSC	101
di-14:0 PC/di-(12',12'-difluoro-14:0) PC (4)	L-M	HSDSC	101
di-14:0 PC/di-14:0 PE (24)	M	DSC	104
di-14:0 PC/di-16:0 PE (∼39)	M	ESR(TEMPO)	31
di-14:0 PC/di-16:1cΔ^9 PE (60)	VL	DSC	26
di-14:0 PC/di-18:1cΔ^9 PE (38)	VL	DSC	26
di-16:0 PC/di-16:0 PE (22)	H	ESR(TEMPO)	31
	H	FP(ParA)	91
di-16:0 PC/di-16:1cΔ^9 PE (78)	VL	DSC	26
di-16:0 PC/di-18:1cΔ^9 PE (57)	VL	DSC	5,26
di-18:0 PC/di-14:0 PE (6)	M-H	HSDSC	25
di-16:1cΔ^9 PC/di-14:0 PE (84)	VL	DSC	26
di-18:1cΔ^9 PC/di-14:0 PE (65)	VL	DSC	5,26
di-18:1cΔ^9 PC/di-16:0 PE (79)	VL	ESR(TEMPO)	45
di-18:1tΔ^9 PC/di-16:0 PE (50)	VL(Fluid immiscibility)	ESR(TEMPO)	45

Table 3 *(Continued)*

Pair (ΔT_c)	Miscibility	Method	Reference
di-14:0 PC/di-14:0 PA·NaH (26)	M	DSC	61
di-14:0 PC/di-14:0 PA·Ca	L-M	DSC	61
di-16:0 PC/di-16:0 PA·NaH (25)	M	DSC	37
di-16:0 PC/di-16:0 PA·Ca,pH 6	No transition <70°	DSC	37
di-16:0 PC/di-16:0 PA·Ca,pH 8	VL-L	DSC	37
di-16:0 PC/di-16:0 PA·H$_2$ (21)	M-H	DSC	37
di-12:0 PC/di-12:0 PG·Na (∿0)	P	DSC	68
di-12:0 PC/di-12:0 PG·1/2Ca (∿75)	H	DSC	68
di-12:0 PC/di-14:0 PG·Na (25)	M-H	DTA	39
di-12:0 PC/di-14:0 PG·1/2Ca (∿87)	M	DTA	39
di-12:0 PC/di-16:0 PG·Na (43)	M	DTA	39
di-12:0 PC/di-16:0 PG·1/2Ca (∿93)	L	DTA	39
di-14:0 PC/di-14:0 PG·Na (0)	P	DTA	39
	H	DSC	61,74
di-14:0 PC/di-14:0 PG·1/2Mg (∿51)	H	DSC	74
di-14:0 PC/di-14:0 PG·1/2Ca (∿61)	H	DTA	39
	H	DSC	74
di-14:0 PC/di-16:0 PG·Na (17)	H	DTA	39
	H	DSC	61
di-14:0 PC/di-16:0 PG·1/2Ca (∿67)	L-M	DTA	39
	M	DSC	61
di-16:0 PC/di-14:0 PG·Na (18)	H	DTA	39
di-16:0 PC/di-14:0 PG·1/2Ca (44)	H	DTA	39
di-16:0 PC/di-16:0 PG·Na (0)	P	DTA	39
di-16:0 PC/di-16:0 PG·1/2Ca (∿50)	H	DTA	39
di-18:0 PC/di-14:0 PG·Na (32)	M	DTA	39
di-18:0 PC/di-16:0 PG·Na (14)	H	DTA	39
di-18:0 PC/di-18:1cΔ⁹ PG·1/2Ca (∿70)	VL	DTA	39
di-18:0 PC/di-18:1tΔ⁹ PG·Na (45)	L,complex	DSC	75
di-18:1cΔ⁹ PC/di-16:0 PG·Na (59)	VL	DSC	75
di-18:1cΔ⁹ PC/di-16:0 PG·1/2Ca (107)	VL	DTA	39
di-18:1cΔ⁹ PC/di-18:1cΔ⁹ PG·Na (∿0)	P	DTA	39
di-18:1cΔ⁹ PC/di-18:1cΔ⁹ PG·1/2Ca (∿0)	H	DTA	39
di-14:0 PC/di-14:0 PS·Na (15)	L	DSC	61
di-14:0 PC/di-14:0 PS·Ca	L	DSC	61
di-14:0 PC/di-16:0 PS·Na (29)	M-H;complex solid mixing	ESR(TEMPO)	79
di-14:0 PC/di-18:1cΔ⁹ PS·Na (∿35)	L-M	DTA	76
di-14:0 PC/di-18:1cΔ⁹ PS·Ca	VL	DTA	76
di-18:0 PC/di-18:1cΔ⁹ PS·Na (65)	VL	DSC	75
di-18:1cΔ⁹ PC/di-14:0 PS·Na (58)	VL	DSC	75
di-18:1cΔ⁹ PC/N-16:0-sphingo-sinephosphorylcholine (60)	VL	DSC	40
(16:0/18:1cΔ⁹)PC/N-18:0-sphingo-sinephosphorylcholine (∿50)	L-M	HSDSC	47
di-16:0 PE/di-18:1cΔ⁹ PE (76)	VL	DSC	26
di-14:0 PG·Na/di-16:0 PG·Na (18)	H	DSC	65
di-16:0 PE/di-18:1tΔ⁹ PG·Na (51)	L,complex	DSC	75

Table 3 *(Continued)*

Pair (ΔT_c)	Miscibility	Method	Reference
di-18:1cΔ^9 PE/di-16:0 PG·Na (63)	L-M	DSC	75
di-18:1cΔ^9 PE/di-14:0 PS·Na (53)	L	DSC	75
di-18:1tΔ^9 PE/di-18:1cΔ^9 PS·Na (45)	L	DSC	75

III.   DATA DESCRIBING THE THERMOTROPIC PHASE BEHAVIOR OF RECONSTITUTED MIXTURES OF PROTEINS AND PURE PHOSPHOLIPIDS

A.   Notes on Table 4

To simplify the presentation of the data, certain special symbols and abbreviations are used. These are listed by the column(s) in which they appear and are explained below.

Protein/Lipid Ratio: (W) indicates a weight ratio, (M) a molar ratio.

Protein Effect on Transition T_c: $\pm x°$ indicates that the presence of the protein raises or lowers the midpoint of the major transition endotherm by $x_{sp}°C$ relative to the pure lipid value at the lipid/protein ratio indicated in the preceding column. NC indicates that the transition midpoint is unchanged, and $\delta\pm$ that the transition is slightly ($< 1°$) shifted up or down in the presence of the protein.

Protein Effect on Transition Amplitude: The amplitude of a transition is measured in various ways by various methods (e.g., an enthalpy of transition in calorimetric experiments, a change in the relative intensities of two Raman scattering bands, or a change in the order parameter determined in a spin-labeling experiment, etc.). Therefore, the effect of a protein on the amplitude of a lipid phase transition is quantified as the percentage change in the amplitude of the transition as measured by the experimental technique listed in the "Method" column, compared to the corresponding transition amplitude, measured in the same way, for the pure lipid. When the transition amplitudes cannot be sufficiently well quantified to calculate a percentage change, the symbols $\pm$ or $\delta\pm$ are used to indicate that the amplitudes are substantially $\gtrsim 20\%$) or slightly $\lesssim 20\%$) changed, respectively, in the presence of protein; NC denotes no significant change.

Protein Effect on Transition Sharpness: The symbols NC, $\delta\pm$, and $\pm$, respectively, indicate that the sharpness of the transition (judged either by its apparent temperature limits or by the rate of change in the experimentally observed index of the extent of transition at the transition midpoint) is unchanged, slightly increased or decreased (by not more than $\sim 50\%$), or substantially increased or decreased, respectively, in the presence of protein.

Protein Effect on Pretransition: NC, Red. and Abol. indicate, respectively, no change, a decrease in amplitude, or an abolition of the pretransition in the presence of protein.

Method: The abbreviations for the techniques used are listed following Table 4.

N: The numbers given indicate the estimated number of lipid molecules whose thermotropic behavior is strongly perturbed by one molecule of protein. The magnitude of such an estimate can, of course, depend on the assumptions employed in its calculation, for details of which the original papers should be consulted.

B. Table 4. Thermotropic Phase Transitions of Reconstituted Lipid-Protein

Protein	Lipid	Protein/Lipid Ratio	Protein Effect on Transition: T_c	Amplitude
Polylysine	Bovine PS	Ppt.[a]	-∿3	
	DMPC	1:3(W)	+∿3	NC
	DMPC	1:3(W)	+3,+6[b]	(-)
	DPPG·Na	0.8:2.1(W)	+∿7	+37%
	DPPG·Na	4:2.1(W)	+4,+8[b]	+50%
	DPPA(pH9)	1:2-1:1(W)	NC,+12 [b]	NC
Poly(Lys,Phe)	Bovine PS	0.25-1.0 Lys/PS(M)[c]	NC	-10-50%
	DPPC	0.6-1.3 Lys/PC(M)[c,d]		-30-60%
Poly(Lys,Tyr)	Bovine PS	0.25-1:1 Lys/PS(M)[c]	NC	-10-50%
Alamethicin	DPPC	1:99(W)	NC	NC
	DPPC	1:99(W)	NC	NC
	DPPC	1:99(W)	NC	+∿30%
Gramicidin A	DMPC	1:33(W)	NC	-∿70%
	DMPC	1:33(W)	δ-	(-)
	DPPC	1:6(M)	-∿1	-∿40%
	DPPG·Na	1:6(M)	δ-	-∿40%
	DPPC	1:150(M)	NC	-∿35%
	DPPC	1:50-1:9(M)	-0-2	-10-55%
	DMPC	1:100-1:6(M)	δ-	-15-85%
	DPPC	1:6(M)	NC	
	DPPC	1:217(M)	NC	-11%
Gramicidin S	DMPC	1:3(W)	-∿8	NC
	DMPC	1:3(W)	-11,-7[b]	-80%
	DMPC	1:2500-1:25(M)	δ-	+0-5%
	DMPC	1:5(M)	-∿5[e]	+39%
	DPPC	1:10-1:5(M)	-∿2	+7-33%
Melittin	DPPC	1:50(M)	NC	-∿15%
	DMPC	1:100(M)	NC	
	DPPC	1:85(M)	NC	
	Porcine PS	1:30(M)	NC	
	DMPC	1:10(M)	-11.5,+7.5[b]	
	DMPC	1:10(M)	-10.5,+11.5[b]	
	DMPC	1:25(M)	-∿1,+8.5[b]	
Melittin fragments: Residues 1-19	DPPC	1:50(M)	NC	NC
	DMPC	1:25(M)	-2.5,+6.5[b]	
Residues 20-26	DPPC	1:50(M)	NC	-∿7%
	DMPC	1:25(M)	+1,+6.5[b]	

Systems

Protein Effect on Transition: Sharpness	Pretransition	N	Method	Reference
δ-			DSC	104
δ-			Raman	107
δ-	Abol.		DSC	107
δ-	Abol.		DSC	108
δ-	Abol.		DSC	108
δ-			FI(Pyrene)	109
δ-		1 PS/2 Lys[c]	DSC	110
		1 PC/2 Lys[c]	DSC	110
δ-		1 PS/2 Lys[c]	DSC	110
δ-			^{1}H-NMR(Bulk Proton T_2)	111
			^{1}H-NMR(Acyl -CH$_3$)	111
			^{1}H-NMR(Choline -CH$_3$)	111
δ-	Abol.		DSC	107
(-)			Raman	107
(-)	Abol.		DSC	108
(-)	Abol.		DSC	108
(-)			DSC	112
(-)	Abol.	6	DSC	113
(-)	Abol.		DSC	113
			DSC	114
(-)	Red.		DSC	115
(-)			Raman	107
δ-	Abol.		DSC	107
δ-			DSC	116
δ-			DSC	116
δ-	Abol.		DSC	116
δ-	Abol.	~10	DSC	117
			FI(Trp)	118
			FI(Trp)	118
			FI(Trp)	118
(-)			Raman[f]	119
(-)			Raman[g]	119
(-)			Raman[f]	119
	Red.		DSC	117
(-)			Raman[f]	120
	NC		DSC	117
NC			Raman[f]	120

Table 4 *(Continued)*

Protein	Lipid	Protein/Lipid Ratio	Protein Effect on Transition: T_c	Amplitude
Polymyxin	DPPA(pH9)	1:20-3:1(M)	NC,-19, -25 [b]	NC
Valinomycin	DMPC	1:10(W)	-4,+2[b]	-26%
	DMPC	1:10(W)	-∿4	(-)
	DMPC	1:100(M)	δ-	-∿30%
	DMPC	1:50(M)	δ-	
	DPPC	1:50(M)	-∿1	
Cytochrome c	Bovine PS	Ppt.[a]	-∿7	
(equine)	DPPG·Na	2:7(W)	-∿5	-47%
Hemoglobin (human)	DPPG·Na	3:1-4:1(W)	-9-11	-50-60%
Lysozyme	Bovine PS	Ppt.[a]	-∿7	
(egg white)	Egg PA	1:1(W)	-∿4	NC
Ribonuclease	DPPG·Na	64:36(W)	NC	+13%
(bovine)	DPPG·Na	3:1(W)	NC	+23%
	DPPC	h	h	h
Spectrin·actin	DMPC	.29:1(W)	NC	-37%
(human red	DMPG·Na	1:100-1:1(W)	NC	-0-55%
cell)	DPPS·Na	1.2:1(W)	-∿5	-43%
Glucagon	DMPC	1:75(M)	NC	NC
(bovine	DMPC	1:15(M)	NC	(-)
-porcine)	DMPC	1:30(M)	-∿6	δ-
	DMPC	1:55-1:35(M)	+∿3	NC
	DMPC	1:20(M)	+∿3	δ+
Myelin proteins:				
Lipophilin[1]	DPPC	1:3-1:1(W)	NC	-25-50%
	DPPG·Na	38:62(W)	NC	-25%
	DPPC	1:1(W)	NC	δ-
	DLPC	46:54(W)	NC	-57%
	DMPC	31:69(W)	NC	-30%
	DPPC	2:3(W)	NC	-54%
	DSPC	3:7(W)	NC	-30%
	DEPC	32:68(W)	NC	-24%
Proteolipid	DMPC	1:340(M)	-13	NC
apolipoprotein	DMPC	1:340(M)	-13.5	δ-
(bovine)	POPC	1:340(M)	-1.5	NC
	POPC	1:340(M)	-1.0 [b,j]	NC
	DMPC	1:15-1:4(W)	NC,+2[b,j]	-10-35%
	DMPC	1:20(W)	NC	δ-

Protein Effect on Transition:				
Sharpness	Pretransition	N	Method	Reference
NC		~3	FP(DPH)	121
(-)			DSC	107
(-)			Raman	107
δ-			DSC	122
(-)			DSC	122
(-)			DSC	122
δ-			DSC	104
(-)			DSC	108
(-)	Abol.		DSC	108
δ-			DSC	104
NC			X-ray	123
δ-	Red.		DSC	108
δ-[h]	Abol.[h]		DSC	108
			DSC	108
δ-	Red.		DSC	124
δ-	Abol.		DSC	124
(-)	Abol.		DSC	125
NC			FI(Pyrene)	126
			FI(Pyrene)	126
NC			Raman	127
(-)			DSC	128
(-)	Abol.		Densitometry	129
(-)	Red.	15	DSC	114
(-)			DSC	114
NC	Abol.		ESR(TEMPO)	130
δ-		17	DSC	131
δ-	Red.	23	DSC	131
δ-	Red.	25	DSC	131
δ-	Abol.	21	DSC	131
δ-		16	DSC	131
(-)			Raman[f]	49
δ-			Raman[g]	49
δ-			Raman[f]	49
δ-			Raman[g]	49
(-)	Abol.	142	DSC	132
(-)			Raman	133

269

Table 4 *(Continued)*

Protein	Lipid	Protein/Lipid Ratio	Protein Effect on Transition: T_c	Amplitude
Basic protein				
(bovine)	DLPG·Na	1:14(W)	+∿15	(-)
	DPPG·Na	1:1(W)	-∿1-2	-35%
	DPPG·Na	3:1(W)	-11	-30%
(human)	DMPA·NaH	1:4(W)	δ-,-5 [b]	
	DPPG·Na	28:72(W)	-∿1	
	DPPG·Na	1:1(W)	-∿5,-∿12[b]	
	DMPC	h	h	h
	DPPC	h	h	h
	Egg PA	1:1(W)	-13	
	Bovine PS	3:7(W)	NC	
	Egg PE	1:3(W)	NC	δ-
Pulmonary surfac-	DPPC	1:300(M)[k]	Transition abolished	
tant apolipopro-	DPPC	1:1100(M)	(-)	(-)
tein (canine)				
Serum apolipoproteins:				
apoA-I (human)	DMPC	1:5-2:1(W)	+∿2	-30-80%
	DMPC	0-1:5(W)	NC	(-)
	DMPC	∿1:2(W)	+∿6	-50%
	DMPC	∿1:2(W)	+∿3	(-)
	DMPC	1:1.9(W)	+4.0	+51%
	DMPC	1:1.9(W)	+2.7	+104%
	DMPC	1:1.9(W)	+5.5	-24%
(bovine)	DMPC	1:90(M)	+∿2.5	(-)
apo-HDL (human)	DMPC	0-1:3(W)	NC	(-)
	DMPC	1:3-3:1(W)	+∿3	-10-70%
	DMPC	3:2(W)	+∿2	-∿70%
	DMPC	9:1(M)	Transition abolished	
	DMPC	1:1.9(W)	+2.8	+21%
	DMPC	1:1.9(W)	+4.7	+70%
	DMPC	1:1.9(W)	+3.1	-16%
(porcine)	DMPC	1:2.5(W)	+∿2	-53%
	DMPC	1:2.5(W)	Transition abolished	
	DMPC	1:2.5(W)	+∿2	NC
	DMPC	1:2.5(W)	+∿2	(-)
	DMPC	1:2.4(W)	Transition abolished	
	DMPC	1:2.4(W)	+∿2	NC
	DMPC	1:2.4(W)	+∿2	NC
	DLPC	1:2.4(W)	+∿2	
apoA-II (human)	DMPC	1:1.9(W)	+3.9	+26%
	DMPC	1:1.9(W)	+4.1	+74%
	DMPC	1:1.9(W)	+3.3	-36%
apoC-I (human)	DMPC	1:1.9(W)	+2.4	+9%
	DMPC	1:1.9(W)	+2.7	+55%
	DMPC	1:1.9(W)	+3.6	-24%

Protein Effect on Transition:

Sharpness	Pretransition	N	Method	Reference
(-)			DSC	68
(-)			DSC	108
(-)			DSC	108
(-)			DSC	134
δ-	Red.		DSC	134
δ-			DSC	134
h	h		DSC	135
h	h		DSC	135
δ-			DSC	135
δ-			DSC	135
(-)			DSC	135
			Turbidity	136
(-)			Turbidity	136
(-)	Abol.	17	DSC	137
NC			DSC	137
(-)	Abol.		DSC	138
(-)			FP(DPH)	139
-56%			ESR(5dS-PC)	140
-17%			ESR(12dS-PC)	140
+81%			ESR(16dS-PC)	140
(-)			FP(DPH)	139
δ-	NC		DSC	137
(-)	Abol.	13	DSC	137
(-)	Abol.		DSC	137
		5-7	DSC	137
-37%			ESR(5dS-PC)	140
+6%			ESR(12dS-PC)	140
+50%			ESR(16dS-PC)	140
(-)	Abol.	25	DSC	141
			ESR(5dS)	141
(-)			ESR(12dS)	141
(-)			ESR(16dS)	141
			ESR(5dS)	142
δ-			ESR(12dS)	142
(-)			FP(12AS,2AP,ANS)	142
(-)			ESR(12dS)	142
-56%			ESR(5dS-PC)	140
-6%			ESR(12dS-PC)	140
+50%			ESR(16dS-PC)	140
-63%			ESR(5dS-PC)	143
-17%			ESR(12dS-PC)	143
-19%			ESR(16dS-PC)	143

Table 4 *(Continued)*

Protein	Lipid	Protein/Lipid Ratio	Protein Effect on Transition: T_c	Amplitude
apoC-III (human)	DMPC	1:1.9(W)	+2.9	+30%
	DMPC	1:1.9(W)	+0.8	+51%
	DMPC	1:1.9(W)	+1.8	-20%
	DMPC	1:50(M)	NC	(-)
	DMPC	1:50(M)	+∿6	(-)
	DMPC	1:50(M)	+∿5	(-)
	DMPC	1:50(M)	+∿4	(-)
	DPPC	1:60(M)	-∿1	NC
apoC (bovine)	DMPC	1:30-1:15(M)	+∿2-4	δ-
apo-ala (human)	DPPC	1:250(M)	NC	NC
	DSPC	1:350(M)	NC	NC
	DMPC	1:20(M)	Transition abolished	
	DMPC	<1:50(M)		(-)
Bacteriorhodopsin	DMPC	1:49(M)	-∿1	(-)
Cytochrome b_5 (bovine)	DPPC	1:44(M)	-∿3	
	DMPC	1:44(M)	NC	
	DPPC	1:40(M)	NC	NC
	DMPC	1:100(M)	-1.7	
	DMPC	1:50(M)	-2.4	
	DMPC	1:22(M)	-0.8	
Cytochrome oxidase (bovine)	POPC	1:200(M)	NC	
(equine)	DOPG·Na	∿1:0.23(W)	+∿3	
	DPPG·Na	∿1:0.19(W)	+∿3	
Glycophorin (human)	DMPC	1:120(M)	-∿1	NC
	DMPC	1:380(M)	NC	-∿20%
	DMPG·Na	1:400(M)	NC	-∿25%
	DMPS·Na	1:310(M)	NC	-∿30%
	DPPC	1:670(M)	-∿16,-6[b]	
	DPPC	1:670(M)	-1.2	
	DPPC	1:280(M)	-∿7	δ-
	DPPC	1:280(M)	Transition abolished	
	DPPC	1:135(M)	-∿23	-∿50%
	DPPC	1:135(M)	Transition abolished	
	d_{62}-DPPC	1:275(M)	-15	NC
	DPPC	1:14(W)	δ-	-∿50%
M13 (fd) Bacteriophage coat protein	DMPC	1:30(M)	NC	-∿40%
	DPPC	1:1(W)	-∿20	
	d_{62}-DPPC	1:1(W)	-∿15	
Rhodopsin (bovine)	DLPC	1:100(M)	NC	
	DMPC	1:100(M)	+4	
	DPPC	1:100(M)	NC	

Protein Effect on Transition:				
Sharpness	Pretransition	N	Method	Reference
-44%			ESR(5dS-PC)	143
-33%			ESR(12dS-PC)	143
NC			ESR(16dS-PC)	143
			FI(Pyrene)	144
(-)			ESR(DTBN)	144
(-)			ESR(5dP-PC,16dS-PC)	144
NC			ESR(12dS-PC)	144
NC	Abol.		FP(2AP,16AP,DPH)	145
$\delta-$			FP(DPH)	146
NC			LS	147
NC			LS	147
			LS	147
(-)			LS	147
(-)			FP(DPH)	148
(-)			FI(Trp)	149
(-)			FI(Trp)	149
NC			FI(DNS-PE,Perylene)	150
			FP(DPH)	151
			FP(DPH)	151
			FP(DPH)	151
			^{2}H-NMR	152
(-)			DSC	153
(-)			DSC	153
$\delta-$	Abol.		ESR(TEMPO)	154
$\delta-$	Abol.	86+14	DSC	155
$\delta-$	Abol.	93∓4	DSC	155
$\delta-$	Abol.	92∓6	DSC	155
$\delta-$			Raman	156
	Abol.		DSC	156
(-)			Raman	156
	Abol.		DSC	156
$\delta-$			Raman	156
	Abol.		DSC	156
NC			Raman	156
(-)	Abol.		DSC	157
NC		$\sim$74	FI(cis-ParA)	158
(-)			Raman	159
(-)			Raman	159
(-)			ESR(6dM-PC)	160
(-)			ESR(6dM-PC)	160
(-)			ESR(6dM-PC)	160

Table 4 *(Continued)*

Protein	Lipid	Protein/Lipid Ratio	Protein Effect on Transition: T_c	Amplitude
Rhodopsin (bovine)	DEPC	1:100(M)	+3	
	DMPC	1:120(M)	NC	
	DPPC	1:120(M)	NC	
	DMPC	1:150(M)	$-\sim1.5$	
	DMPC	1:50(M)	$-\sim3$	
Sarcoplasmic reticular Ca^{2+}-ATPase (rabbit)	DMPC	1:9(W)	$+\sim1$	$\delta-$
	DMPC	3:7(W)	$+\sim1.5$	$\delta-$
	DMPC	1:237–1:66(M)	NC	
	DMPC	1:250–1:70(M)	NC	$-\sim5$–70%
	DPPC	1:260–1:80(M)	NC	$-\sim20$–50%
	DPPC	1:67(M)	NC	
	DPPC	1:29(M)	Transition abolished[1]	
	DPPC	1:7(W)	$-\sim12^{m}$	
Vesicular Stomatitis Virus Coat Protein	DPPC	1:3000(M)	-0.7	-12%
	DPPC	1:1000(M)	-1.5	-19%
	DPPC	1.5:1000(M)	-2.0	-41%

```
Protein Effect on Transition:
Sharpness    Pretransition        N        Method              Reference
```

Sharpness	Pretransition	N	Method	Reference
(-)			ESR(6dM-PC)	160
(-)			ESR(16dS-PC)	161
(-)			ESR(16dS-PC)	161
δ-			^{2}H-NMR	162
(-)			^{2}H-NMR	162
δ-			ESR(TEMPO)	163
(-)			ESR(TEMPO)	163
(-)			DSC,FP(DPH)	164
(-)	Abol.	45	DSC	165
(-)	Abol.	42	DSC	165
δ-	Red.		ESR(TEMPO)	166
			ESR(TEMPO)	166
			ESR(12dS,16dS)	167
δ-		270+150	HSDSC	169
(-)			HSDSC	169
(-)			HSDSC	169

Footnotes for Table 4

[a]The lipid was sedimented as a complex with an excess of protein.

[b]Multiple transitions were observed; the numbers given refer to the midpoints of the transitions.

[c]Expressed as mol of phospholipid per mol of lysine residues (N) or vice versa (Protein-Lipid Ratio).

[d]A very concentrated mixture was obtained by evaporation of excess water.

[e]The overall transition shape is complex; the T_c value given refers to the peak of the calorimetric transition.

[f]The transition was monitored as the temperature-dependent change in the ratio of the intensities of the Raman C—C stretching bands at 1090 and 1126 cm^{-1}.

[g]The transition was monitored as the temperature-dependent change in the ratio of the intensities of the Raman C—H stretching bands at 2940 and 2880 cm^{-1}.

[h]The protein was reported not to bind to the lipid.

[i]Called "myelin N-2 protein" in the earlier literature.

[j]In a recent paper, Boggs et al. (170) have reported that a double transition is not ordinarily found in the reconstituted DMPC–bovine myelin proteolipid apoprotein system unless small amounts of lipid degradation products are present.

[k]This is slightly higher than the lipid/protein ratio in vivo.

[l]The authors do not interpret the measured transition in this sample (which is in fact merely a nonlinearity in the TEMPO partition parameter's variation with temperature) as a true phase transition.

[m]The measured transition may not be a property of the lipid phase alone but may represent in part a protein conformational change (see Ref. 168).

ACKNOWLEDGMENTS

I thank Drs. O. H. Griffith, Patricia C. Jost, and Thomas E. Thompson for valuable suggestions regarding the content and format of this chapter, and Drs. Richard M. Epand, Jean François Faucon, Martin P. Heyn, Mahendra K. Jain, Ana Jonas, Kevin M. W. Keough, Ira W. Levin, Richard Mendelsohn, P. Nuhn, Frank Sixl, N. Z. Stanacev, Julian M. Sturtevant, Thomas E. Thompson, Jean-François Tocanne, Pieter W. M. van Dijck, Winchil L. Vaz, and Anthony Watts for providing me with results before publication. I also thank many of the other cited authors for checking for accuracy the tabular entries that were derived from their studies. Finally, I gratefully acknowledge the financial support of the Medical Research Council of Canada in the form of a fellowship award during the course of this work.

REFERENCES

(1) Ladbrooke, B. D., & Chapman, D. (1969). *Chem. Phys. Lipids* **3**, 304–364.

(2) Levine, Y. K. (1972). *Prog. Biophys. Mol. Biol.* **24**, 1–74.

(3) Melchior, D. L., & Steim, J. M. (1976). *Annu. Rev. Biophys. Bioeng.* **5**, 205–238.

(4) Cronan, J. E., Jr., & Gelmann, E. P. (1975). *Bacteriol. Rev.* **39**, 232–256.

(5) Cullis, P. R., van Dijck, P. W. M., de Kruijff, B., & de Gier, J. (1978). *Biochim. Biophys. Acta* **513**, 21–30.

(6) Cullis, P. R., & de Kruijff, B. D. (1978). *Biochim. Biophys. Acta* **507**, 207–218.

(7) De Kruijff, B., van den Besselaar, A. M. H. P., Cullis, P. R., van den Bosch, H., & van Deenen, L. L. M. (1978). *Biochim. Biophys. Acta* **514**, 1–8.

(8) Cullis, P. R. & de Kruijff, B. D. (1979). *Biochim. Biophys. Acta* **559**, 399–420.

(9) Albon, N., & Sturtevant, J. M. (1978). *Proc. Natl. Acad. Sci. U.S.A.* **75**, 2258–2260.

(10) Lee, A. G. (1977). *Biochim. Biophys. Acta* **472**, 237–281.

(11) Rand, R. P., Chapman, D., & Larrson, K. (1975). *Biophys. J.* **15**, 1117–1124.

(12) Janiak, M. J., Small, D. M., & Shipley, G. G. (1976). *Biochemistry* **15**, 4575–4580.

(13) Janiak, M. J., Small, D. M., & Shipley, G. G. (1979). *J. Biol. Chem.* **254**, 6068–6078.

(14) Lentz, B. R., Freire, E., & Biltonen, R. L. (1978). *Biochemistry* **17**, 4475–4480.

(15) Lee, A. G. (1977). *Biochim. Biophys. Acta* **472**, 285–344.

(16) Van Dijck, P. W. M., Kaper, A. J., Oonk, M. A. J., & de Gier, J. (1977). *Biochim. Biophys. Acta* **470**, 58–69.

(17) Marčeljà, S. (1976). *Biochim. Biophys. Acta* **455**, 1–7.

(18) Owicki, J. C., Springate, M. W., & McConnell, H. M. (1978). *Proc. Natl. Acad. Sci. U.S.A.* **75**, 1616–1619.

(19) Owicki, J. C., & McConnell, H. M. (1979). *Proc. Natl. Acad. Sci. U.S.A.* **76**, 4750–4754.

(20) Suurkuusk, J., Lentz, B. R., Barenholz, Y., Biltonen, R. L., & Thompson, T. E. (1976). *Biochemistry* **15**, 1393–1401.

(21) Van Dijck, P. W. M., de Kruijff, B., Aarts, P. A. M. M., Verkleij, A. J., & de Gier, J. (1978). *Biochim. Biophys. Acta* **506**, 183–191.

(22) Van Dijck, P. W. M. (1977). Thesis, State University of Utrecht, Utrecht, the Netherlands.

(23) Cater, B. A., Chapman, D., Hawes, S., & Saville, J. (1974). *Biochim. Biophys. Acta* **363**, 54–69.

(24) Mabrey, S., Mateo, P. L., & Sturtevant, J. M. (1978). *Biochemistry* **17**, 2464–2468.

(25) Mabrey, S., & Sturtevant, J. M. (1976). *Proc. Natl. Acad. Sci. U.S.A.* **73**, 3862–3866.

(26) Van Dijck, P. W. M., de Kruijff, B., van Deenen, L. L. M., de Gier, J., & Demel, R. A. (1976). *Biochim. Biophys. Acta* **455**, 576–587.

(27) Silvius, J. R., Read, B. D., & McElhaney, R. N. (1979). *Biochim. Biophys. Acta* **555**, 175–178.

(28) Nagle, J. F., & Wilkinson, D. A. (1978). *Biophys. J.* **23**, 159–175.

(29) Lentz, B. R., Barenholz, Y., & Thompson, T. E. (1976). *Biochemistry* **15**, 4521–4528.

(30) Chen, S. C., & Sturtevant, J. M. (1981). *Biochemistry* **20**, 713–718.

(31) Shimshick, E. J., & McConnell, H. M. (1973). *Biochemistry* **12**, 2351–2359.

(32) Marsh, D., Watts, A., & Knowles, P. F. (1977). *Biochim. Biophys. Acta* **465**, 500–514.

(33) Estep, T. N., Mountcastle, D. B., Barenholz, Y., Biltonen, R. L., & Thompson, T. E. (1979). *Biochemistry* **18**, 2112–2117.

(34) Petersen, N. O., Kroon, P. A., Kainosho, M., & Chan, S. I. (1975). *Chem. Phys. Lipids* **14**, 343–349.

(35) Vaughan, D. J., & Keough, K. M. (1974). *FEBS Lett.* **47**, 158–161.

(36) DeKruyff, B., Demel, R. A., Slotboom, A. J., van Deenen, L. L. M., & Rosenthal, A. F. (1973). *Biochim. Biophys. Acta* **307**, 1–19.

(37) Jacobson, K., & Papahadjopoulos, D. (1975). *Biochemistry* **14**, 152–161.

(38) Salvati, S., Serlupi-Crescenzi, G., & de Gier, J. (1979). *Chem. Phys. Lipids* **24**, 85–89.

(39) Findlay, E. J., & Barton, P. G. (1978). *Biochemistry* **17**, 2400–2405.

(40) Demel, R. A., Jansen, J. W. C. M., van Dijck, P. W. M., & van Deenen, L. L. M. (1977). *Biochim. Biophys. Acta* **465**, 1-10.

(41) Silvius, J. R., & McElhaney, R. N. (1979). *Chem. Phys. Lipids* **25**, 125–134.

(42) Barton, P. G., & Gunstone, F. D. (1975). *J. Biol. Chem.* **250**, 4470–4476.

(43) Op den Kamp, J. A. F., Kauerz, M. T., & van Deenen, L. L. M. (1975). *Biochim. Biophys. Acta* **406**, 169–177.

(44) Van Echteld, C. J. A., de Kruijff, B., & de Gier, J. (1980). *Biochim. Biophys. Acta* **595**, 71–81.

(45) Wu, S. H., & McConnell, H. M. (1975). *Biochemistry* **14**, 847–854.

(46) Keough, K. M. W., & Davis, P. J. (1979). *Biochemistry* **18**, 1453–1459.

(47) Barenholz, Y., Suurkuusk, J., Mountcastle, D., Thompson, T. E., & Biltonen, R. L. (1976). *Biochemistry* **15**, 2441–2447.

(48) Davis, P. J., Coolbear, K. P., & Keough, K. M. W. (1980). *Can. J. Biochem.* **58**, 851–858.

(49) Lavialle, F., & Levin, I. W. (1980). *Biochemistry* **19**, 6044–6050.

(50) De Kruyff, B., Demel, R. A., & van Deenen, L. L. M. (1972). *Biochim. Biophys. Acta* **255**, 331–347.

(51) Sklar, L. A., Miljanich, G. P., & Dratz, E. A. (1979). *Biochemistry* **18**, 1707–1716.

(52) Silvius, J. R., & McElhaney, R. N. (1979). *Chem. Phys. Lipids* **24**, 287–296.

(53) Silvius, J. R., & McElhaney, R. N. (1980). *Chem. Phys. Lipids* **26**, 67–77.

(54) Roseman, M. A., Lentz, B. R., Sears, B., Gibbes, D., & Thompson, T. E. (1978). *Chem. Phys. Lipids* **21**, 205–222.

(55) Mendelsohn, R., & Taraschi, T. (1978). *Biochemistry* **17**, 3944–3949.

(56) Blume, A. (1976). Thesis, Universität Freiburg, Freiburg, Germany.

(57) Yang, R. D., Patel, K. M., Pownall, H. J., Knapp, R. D., Sklar, L. A., Crawford, R. B., & Morrissett, J. D. (1979). *J. Biol. Chem.* **254**, 8256–8262.

(58) Jackson, M. B., & Sturtevant, J. M. (1977). *J. Biol. Chem.* **252**, 4749–4751.

(59) Eibl, H., & Wooley, P. (1979). *Biophys. Chem.* **10**, 261–271.

(60) Stoffel, W., & Michaelis, G. (1976). *Hoppe-Seyler's Z. Physiol. Chem.* **357**, 21–33.

(61) Van Dijck, P. W. M., de Kruijff, B., Verkleij, A. J., van Deenen, L. L. M., & de Gier, J. (1978). *Biochim. Biophys. Acta* **512**, 84–96.

(62) Van Dijk, P. W. M. Personal communication.

(63) Eibl, H., & Blume, A. (1979). *Biochim. Biophys. Acta* **553**, 476–488.

(64) Träuble, H., & Eibl, H. (1974). *Proc. Natl. Acad. Sci. U.S.A.* **73**, 214–219.

(65) Papahadjopoulos, D., Vail, W. J., Panborn, W. A., & Poste, G. (1976). *Biochim. Biophys. Acta* **448**, 265–283.

(66) Blume, A., & Eibl, H. (1979). *Biochim. Biophys. Acta* **558**, 13–21.

(67) Sacré, M.-M., Hoffmann, W., Turner, M., Tocanne, J.-F., & Chapman, D. (1979). *Chem. Phys. Lipids* **25**, 69–83.

(68) Verkleij, A. J., de Kruyff, B., Ververgaert, P. H. J. T., Tocanne, J.-F., & van Deenen, L. L. M. (1974). *Biochim. Biophys. Acta* **339**, 432–437.

(69) Ververgaert, P. H. J. T., de Kruyff, B., Verkleij, A. J., Tocanne, J.-F., & van Deenen, L. L. M. (1975). *Chem. Phys. Lipids* **14**, 97–101.

(70) Watts, A., Harlos, K., Maschke, W., & Marsh, D. (1978). *Biochim. Biophys. Acta* **510**, 63–74.

(71) Watts, A. Personal communication.

(72) Faucon, J.-F., & Tocanne, J.-F. Personal communication.

(73) Findlay, E. J. (1978). Ph.D. thesis, University of Alberta, Edmonton, Alberta, Canada.

(74) Van Dijck, P. W. M., Ververgaert, P. M. J. T., Verkleij, A. J., van Deenen, L. L. M., & de Gier, J. (1975). *Biochim. Biophys. Acta* **406**, 465–478.

(75) Van Dijck, P. W. M. (1979). *Biochim. Biophys. Acta* **555**, 89–101.

(76) Silvius, J. R., & Poznansky, M. Unpublished results.

(77) Browning, J. L., & Seelig, J. (1980). *Biochemistry* **19**, 1262–1270.

(78) MacDonald, R. C., Simon, S. A., & Baer, E. (1976). *Biochemistry* **15**, 885–891.

(79) Luna, E. J., & McConnell, H. M. (1977). *Biochim. Biophys. Acta* **470**, 303–316.

(80) Comfurius, P., & Zwaal, R. F. A. (1977). *Biochim. Biophys. Acta* **488**, 36–42.

(81) Estep, T. N., Calhoun, W. I., Barenholz, Y., Biltonen, R. L., Shipley, G. G., & Thompson, T. E. (1980). *Biochemistry* **19**, 20–24.

(82) Skarjune, R., & Oldfield, E. (1979). *Biochim. Biophys. Acta* **556**, 208–218.

(83) Rainier, S., Jain, M. K., Ramirez, F., Ioannou, P. V., Marecek, J. F., & Wagner, R. (1979). *Biochim. Biophys. Acta* **558**, 187–198.

(84) Harlos, K., Stumpel, J., & Eibl, H. (1979). *Biochim. Biophys. Acta* **555**, 409–416.

(85) Harlos, K., & Eibl, H. (1980). *Biochemistry* **19**, 895–899.

(86) Gawrisch, K., Arnold, K., Ruger, H.-J., Kertscher, P., & Nuhn, P. (1977). *Chem. Phys. Lipids* **20**, 285–293.

(87) Jain, M. K., & Wu, N. M. (1977). *J. Membrane Biol.* **34**, 157–201.

(88) Ruger, H.-J., Kertscher, P., Arnold, K., Gawrisch, K., & Nuhn, P. (1981). *Chem. Phys. Lipids* (in press).

(89) Lee, T.-C., & Fitzgerald, V. (1980). *Biochim. Biophys. Acta* **598**, 189–192.

(90) Seelig, J., Kijkman, R., & de Haas, G. H. (1980). *Biochemistry* **19**, 2215–22219.

(91) Sklar, L. A., Hudson, B. S., & Simoni, R. D. (1977). *Biochemistry* **16**, 819–828.

(92) Bach, D., Bursuker, I., Eibl, H., & Miller, I. R. (1978). *Biochim. Biophys. Acta* **514**, 310–319.

(93) Turcotte, J. G., Sacco, A. M., Steim, J. M., Tabak, S. A., & Notter, R. H. (1977). *Biochim. Biophys. Acta* **488**, 235–248.

 (94) Sunder, S., Bernstein, H. J., & Paltauf, F. (1978). *Chem. Phys. Lipids* **22**, 279–283.

 (95) Lindsey, H., Petersen, N. O., & Chan, S. I. (1979). *Biochim. Biophys. Acta* **555**, 147–167.

 (96) Mendelsohn, R., & Maisano, J. (1978). *Biochim. Biophys. Acta* **506**, 192–201.

 (97) Sunder, S., Cameron, D., Mantsch, H. H., & Bernstein, H. J. (1978). *Can. J. Chem.* **56**, 2121–2126.

 (98) Gaber, B. P., Yager, P., & Peticolas, W. L. (1978). *Biophys. J.* **22**, 191–207.

 (99) Mendelsohn, R. Personal communication.

(100) Dahlquist, F. W., Muchmore, D. C., Davis, J. H., & Bloom, M. (1977). *Proc. Natl. Acad. Sci. U.S.A.* **74**, 5435–5439.

(101) Sturtevant, J. M., Ho, C., & Reimann, A. (1979). *Proc. Natl. Acad. Sci. U.S.A.* **76**, 2239–2243.

(102) Longmuir, K. J., Capaldi, R. A., & Dahlquist, F. W. (1977). *Biochemistry* **16**, 5746–5755.

(103) Chen, S.-C., & Gaffney, B. J. (1978). *J. Magn. Reson.* **29**, 341–353.

(104) Chapman, D., Urbina, J., & Keough, K. M. (1974). *J. Biol. Chem.* **249**, 2512–2521.

(105) Lentz, B. R., Barenholz, Y., & Thompson, T. E. (1976). *Biochemistry* **15**, 4529–4537.

(106) Wilkinson, D. A., & Nagle, J. F. (1979). *Biochemistry* **18**, 4244–4249.

(107) Susi, H., Sampugna, J., Hampson, J. W., & Ard, J. S. (1979). *Biochemistry* **18**, 297–301.

(108) Papahadjopoulos, D., Moscarello, M., Eylar, E. H., & Isac, T. (1975). *Biochim. Biophys. Acta* **401**, 317–335.

(109) Hartmann, W., & Galla, H.-J. (1978). *Biochim. Biophys. Acta* **509**, 474–490.

(110) Bach, D., & Miller, I. R. (1976). *Biochim. Biophys. Acta* **433**, 13–19.

(111) Lau, A. L. Y., & Chan, S. I. (1974). *Biochemistry* **13**, 4942–4948.

(112) Weidekamm, E., Bamberg, E., Brdiczka, D., Wildermuth, G., Macco, F., Lehmann, W., & Weber, R. (1977). *Biochim. Biophys. Acta* **464**, 442–447.

(113) Chapman, D., Cornell, B. A., Eliasz, A. W., & Perry, A. (1977). *J. Mol. Biol.* **113**, 517–538.

(114) Papahadjopoulos, D., Vail, W. J., & Moscarello, M. (1975). *J. Membrane Biol.* **22**, 143–164.

(115) Boehler, B. A., de Gier, J., & van Deenen, L. L. M. (1978). *Biochim. Biophys. Acta* **512**, 480–488.

(116) Wu, E.-S., Jacobson, K., Szoka, F., & Portis, A., Jr. (1978). *Biochemistry* **17**, 5543–5550.

(117) Mollay, C. (1976). *FEBS Lett.* **64**, 65–68.

(118) Dufourcq, J., & Faucon, J.-F. (1977). *Biochim. Biophys. Acta* **467**, 1–11.

(119) Lavialle, F., Levin, I. W., & Mollay, C. (1980). *Biochim. Biophys. Acta* **600**, 62–71.

(120) Levin, I. W., & Lavialle, F. Personal communication.

(121) Sixl, F., & Galla, H.-J. (1979). *Biochim. Biophys. Acta* **557**, 320–330.

(122) Hsu, M., & Chan, S. I. (1973). *Biochemistry* **12**, 3872–3876.

(123) Mateu, L., Caron, F., Luzzati, V., & Billecocq, A. (1978). *Biochim. Biophys. Acta* **508**, 109–121.

(124) Mombers, C., van Dijck, P. W. M., van Deenen, L. L. M., de Gier, J., & Verkleij, A. J. (1977). *Biochim. Biophys. Acta* **470**, 152–160.

(125) Mombers, C., Verkleij, A. J., de Gier, J., & van Deenen, L. L. M. (1979). *Biochim. Biophys. Acta* **551**, 271–281.

(126) Epand, R. M., Jones, A. J. S., & Schreier, S. (1977). *Biochim. Biophys. Acta* **491**, 296–304.

(127) Taraschi, T., & Mendelsohn, R. (1979). *J. Am. Chem. Soc.* **101**, 1050–1052.

(128) Epand, R. M., Jones, A. J. S., & Sayer, B. (1977). *Biochemistry* **16**, 4360–4368.

(129) Epand, R. M., & Epand, R. F. (1981). *Biochim. Biophys. Acta*. Submitted for publication.

(130) Boggs, J. M., Vail, W. J., & Moscarello, M. A. (1976). *Biochim. Biophys. Acta* **448**, 517–530.

(131) Boggs, J. M., & Moscarello, M. A. (1978). *Biochemistry* **17**, 5734–5739.

(132) Curatolo, W., Sakura, J. D., Small, D. M., & Shipley, G. G. (1977). *Biochemistry* **16**, 2313–2319.

(133) Curatolo, W., Verma, S. P., Sakura, J. D., Small, D. M., Shipley, G. G., & Wallach, D. F. H. (1978). *Biochemistry* **17**, 1802–1807.

(134) Boggs, J. M., Moscarello, M. A., & Papahadjopoulos, D. (1977). *Biochemistry* **16**, 5420–5426.

(135) Boggs, J. M., & Moscarello, M. A. (1978). *J. Membrane Biol.* **39**, 75–96.

(136) King, R. J., & Macbeth, M. C. (1979). *Biochim. Biophys. Acta* **557**, 86–101.

(137) Tall, A. R., Small, D. M., Deckelbaum, R. J., & Shipley, G. G. (1977). *J. Biol. Chem.* **252**, 4701–4711.

(138) Tall, A. R., & Lange, Y. (1978). *Biochim. Biophys. Acta* **513**, 185–197.

(139) Jonas, A., Krajnovich, D. J., & Patterson, B. W. (1977). *J. Biol. Chem.* **252**, 2200–2205.

(140) Vaughan, D. J., Breckenridge, W. C., & Stanacev, N. Z. (1980). *Can. J. Biochem.* **58**, 581–591.

(141) Andrews, A. L., Atkinson, D., Barratt, M. D., Finer, E. G., Hauser, H., Henry, R., Leslie, R. B., Owens, N. L., Phillips, M. C., & Robertson, R. N. (1976). *Eur. J. Biochem.* **64**, 549–563.

(142) Barratt, M. D., Badley, R. A., Leslie, R. B., Morgan, C. G., & Radda, G. K. (1974). *Eur. J. Biochem.* **48**. 595–601.

(143) Vaughan, D. J., Breckenridge, W. C., & Stanacev, N. Z. (1980). *Can. J. Biochem.* **58**, 592–598.

(144) Novosad, Z., Knapp, R. D., Gotto, A. M., Pownall, H. J., & Morrisett, J. D. (1976). *Biochemistry* **15**, 3176–3183.

(145) Vaz, W. L. C., Jacobson, K. L., Wu, E.-S., & Derzko, Z. (1979). *Proc. Natl. Acad. Sci.* **76**, 5645–5649.

(146) Patterson, B. W., & Jonas, A. (1980). *Biochim. Biophys. Acta* **619**, 587–603.

(147) Träuble, H., Middelhoff, G., & Brown, V. W. (1974). *FEBS Lett.* **49**, 269–275.

(148) Heyn, M. P. (1979). *FEBS Lett.* **108**, 359–364.

(149) Dufourcq, J., Faucon, J.-F., Lussan, C., & Bernon, R. (1975). *FEBS Lett.* **57**, 112–116.

(150) Faucon, J.-F., Dufourcq, J., Lussan, C., & Bernon, R. (1976). *Biochim. Biophys. Acta* **436**, 283–294.

(151) Vaz, W. L. C., Austin, R. H., & Vogel, H. (1979). *Biophys. J.* **26**, 415–426.

(152) Seelig, A., & Seelig, J. (1978). *Hoppe-Seyler's Z. Physiol. Chem.* **359**, 1747–1756.

(153) Denes, A. S., & Stanacev, N. Z. (1979). *Can. J. Biochem.* **57**, 238–249.

(154) Grant, C. W. M., & McConnell, H. M. (1974). *Proc. Natl. Acad. Sci. U.S.A.* **71**, 4653–4657.

(155) Van Zoelen, E. J. J., van Dijck, P. W. M., de Kruijff, B., Verkleij, A. J., & van Deenen, L. L. M. (1978). *Biochim. Biophys. Acta* **514**, 9–24.

(156) Taraschi, T., & Mendelsohn, R. (1980). *Proc. Natl. Acad. Sci. U.S.A.* **77**, 2362–2366.

(157) MacDonald, R. I., & MacDonald, R. C. (1975). *J. Biol. Chem.* **250**, 9206–9214.

(158) Kimelman, D., Tecoma, E. S., Wolber, P. K., Hudson, B. S., Wickner, W. T., & Simoni, R. D. (1979). *Biochemistry* **18**, 5874–5880.

(159) Dunker, A. K., Williams, R. W., Gaber, B. P., & Peticolas, W. L. (1979). *Biochim. Biophys. Acta* **553**, 351–357.

(160) Chen, Y. S., & Hubbell, W. L. (1973). *Exp. Eye Res.* **17**, 517–532.

(161) Davoust, J., Bienvenue, A., Fellmann, P., & Devaux, P. F. (1980). *Biochim. Biophys. Acta* **596**, 28–42.

(162) Bienvenue, A., Devaux, P. F., Davis, J. M., & Bloom, M. (1980). *FEBS Lett.* Submitted for publication.

(163) Kleemann, W., & McConnell, H. M. (1976). *Biochim. Biophys. Acta* **419,** 206–222.

(164) Gomez-Fernandez, J. C., Goñi, F. M., Bach, D., Restall, C., & Chapman, D. (1979). *FEBS Lett.* **98,** 224–228.

(165) Gomez-Fernandez, J. C., Goñi, F. M., Bach, D., Restall, C. J., & Chapman, D. (1980). *Biochim. Biophys. Acta* **598,** 502–516.

(166) Hesketh, T. R., Smith, G. A., Houslay, M. D., McGill, K. A., Birdsall, N. J. M., Metcalfe, J. C., & Warren, G. B. (1976). *Biochemistry* **15,** 4145–4151.

(167) Hidalgo, C., Ikemoto, N., & Gergely, J. (1976). *J. Biol. Chem.* **251,** 4224–4232.

(168) Dean, W. L., & Tanford, C. (1978). *Biochemistry* **17,** 1683–1690.

(169) Petri, W. A., Jr., Estep, T. N., Pal, R., Thompson, T. E., Biltonen, R. L., & Wagner, R. R. (1980). *Biochemistry* **19,** 3088–3091.

(170) Boggs, J. M., Clement, I. R., & Moscarello, M. A. (1980). *Biochim. Biophys. Acta* **601,** 134–151.

(171) Keough, K. M. W. Personal communication.

(172) Patel, K. M., Morrisett, J. D., & Sparrow, J. T. (1979). *J. Lipid Res.* **20,** 674–677.

(173) Lentz, B. R., Alford, D. R., & Dombrose, F. A. (1980). *Biochemistry* **19,** 2555–2559.

(174) Stümpel, J., Harlos, K., & Eibl, H. (1980). *Biochim. Biophys. Acta* **599,** 464–472.

(175) Steim, J. M. (1974). *Methods Enzymol.* **32,** 262–272.

(176) Mabrey, S., & Sturtevant, S. (1978). In *Methods in Membrane Biology* (E. D. Korn, Ed.), Vol. 9, pp. 237–274, Plenum Press, New York.

(177) Yguerabide, J., & Foster, M. C. (1979). *J. Membrane Biol.* **45,** 109–123.

(178) Lentz, B. R., Barenholz, Y., & Thompson, T. E. (1976). *Biochemistry* **15,** 4529–4537.

(179) Wilkinson, D. A., & Nagle, J. F. (1978). *Anal. Biochem.* **84,** 263–271.

(180) Gaffney, B. J., & Chen, S.-C. (1977). In *Methods in Membrane Biology* (E. D. Korn, Ed.), Vol. 8, pp. 291–358, Plenum Press, New York.

Author Index

Numbers in parentheses refer to reference numbers. Numbers in *italics* refer to pages where full reference appear.

Subject Index